SUNIL V. HATTANGADY

Gallium Arsenide Processing Techniques

The Artech House Microwave Library

Introduction to Microwaves by Fred E. Gardiol
Receiving Systems Design by Stephen J. Erst
Applications of GaAs MESFETs by R.A. Soares, J. Graffeuil, and J. Obregon
GaAs Processing Techniques by R.E. Williams
GaAs FET Principles and Technology, J.V. DiLorenzo and D. Khandelwal, editors
Modern Spectrum Analyzer Theory and Applications by Morris Engelson
Microwave Materials and Fabrication Techniques by Thomas S. Laverghetta
Handbook of Microwave Testing by Thomas S. Laverghetta
Microwave Measurements and Techniques by Thomas S. Laverghetta
Principles of Electromagnetic Compatibility by Bernhard E. Keiser
Microwave Filters, Impedance Matching Networks, and Coupling Structures by G.L. Matthaei, Leo Young, and E.M.T. Jones
Microwave Engineer's Handbook, 2 vol., Theodore Saad, ed.
Computer-Aided Design of Microwave Circuits by K.C. Gupta, R. Garg, and R. Chadha
Microstrip Lines and Slotlines by K.C. Gupta, R. Garg, and I.J. Bahl
Microstrip Antennas by I.J. Bahl and P. Bhartia
Microwave Circuit Design Using Programmable Calculators by J. Lamar Allen and Max Medley, Jr.
Stripline Circuit Design by Harlan Howe, Jr.
Microwave Transmission Line Filters by J.A.G. Malherbe
Electrical Characteristics of Transmission Lines by W. Hilberg
Multiconductor Transmission Line Analysis by Sidney Frankel
Microwave Diode Control Devices by Robert V. Garver
A Practical Introduction to Impedance Matching by Robert Thomas
Active Filter Design by A.B. Williams
Adaptive Electronics by Wolfgang Gaertner
Laser Applications by W.V. Smith
Electronic Information Processing by W.V. Smith
Logarithmic Video Amplifiers by Richard S. Hughes
Avalanche Transit-Time Devices, George Haddad, ed.
Gallium Arsenide Bulk and Transit-Time Devices, Lester Eastman, ed.
Ferrite Control Components, 2 vol., Lawrence Whicker, ed.

Gallium Arsenide Processing Techniques

Ralph E. Williams

Copyright © 1984
ARTECH HOUSE, INC.
610 Washington Street
Dedham, MA 02026

All rights reserved. Printed and bound in the United States of America. No part of this book may be reproduced or utilized in any form or by any means, electronic or mechanical, including photocopying, recording, or by any information storage and retrieval system, without permission in writing from the publisher.

International Standard Book Number: 0-89006-152-1
Library of Congress Catalog Card Number: 84-071257

Second Printing, April, 1985.

TABLE OF CONTENTS

PREFACE ix
ACKNOWLEDGMENTS xiii

PART I. PRELIMINARY CONCEPTS

1. Introduction 1
   1.1 Why Gallium Arsenide? 1
   1.2 Fundamental Concepts and Definitions 2
   1.3 Example Process Flows 6
2. GaAs Material and Crystal Properties: A Tutorial 17
   2.1 General Concepts 18
   2.2 Physical and Electrical Characteristics 21
   2.3 Bulk Crystal Growth and Slice Generation 35
   2.4 Epitaxy 42
   2.5 Ion Implantation 47
3. GaAs Devices: A Tutorial 57
   3.1 Schottky Diodes and VARACTORs 58
   3.2 Field Effect Transistor 61
   3.3 Microwave Diodes (IMPATT & GUNN) 74
   3.4 Heterojunction Devices 79

PART II. GENERAL PROCESS TECHNIQUES

4. Cleaning and Cleanliness 85
   4.1 Introduction 85
   4.2 Environment and Handling (Cleanliness) 87
   4.3 Cleaning Techniques 93

5. Wet Etching 101
   5.1 Introduction 101
   5.2 Basic Considerations in GaAs Etching 102
   5.3 GaAs Etchants 108
   5.4 Wet Etching Other Materials 122
6. Photolithography 125
   6.1 Introduction 125
   6.2 Types of Photolithography and Equipment 126
   6.3 Resist Properties 135
   6.4 Edge Profile and Multilevel Techniques 143
7. Non-Optical Lithography 149
   7.1 Introduction 149
   7.2 Electron Beam Lithography 150
   7.3 X-ray Lithography 158
   7.4 Ion Beam Lithography 162
8. Plasma Assisted Deposition 165
   8.1 Introduction 165
   8.2 Equipment Considerations 167
   8.3 Thin Film Deposition 171
9. Dry Etching — Plasma, RIE, RIBE, Ion Milling 183
   9.1 Introduction 183
   9.2 Plasma Etching 187
   9.3 Reactive Ion Etching 192
   9.4 Reactive Ion Beam Etching 197
   9.5 Ion Milling 198

PART III. SPECIFIC PROCESS STEPS

10. Device Isolation 211
    10.1 Purpose 211
    10.2 Isolation by Etching — Slice Orientation 214
    10.3 Ion Implant Isolation 218
    10.4 Selective Implantation 221
11. Ohmic Contacts 225
    11.1 Introduction 225
    11.2 Theoretical Basis of Ohmic Contacts 228
    11.3 Fabrication and Structure of Ohmic Contacts ... 232
    11.4 Measurement of Contact Resistance Parameters .. 241
12. Schottky Barriers and Gate Formation 259
    12.1 Introduction 259
    12.2 Measurement of Schottky Barrier Parameters ... 267
    12.3 Gate Fabrication 270

13. First-Level Metal, Dielectric Formation,
    Second-Level Metal 285
    13.1 Introduction 285
    13.2 First-Level Metal 294
    13.3 Dielectric/Formation 297
    13.4 Second-Level Metal 300
14. Capacitors, Inductors, and Resistors 303
    14.1 Introduction 303
    14.2 Capacitors 306
    14.3 Inductors 315
    14.4 Resistors 319
15. Plating and Bridge Interconnects 327
    15.1 Introduction 327
    15.2 Gold Plating 329
    15.3 Bridge Formation 334
16. Back Side Processing 341
    16.1 Introduction 341
    16.2 Wafer Thinning 343
    16.3 Via Formation 346
    16.4 Die Separation 351
17. Process Integration and Control 355
    17.1 Introduction 355
    17.2 Interrelationship of Process Steps 356
    17.3 In-process Controls 358

PART IV. CHARACTERIZATION AND MEASUREMENTS

18. Electrical Characterization 363
    18.1 Introduction 363
    18.2 Characterization of Materials 364
    18.3 Device Characterization 373
19. Diagnostic Techniques 385
    19.1 Introduction 385
    19.2 Surface Analysis 393
    19.3 Bulk Analysis 398

INDEX 403

PREFACE

Purpose of the Book

Gallium arsenide (GaAs) has emerged from the laboratory into production. The years, even decades, of laboratory development of GaAs devices have culminated in a body of fabrication expertise specialized to the GaAs arena. This expertise currently resides in years of separate journal articles, in conference proceedings, and with a few individuals engaged in the relevant research. The growing production effort is already resulting in substantial increases in the number of people concerned with GaAs, either directly in the production of devices and monolithic ICs, or indirectly in device use or circuit design. Many, if not most, of these people are relatively new to GaAs topics and are not familiar with all the relevant information. This situation makes it useful to gather together a compendium of GaAs fabrication techniques directed toward people inexperienced with these concepts. This book attempts to fulfill that goal. It addresses all the major GaAs processing techniques and indicates the major references for further or more detailed study.

Gallium arsenide processing differs substantially from the well-established silicon processing, and it is not the intent of this book to dwell on techniques well-known to silicon process engineers. Of course, a certain amount of such material must be included for completeness, but GaAs processing is sufficiently different from silicon processing that almost every process step has features unique to GaAs.

Few colleges or universities have programs that produce students already knowledgable in most of these topics. Further, the rapid expansion of GaAs production efforts is resulting in inclusion of personnel who, while perhaps holding graduate degrees in some field, are not expert in all topics pertaining to GaAs. These may be persons with backgrounds in silicon production, or physicists, chemists, and electrical engineers lacking familiarity with GaAs. I have spent many hours responding to questions from such persons on exactly the material contained in this book. I have also been frankly surprised at the interest and many questions from circuit designers involved in GaAs FET use and/or GaAs microwave monolithic IC design. In fact, it is precisely these many discussions over several years that convinced me of the necessity and usefulness of this book.

It should also be noted that many companies in private industry consider it healthy to hire bright people of various backgrounds regardless of whether or not their formal training matches their initial assignments. The present relative obscurity of GaAs topics in academia means that the GaAs arena will be particularly suited to such action for some time. The growing importance of GaAs devices (Chapter 1), however, will almost certainly result in increased academic attention and this book should prove useful to those students.

Because the intended audience is somewhat diverse, some material will not apply to all readers. Persons with a background in solid-state physics may have little interest in the elementary review of the energy band structure of semiconductors contained in Chapter 2. But they may not be familiar with other GaAs material properties and processing techniques. Electrical engineers involved in circuit design may find much of the material useful, including the tutorial review of GaAs material properties (Chapter 2). Persons holding graduate degrees in disciplines such as chemistry may find virtually all of the material useful.

GaAs processing has not achieved the maturity that is characteristic of silicon processing, and research continues in virtually all aspects of device fabrication. Many, if not most, of these research efforts will not yield processes having the necessary reliability, reproducibility, and yield required for production. Consistent with its intention of addressing production-worthy techniques, this book will generally restrict itself to those processes and procedures that have proven viable in a production environment.

Some topics related to GaAs device fabrication have an exhaustive research history and are well-represented in the technical literature.

Two examples are ohmic contacts and plasma processing. In these cases, the purpose of the book is to distill that information and present it in a format useful to those unfamiliar with these subjects. Other topics, however, are less well publicized. Examples are cleaning and cleanliness, and backside processing. Treatment of these topics is less exhaustive, but may provide information very difficult to find or unavailable to beginning process engineers.

Discussions in the book are often in the context of field effect transistors (FETs) rather than other GaAs devices such as microwave diodes or bipolar transistors. This is because the FET is the dominant device for GaAs microwave applications and for many digital applications. Nevertheless, all techniques (lithography, etching, ohmics, isolation, etc.) apply generally to all GaAs fabrication.

Organization of the Book

The book is divided into four parts, each of which is summarized below.

Part I, Preliminary Concepts, contains general information that provides a foundation for the remainder of the book. Chapter 1 is a general introduction. It details the motivation for using GaAs instead of other materials such as silicon. It reviews some fundamental concepts and definitions and contains example step-by-step process flows for several types of devices. Although this is not a book on GaAs materials science, GaAs devices, or device physics, a minimal understanding of these topics is necessary. Many of the process techniques are best understood in relation to device design and function and are directly motivated by GaAs crystal properties. For this reason, a tutorial review of GaAs material properties and growth techniques is contained in Chapter 2. A similar tutorial review of the major GaAs devices and their operation is contained in Chapter 3.

Part II, General Process Techniques, contains descriptions of techniques that are generally applicable to many different fabrication steps. Each chapter treats one of the following major topics: cleaning and cleanliness; wet etching (of GaAs and other materials); photolithography (emphasizing the special requirements necessitated by liftoff procedures); electron beam lithography (with brief mention of ion beam and x-ray lithography); plasma enhanced deposition; and dry etching techniques (plasma assisted etching, reactive ion etching, reactive ion beam etching, and ion milling).

Part III, Specific Process Steps, contains descriptions of specific fabrication steps used in GaAs processing. These include device isolation;

ohmic contact formation; gate formation; first-level metalization, dielectric coatings, and second-level metalization; fabrication of capacitors, inductors, and resistors; plating and bridge interconnects; and backside processing, including via formation. Finally, Chapter 17 treats the interrelationship of these processes.

Part IV, Characterization and Measurements, treats the diagnostic and measurement techniques used for in-process evaluation and for diagnosing process problems. A portion of this information is included in the appropriate chapter in Part III (contact resistance in Chapter 11; Schottky barrier parameters in Chapter 12). But most of the in-process measurement techniques are discussed in Chapter 18. Chapter 19 reviews the diagnostic and analytical techniques that have proven useful for diagnosing process problems.

ACKNOWLEDGMENTS

Let me state the obvious: I was not born with knowledge of this book's contents, nor did I independently discover most of it. I have been fortunate to have contributed to the field, both by research and subsequently by aiding initiation of GaAs production facilities. But the overwhelming majority of the book's contents is entirely the result of others' work. It would be impossible to list them all and you wouldn't read it anyway. To attempt to list only ten or twenty would (rightly) offend the others by their omission. I therefore will limit myself to listing very few, all but two of which directly aided preparation of the book with reviews and suggestions.

I thank H. Michael (Mike) Macksey who first introduced me to many of the aspects of GaAs processing and device operation.

I thank David Seymour for many instructive conversations over the years. His novel and effective approaches to many topics, mostly unpublished, have proven quite useful.

The remainder of the acknowledgements go to those below, who have reviewed all or parts of the book and made numerous helpful suggestions. For various reasons, I have not incorporated all of their suggestions, and so any faults in the text are likely my own. I appreciate the efforts of my colleagues David Rhine and Garry Boggan for reviewing each chapter in the book and making innumerable, valuable suggestions. Fred Doerbeck, Jim Latham, Steve Nelson, and Burhan Bayraktaroglu reviewed specific portions of the material. Ruth Williams helped in preparation of figures. I also greatly appreciate the support of the editors at Artech House, especially Dennis N. Ricci and Barbara W. Modelski, during the course of the book's preparation.

Dallas, 1984

PART I.
PRELIMINARY CONCEPTS

CHAPTER 1

INTRODUCTION

1.1 WHY GALLIUM ARSENIDE?

Gallium arsenide (GaAs) has been a subject of research in many industrial and university laboratories for two decades. The high mobility and high saturated drift velocity (compared to silicon) have meant that GaAs semiconductor devices could operate at microwave frequencies where silicon devices are unable to function. The ability to produce semi-insulating substrates allows low parasitics and true monolithic circuit implementation. GaAs devices are more radiation tolerant than silicon MOS devices. (These properties will be discussed further in Chapter 2.)

In spite of its promised performance, GaAs did not evolve rapidly into the commercial market (with the exception of GaAs light emitting diodes). In fact, it was a long-standing joke within the semiconductor community that "GaAs is the material of the future, and it always will be." This is no longer true; the future has arrived. Proof may be found in the large number of major semiconductor companies and microwave houses that have established prototype or full-scale GaAs production facilities.

This maturity of GaAs has resulted from two forces. First, the years of laboratory work resulted in viable processes and devices that exhibit high performance and reliability. Second, there is a rapidly growing demand for such devices in communication, radar, and electronic warfare. A commercial market also exists for direct satellite broadcast electronics and high-speed computers. All these markets may be characterized by the need for large numbers of economical components operating above 1 GHz, and often above 10 GHz.

A prime example is phased array radar. It is currently one of the major driving forces encouraging GaAs production. In conventional radars, the signal is transmitted and received from a dish antenna that is physically moved to direct the beam. Only one transmitter and one receiver are required. A phased array radar consists of a plane of many separate modules, each of which can transmit or receive a radar signal. Each can also adjust the phase of this signal. The beam can be steered by adjusting the relative phases of all modules; no physical movement of the antenna is required. This allows almost instantaneous redirection of the beam. Further, it opens the possibility of more exotic procedures, such as simultaneously transmitting two beams in different directions from the same antenna. Obviously the phased array radar has many advantages. It has one major disadvantage: the large number of microwave modules it requires. Even a relatively small radar may require thousands. Large radars, such as those considered for use in the earth orbit, might use approximately one million modules. Cost, size, weight, and reliability all demand the use of solid-state devices. It is these kinds of considerations that promise a gigantic market for GaAs technology.

Direct broadcast satellite (DBS) applications also promise a large market for microwave receivers operating well above 1 GHz and requiring low noise operation. Many companies are currently working on GaAs receivers having a gain and noise figure sufficient to require only a small dish antenna to receive DBS signals.

High-speed logic is needed for electronic warfare applications and high-speed computers, both military and civilian. Many laboratories are actively exploring GaAs logic. GaAs devices can provide memory access times and logical operations at speeds several times greater than silicon can provide (Chapter 3). GaAs-based devices such as the *high electron mobility transistor* (HEMT) promise performance equivalent to that expected from the far more complex superconducting Josephson junction technology.

In summary, GaAs has established a firm foothold in high-speed electronics and will certainly play a greatly expanded role in the near future.

1.2 FUNDAMENTAL CONCEPTS AND DEFINITIONS

As indicated in the Preface, many people are becoming involved with GaAs without formal training in all relevant topics. This section reviews some of the concepts and definitions assumed throughout the rest of the book. This information is generally below the technical level of the book, but some readers may be unfamiliar with these concepts.

The *resistivity*, ρ, is a fundamental bulk property of any homogeneous material. As indicated in Figure 1.1, a solid slab having length L and cross-sectional area A will have a resistance (measured between the two ends) of

$$R = \rho (L/A) = \rho (L/tW) \tag{1.1}$$

where the end area, A, is the product of width, W, and thickness, t. Note that ρ has dimensions of ohm-cm. (In this and all future cases, we shall use the units most commonly employed in the field — in this instance, cgs units.) In semiconductor applications, one often deals with thin layers of conductive material. These may be thin-film metals or active semiconductor layers formed on semi-insulating substrates. It is then useful to define a quantity called the *sheet resistance* as follows:

$$R_s = \rho/t$$

and therefore equation (1.1) becomes

$$R = R_s (L/W)$$

In the strictest sense, R_s has the dimension of ohms. But it is universally given the units "ohms per square," sometimes written $\Omega/\square$. The reason is indicated in Figure 1.2. A "square" of material ($L=W$) has an end-to-end resistance of R_s; three "squares" of material ($L=3W$) have an end-to-end resistance of $3R_s$; and so on. Sheet resistance can be measured (Chapters 11 and 18) and usefully employed in calculations, all without any knowledge of the exact thickness of the layer (which may be hard to measure

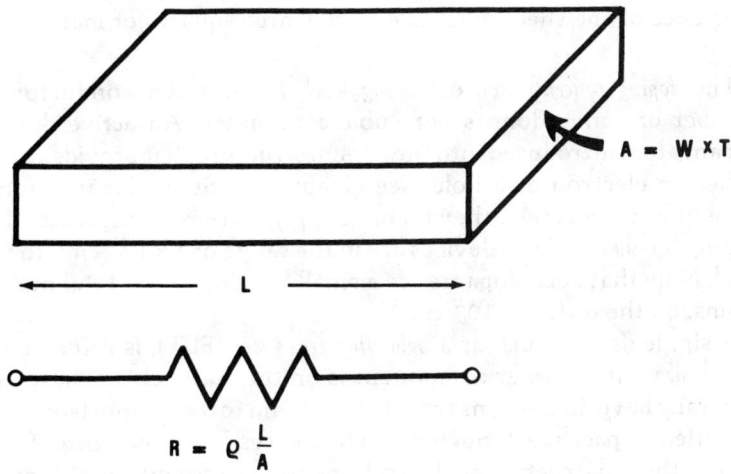

Figure 1.1 Resistance of a homogeneous solid as a function of its resistivity and shape.

accurately). More importantly, sheet resistance is a valid concept even when the resistivity of the conductive layer is not constant (but is some unknown function of depth), and even when there is no sharp boundary between conductive and insulating material.

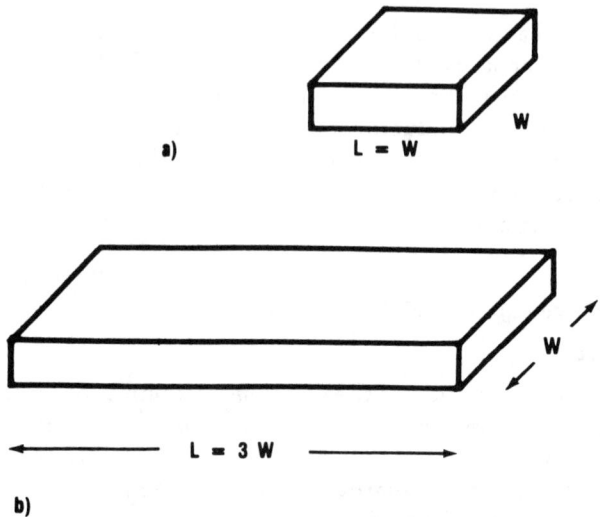

Figure 1.2 "Squares" of material, as used in the context of sheet resistance. If the sheet resistance is R_s ohms/square, the horizonal end-to-end resistance of the material in (a) is $R = R_s$ (one square of material). The resistance of the sheet in (b) is $R = 3R_s$ (three squares of material).

The *doping concentration* or *doping level*, N, of a semiconductor is the number of active donors per cubic centimeter. An active donor is a dopant atom introduced into the crystal structure that provides a carrier, either an electron or a hole (see Chapter 2). Some dopant atoms are present but not active: they do not supply carriers (Chapter 2). Typical doping levels for GaAs devices are in the range of 1×10^{16} cm^{-3} to 1×10^{18} cm^{-3}. Note that such dopants are a small fraction of the total number of atoms, on the order of 10^{22} cm^{-3}.

A single device, such as a *field effect transistor* (FET), is referred to as a *discrete device* if no other component is on the GaAs chip. These devices generally have dimensions from 0.3 x 0.3 mm to 0.6 x 1 mm (some may be mounted in packages). Such discrete devices have been used for some time in the construction of "hybrid" microwave circuits. In this process a substrate material such as silica (quartz) or alumina is used. Transmission lines to carry microwave signals are fabricated (usually photolithographically) on the substrate. Then other components, such as discrete GaAs FETs and very small "chip" capacitors, are mounted onto the substrate. These components are interconnected using gold wire (usually 0.001-inch in diameter). These wires have appreciable inductance at microwave frequencies, are an active part of the circuit, and must be considered in the circuit design. Such a hybrid circuit is designated a *microwave integrated cirucit* or MIC. It is an integrated circuit in the sense that all components exist on the same small substrate. Each hybrid circuit may be individually tuned by adjusting bond wires, placing small metal or dielectric "tuning chips" near transmission lines, or by other methods. Such individual tuning is usually necessary because of the difficulty of making all MICs identical. For example, minor differences in the length or shape of the wire bonds can cause such variations. Fabrication and tuning of MICs is labor intensive and very expensive.

It has become possible and desirable to fabricate these microwave circuits in a truly monolithic manner: all components (FETs, transmission lines, inductors, resistors, capacitors) are fabricated on the same GaAs substrate. No bonding is required except to connect the chip to the outside world. This monolithic approach promises high uniformity in chip-to-chip performance. Such uniformity and elimination of the assembly required by MICs promise low cost. It would have been tempting to call such a device a monolithic integrated circuit and to use the acronym MIC, but that designation was already in use as described above. Such a chip has come to be called a *monolithic microwave integrated circuit* or MMIC. Hence, MICs are hybrid circuits; MMICs are monolithic circuits.

The terms *first-level* and *second-level metal* distinguish metalization placed below or above a dielectric spacer. First-level metal is metal placed directly on the slice and/or on other metal directly on the slice. In this sense, ohmic contact metal and any overlay metal (to improve conductivity) would jointly constitute the first-level metalization. However, common usage often does not include the alloyed, ohmic metal but refers only to the overlay metal placed on the slice before applying the dielectric. A dielectric (insulating) layer is generally employed over the first-level metal. Holes are etched in this dielectric to allow electrical contact to the first-level metal at desired locations. Second-level metal may then be formed on top of the dielectric in such a manner that it contacts the first-level metal at the openings. This is the typical method to allow electrical *crossovers* in digital logic circuits and in some discrete GaAs devices. In other cases, such crossovers are accomplished by an *air bridge* technique (Chapter 15).

6   GaAs Processing Techniques

The term *wafer* and *slice* will be used interchangeably. Similarly, the terms *chip*, *bar*, and *die* will all be used to designate the unit that results from sawing or scribing the slice into separate pieces.

Finally, a few words of philosophy: semiconductor processing, unlike cooking or building a model airplane, does not consist of blindly following well-specified instructions. In those cases, clear and straightforward instructions (almost) always lead to successful results. Semiconductor process results are inordinately sensitive to many complex parameters and process engineers must fine tune each process. Even so, day-to-day experience in semiconductor production is a constant struggle to maintain process integrity and high yield. The situation is not quite as demanding in a research environment where only a few good devices are needed from a slice. But the point is this: the inexperienced process engineer should not be discouraged by initial difficulty or failure when attempting to establish a new process. About two-thirds of process engineering is finding out what is going wrong. A firm understanding of the underlying principles and knowledge of which processes are proven techniques aid in such difficulties.

1.3 EXAMPLE PROCESS FLOWS

A *process flow* is a list of process steps used to fabricate a given device. It is useful to present several example process flows to provide a context for the specific descriptions addressed in the remainder of the book. There are always many ways to fabricate a given device. The multiplicity of choices is considered in Chapter 17, Process Integration. The following process flows are not intended to represent any one manufacturer's or laboratory's exact method. Indeed, an attempt has been made to assure they don't. They are intended to be "generic" process flows and serve only as examples. They omit many of the detailed steps that a complete process flow would contain. They also omit in-process inspections and electrical characterizations. Three examples are given: an unplated analog FET such as might be used for low noise applications, a digital FET (assumed part of a larger logic array), and an MMIC which includes FETs, capacitors, inductors or transmission lines, via holes for grounding, and a plated heat sink. It should be emphasized that the figures accompanying these process flows are significantly out of scale. This is necessary for clarity, but sometimes may leave wrong impressions. Figure 1.4 shows a correctly scaled cross section of the analog FET device presented in Figure 1.3. On this scale, the completed slice would be approximately five feet thick and the ohmic contacts might extend a foot further to the left and right.

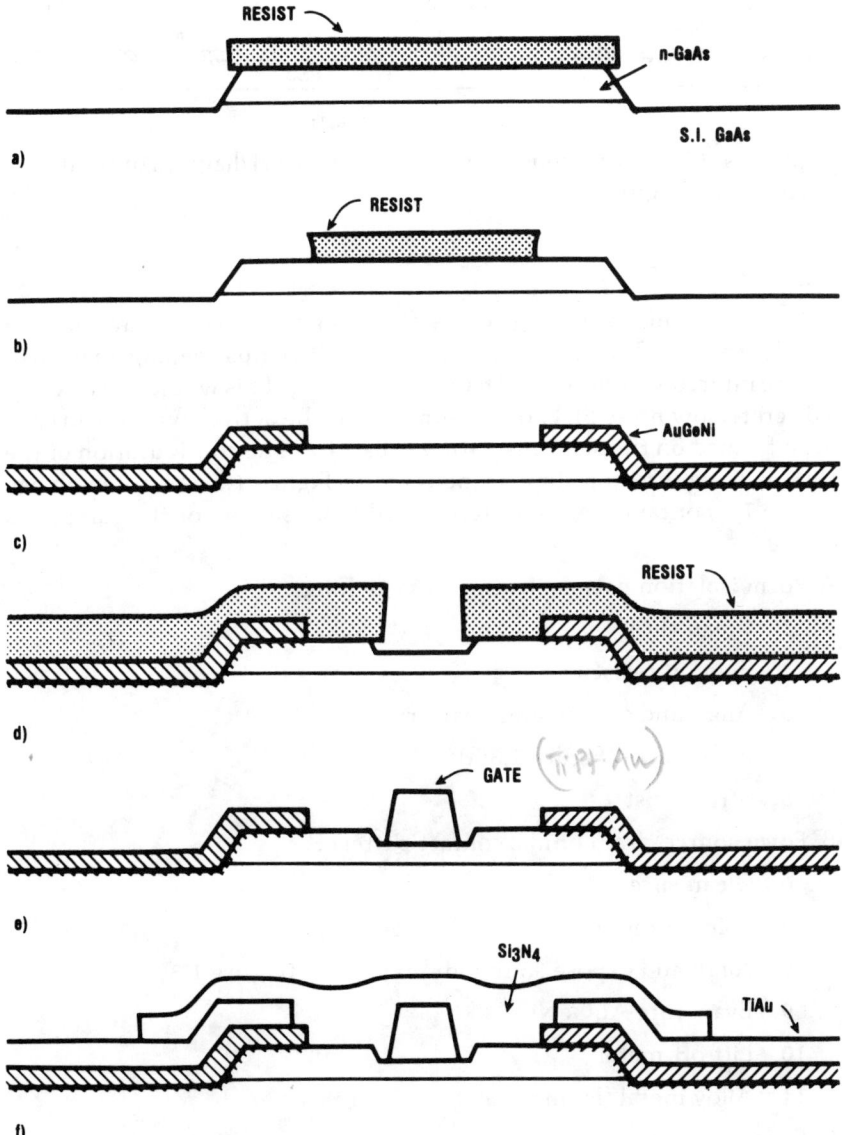

Figure 1.3 Stages in the fabrication of a typical low noise, discrete GaAs field effect transistor. See text for details.

8   GaAs Processing Techniques

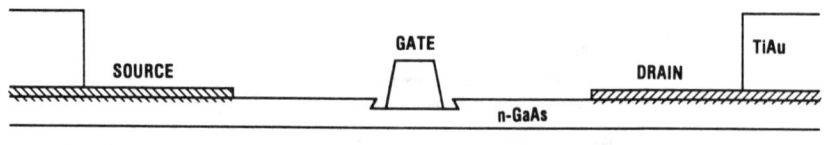

Figure 1.4 Correctly proportioned cross-sectional diagram of the device illustrated in Figure 1.3.

1.3.1 Analog FET

The following example process flow might be used to fabricate a simple analog FET having small total gate width (thus needing no form of source interconnections) and having no plating. This would be typical of a discrete, low noise FET. It is assumed that the active layer has already been formed on the semi-insulating substrate. The configuration of the slice at several major steps is indicated in Figure 1.3. As noted above, Figure 1.4 contains an accurately scaled cross section of the completed device.

A. Form isolation pattern (using mesa etching):
  1. Clean slice
  2. Spin on resist
  3. Align and expose mesa pattern
  4. Etch mesas (Figure 1.3(a))
  5. Strip resist

B. Form source-drain ohmic contact pattern:
  6. Clean slice
  7. Spin on resist
  8. Align and expose source-drain pattern (Figure 1.3(b))
  9. Evaporate AuGeNi
  10. Lift off metal
  11. Alloy metal (Figure 1.3(c))

C. Fabricate gate:
  12. Clean slice
  13. Spin on resist
  14. Expose gate pattern (e-beam)
  15. Recess gates (etch) (Figure 1.3(d))

16. Evaporate TiPtAu
   17. Lift off metal (Figure 1.3(e))
D. Form first metalization:
   18. Clean slice
   19. Spin on resist
   20. Align and expose first metalization pattern
   21. Evaporate TiAu
   22. Lift off metal
E. Dielectric protection:
   23. Clean slice
   24. Grow silicon nitride (protection)
   25. Clean slice
   26. Spin on resist
   27. Align and expose bonding pad pattern
   28. Plasma etch exposed silicon nitride
   29. Strip resist (Figure 1.3(f))
F. Back side processing:
   30. Thin slice to 125 μm
   31. Metalize back of slice (TiAu)
   32. Scribe and separate

1.3.2 Typical Process Flow for a Digital FET

The following example process flow might be used to fabricate an enhancement mode FET digital logic circuit. The FET is assumed part of a larger circuit that is being fabricated at the same time. The device configuration at various steps in the process is indicated in Figure 1.5. It is assumed that the slice does not already have an active layer formed on it. An active layer and an N^+, self-aligned contact layer will be formed during the process.

A. Ion implant active layer:
   1. Clean slice
   2. Spin on resist
   3. Align and expose active layer pattern
   4. Implant Si into GaAs to form active layer (Figure 1.5(a))
   5. Strip resist pattern

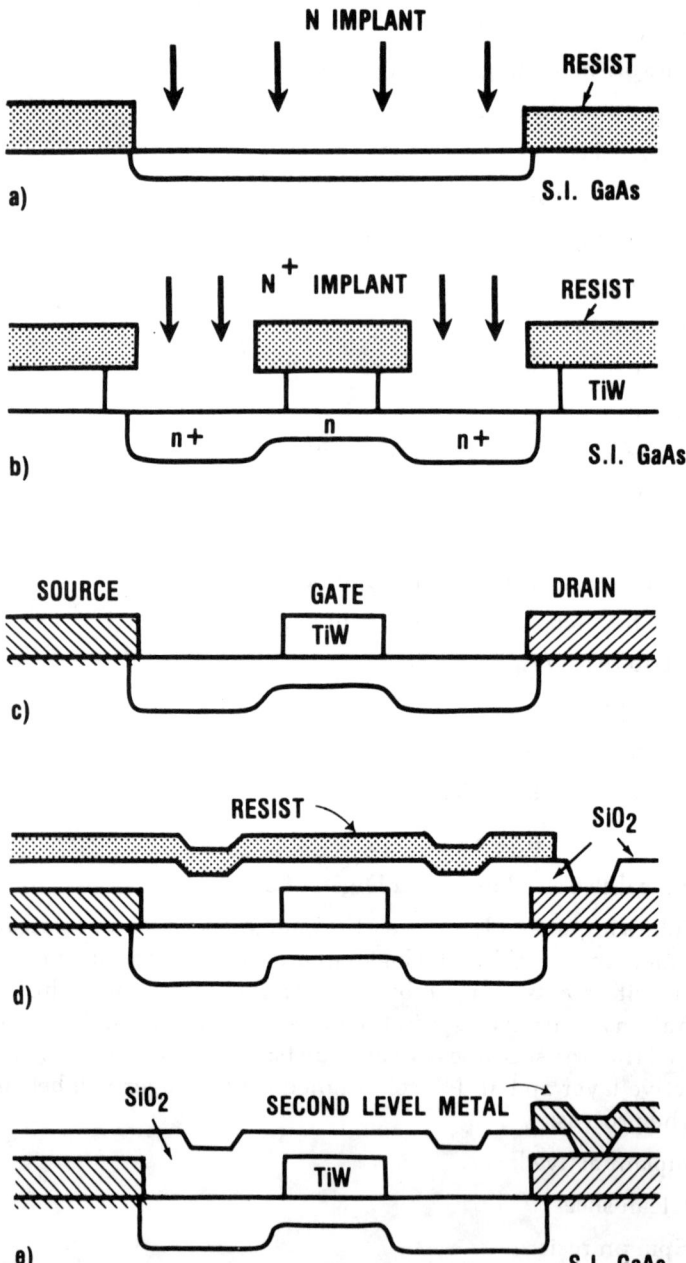

Figure 1.5 Stages in the fabrication of a typical GaAs field effect transistor for use in a digital logic circuit. See text for details.

B. Form TiW gate:
   6. Clean slice
   7. Sputter TiW
   8. Clean slice
   9. Spin on resist
   10. Align and expose gate pattern
   11. Etch TiW (undercutting resist)
C. Ion Implant N^+ contact layer (self-aligned to gate pattern):
   12. Ion implant N^+ dose (Figure 1.5(b))
   13. Strip resist
D. Remove extraneous TiW and anneal implant:
   14. Clean slice
   15. Spin on resist
   16. Align and expose pattern (to protect TiW gates)
   17. Etch away all exposed TiW
   18. Strip resist and clean
   19. Anneal to activate implanted dose
E. Fabricate source-drain ohmic contacts:
   20. Clean slice
   21. Spin on resist
   22. Evaporate AuGeNiAu
   23. Lift off metal
   24. Alloy metal (Figure 1.5(c))
F. Form crossover dielectric and pattern for vias:
   25. Clean slice
   26. Grow silicon dioxide
   27. Clean slice
   28. Spin on resist
   29. Align and expose via pattern
   30. Etch SiO_2
   31. Strip resist

12 GaAs Processing Techniques

G. Second metalization:
   32. Clean slice
   33. Spin on resist
   34. Align and expose second metalization pattern (Figure 1.5(d))
   35. Evaporate TiAu
   36. Lift off metal (Figure 1.5(e))

H. Back side processing:
   37. Lap to 200 microns
   38. Metalize back
   39. Scribe

1.3.3 Typical Process Flow for MMIC

The following example process flow might be used to fabricate a monolithic microwave integrated circuit (MMIC). The circuit includes resistors, capacitors, inductors, air-bridge interconnects, and via holes through the slice. In this particular example, the resistors are formed using conductive GaAs material. There are other options (Chapter 14). The configuration of various elements at several major steps is shown in Figure 1.6. It is assumed that the slice already has an active layer formed on a semi-insulating substrate.

A. Form isolation pattern (using unannealed ion implant):
   1. Clean slice
   2. Spin on resist
   3. Align and expose isolation pattern
   4. Etch slightly (for subsequent alignment)
   5. Ion implant boron (Figure 1.6(a))
   6. Strip resist

B. Fabricate source-drain ohmic contacts:
   7. Clean slice
   8. Spin on resist
   9. Align and expose source-drain pattern
   10. Evaporate AuGeNiAu
   11. Lift off metal
   12. Alloy metalization (Figure 1.6(b))

C. Fabricate gate:
   13. Clean slice
   14. Spin on resist
   15. E-beam expose gate pattern
   16. Recess gates (etch)
   17. Evaporate TiPtAu
   18. Lift off metal (Figure 1.6(c))
D. Form first metalization (includes inductors):
   19. Clean slice
   20. Spin on resist
   21. Align and expose first metalization pattern (includes inductors)
   22. Evaporate TiAu
   23. Lift off metal (Figure 1.6(d))
E. Capacitor formation:
   24. Clean slice
   25. Grow dielectric layer
   26. Clean slice
   27. Spin on resist
   28. Align and expose capacitor top plate pattern
   29. Evaporate TiAu
   30. Lift off metal (Figure 1.6(e))
F. Plating sequence (forms air bridges):
   31. Clean slice
   32. Spin on resist
   33. Align and expose plating pattern
   34. Etch open dielectric
   35. **Sputter 100Å TiAu (Figure 1.6(f))**
   36. Spin on resist
   37. Align and expose air bridge pattern (Figure 1.6(g))
   38. Gold plate
   39. Lift off (Figure 1.6(h))

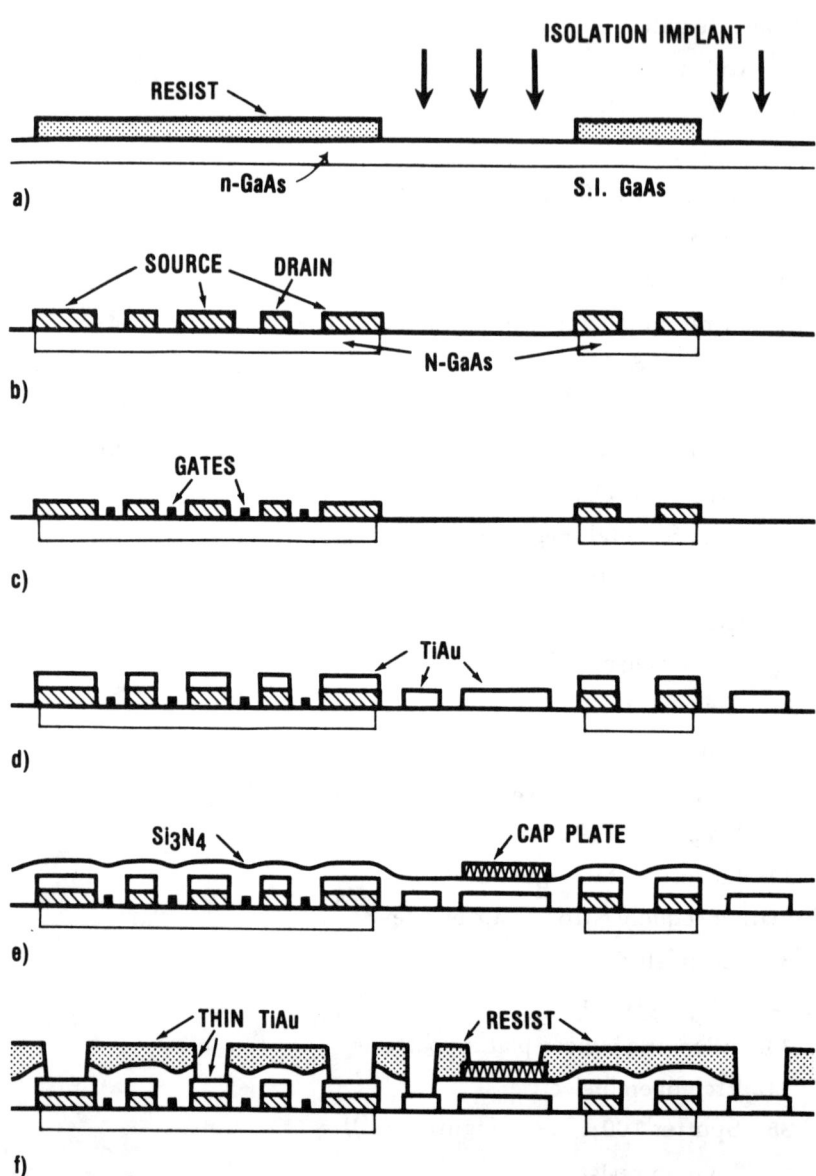

Figure 1.6 Stages in the fabrication of a typical GaAs MMIC incorporating field effect transistors, air bridges, capacitors, resistors, and via holes. See text for details.

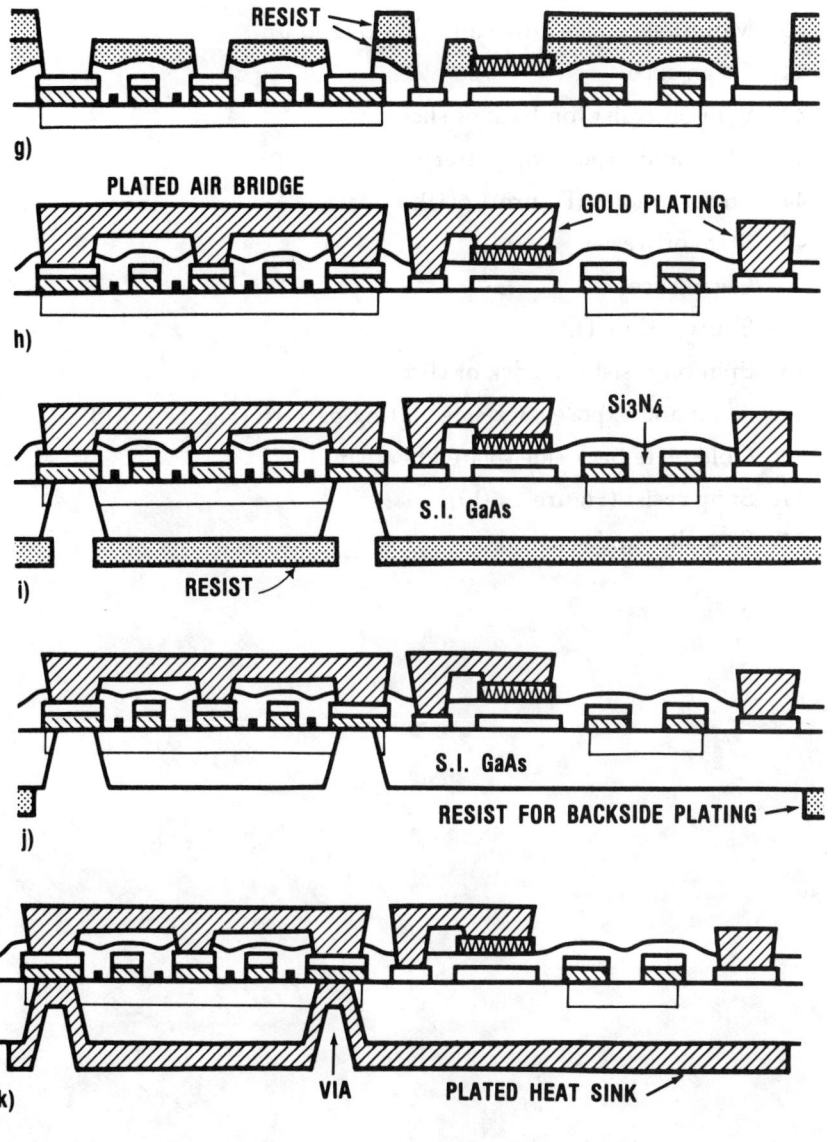

G. Back side processing (forms vias):
   40. Mount slice face down and thin to 100 μm
   41. Clean slice
   42. Spin on resist (on back of slice)
   43. Align and expose via pattern
   44. Etch via holes (Figure 1.6(i))
   45. Strip off resist
   46. Clean slice
   47. Sputter thin TiAu
   48. Spin on resist (on back of slice)
   49. Align and expose back side plating pattern (Figure 1.6(j))
   50. Gold plate back side (plated heat sink)
   51. Strip resist (Figure 1.6(k))
   52. Saw slice and separate

CHAPTER 2

GaAs MATERIAL
AND
CRYSTAL PROPERTIES:
A TUTORIAL

The motivation for using gallium arsenide (GaAs) instead of silicon was summarized in Chapter 1. These advantageous properties will be considered in more detail in this Chapter, which serves as a tutorial review of GaAs material properties. Tutorial reviews have the advantage of presenting major elements of a topic without the many detailed complexities. In doing so, there is danger of creating an impression of simplicity. It should be emphasized that GaAs materials technology began as almost a "black art" and has reached its present status through an enormous amount of scientific and technological development. This is reflected in the inclusion of materials people as authors on many of the device papers that have been published over the years. Their work has not been trivial.

Gallium arsenide is a III-V compound semiconductor in that it is composed of an element (Ga) from Column III of the periodic chart and an element (As) from Column V of the periodic chart. GaAs was first created by Goldschmidt [1] in the 1920s. The first published article on the electronic properties of III-V compounds (as semiconductors) appeared in 1952 [2]. Since that time, the importance of GaAs as a III-V semiconductor to both solid-state physics in general and device physics in particular has resulted in literally thousands of articles on this material. A recent review article on the intrinsic, major properties of GaAs [3] that intentionally omits any discussion of impurity phenomena or recombination phenomena lists over four hundred references. Other reviews, some addressing the entire III-V family, are listed in references 4-11. Further information may be found in the set of volumes from the Institute of Physics containing the proceedings of the biennial Conference on Gallium Arsenide and Related Compounds [12]. The series of yearly volumes entitled *Semiconductors and Semimetals* from Academic Press [13] also contains much information about GaAs.

Section 2.1 describes the advantages of GaAs. Section 2.2 reviews the physical and electrical properties of GaAs, including the troublesome determination of the dielectric constant. Section 2.3 reviews GaAs bulk crystal growth and individual slice preparation. Section 2.4 reviews the epitaxial methods used to grow active layers on GaAs substrates including liquid phase epitaxy (LPE), vapor phase epitaxy (VPE), and molecular beam epitaxy (MBE). Section 2.5 describes the use of annealed ion implants to form active layers.

2.1 GENERAL CONCEPTS

Gallium arsenide has two principal advantages over silicon for microwave device use. First, its higher mobility and saturated drift velocity (discussed below) mean that electrons can move faster in GaAs than in silicon, and therefore, devices can operate at higher frequencies. Second, GaAs can be readily produced in a semi-insulating substrate form. This semi-insulating substrate greatly reduces parasitic capacitances and allows the use of true monolithic integrated circuits operating at high speed (above 1 GHz). Other advantages of GaAs will be considered below.

The speed advantage is illustrated in Figure 2.1. When a carrier (electrons will be considered the carriers for now) in a semiconductor is subjected to an electric field, it rapidly (about 10^{-12} seconds) achieves a velocity that is a function of the electric field strength. At low field strengths this relationship is linear and the constant of proportionality is called the mobility:

$$v = \mu E$$

where v is the electron velocity (cm/sec), E is the electric field strength (volts/cm) and μ is the mobility and has units of cm^2/V-sec. (Note that capital V is used for voltage; lower case v is used for velocity.) This equilibrium velocity is the result of two opposing forces: the electric field which tends to increase velocity, and electron scattering with the crystal lattice which tends to limit the velocity. The scattering time is on the order of 10^{-13} seconds. However, as electric field strength continues to increase, the linear relationship no longer holds (Figure 2.1). The reasons have to do with the energy band structure of the semiconductor, and will be considered in the next section. The maximum velocity is called the *saturated drift velocity*. Drift velocities are those caused by electric fields as opposed to electron movement driven by concentration gradients.

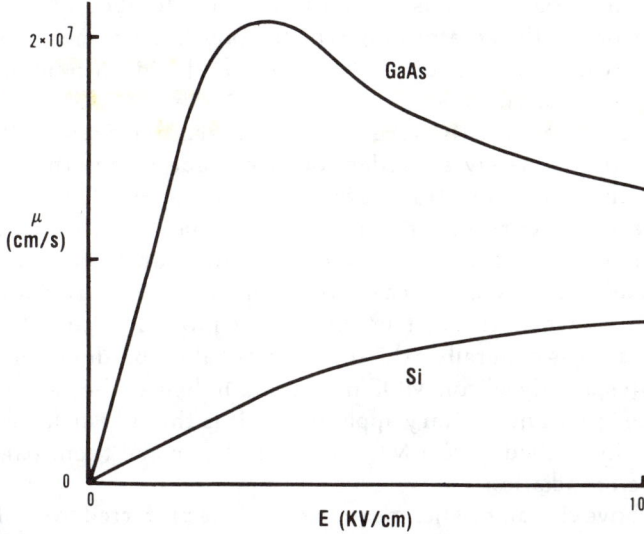

Figure 2.1 Drift velocity of electrons in GaAs and Si as a function of the electric field.

The electron mobility in the linear region is a function of temperature and impurity concentration (see section 2.2), but is approximately six times greater for GaAs than for silicon. The maximum GaAs drift velocity is at least twice that of silicon for field strengths less than 2×10^4 V/cm. At typical field strengths the advantage of GaAs may be much greater than a factor of two (Figure 2.1).

These advantages simply mean that appropriate GaAs devices and circuits can work at higher frequencies than silicon devices. The exact increase in speed depends on many factors. For example, comparable devices do not always exist in both mediums: MOSFETs (*metal oxide semiconductor field effect transistors*) cannot be constructed on GaAs because of the lack of a suitable native oxide. Further, total speed for digital logic circuits also depends on the circuit capacitances that must be driven and the electric field regime in which the device operates (whether in the linear or the saturated region). The consensus of recent analysis [14-17] is that GaAs digital circuits should run two to five times faster than Si circuits and/or have lower power dissipation. For analog applications, silicon transistors are hard pressed to operate above 1 GHz. GaAs FETs work easily into the 20 GHz range and laboratory devices have been operated near 60 GHz [18].

The second major advantage of GaAs over silicon that was mentioned above was the availability of semi-insulating substrates. The bulk resistivity of materials extends from 10^{-6} ohm-cm to about 10^{22} ohm-cm, certainly one of the greater ranges of any physical parameter. Semiconductors are in the range 10^{-2} to 10^{9} ohm-cm [19]. Silicon generally can be made with resistivities of approximately 100 ohm-cm. GaAs, however, can be made with resistivities above 10^{8} ohm-cm. This resistivity difference of approximately six orders of magnitude means that, speaking colloquially, silicon substrates are not insulating, but GaAs substrates are. The active (current carrying) GaAs for many devices consists of a thin layer on the surface of the semi-insulating substrate. Such a situation makes device isolation easy. More importantly, it provides an ideal substrate on which to construct monolithic integrated circuits.

GaAs devices generally exhibit higher radiation hardness than silicon devices (especially silicon MOS devices) [20]; hence, they are attractive for some space and military applications. It is the thin dielectric layers, such as SiO_2, used in the MOS devices that make them particularly sensitive to radiation.

The above characteristics mean that GaAs is preferred over silicon for high speed operation. Nevertheless, GaAs does have some problems. First, it is much more susceptible to breakage than is silicon. Dropping a GaAs wafer onto a hard surface from even a few inches can result in breakage, or even total shattering of the slice into small pieces. Thus, handling GaAs is more difficult than silicon. Second, GaAs does not have a stable native oxide that can "passivate" the slice (see section 2.2) in the manner that SiO_2 does for silicon. This means that *metal-oxide-semiconductor* (MOS) devices that are a mainstay of silicon device technology cannot be used on GaAs. The dominant GaAs device is the *metal-semiconductor field effect transistor* (MESFET) (see Chapter 3).

One final point should be made about the importance of GaAs. Unlike silicon, GaAs is a direct bandgap semiconductor (see section 2.2) and so is useful for fabricating optical components such as light emitting diodes (LEDs) and lasers [8]. Although a discussion of such devices is beyond the scope of this book, the ability to fabricate those devices on the same substrate as other electronic circuitry opens up the potentially vast field of integrated opto-electronics. Such work is being actively pursued in the laboratory.

2.2 PHYSICAL AND ELECTRICAL CHARACTERISTICS

The following subsections consider geometric, lattice, and thermal properties; the dielectric constant; the energy band structure of GaAs; and electron and hole transport.

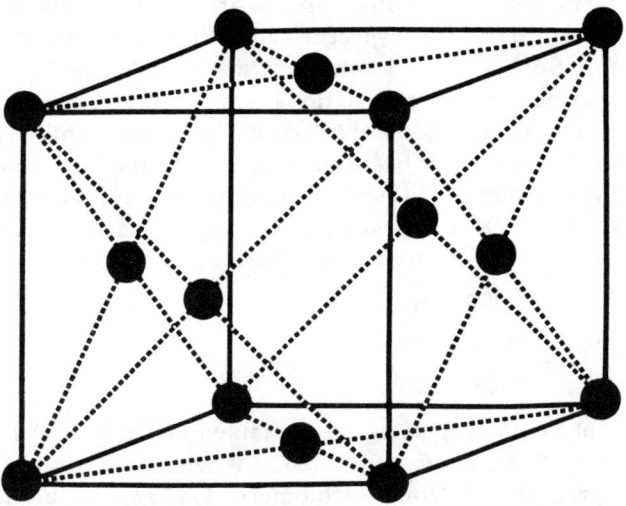

Figure 2.2 Face-centered cubic (fcc) crystal structure: atoms are at the corners and in the middle of all faces of a cube. The gallium and arsenic atoms each form a fcc sublattice within the GaAs lattice.

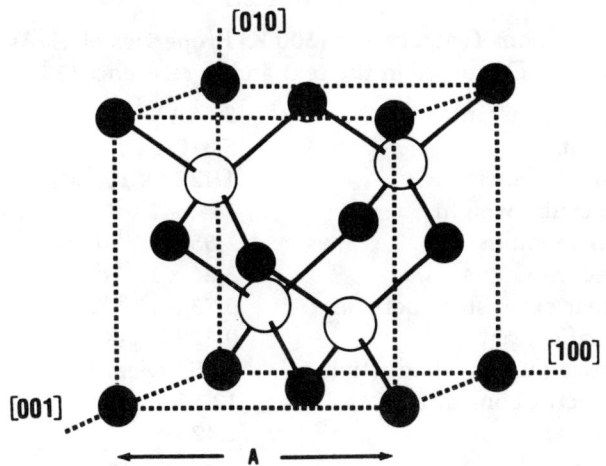

Figure 2.3 Unit cube of GaAs crystal lattice (after reference [3]).

2.2.1 Geometric and Lattice Properties

The GaAs crystal is composed of two sublattices, each *face-centered cubic* (fcc). The fcc structure is shown in Figure 2.2 and consists of atoms at the corners of a cube and also at the center of each face of the cube. The Ga atoms are arranged in such a pattern, as are the As atoms. These two sublattices are offset with respect to each other by half the diagonal of the fcc cube. Such a crystal configuration is called *cubic sphalerite* or *zincblende*, and has the fcc translational symmetry. The unit cube for GaAs, showing the arrangement of Ga and As atoms, is shown in Figure 2.3. Each Ga (or As) atom has four neighbor As (or Ga) atoms. If the length of the unit cube is A (Figure 2.3), the bond length between these neighbors is $r_o = 3A/4$ and these bonds are separated by the tetrahedral bond angle of 109.47°. Undoped GaAs has a unit cell size at 300 K of [21]

$$A = 5.65325 \pm 0.00002 \text{ Å}$$

and a density at 300 K of [22]

$$d = 5.3174 \pm 0.0026 \text{ g/cm}^3$$

(A number of GaAs properties are contained in Table 2.1.) The GaAs lattice constant A can increase up to 0.02% when a large concentration of a dopant is present [21]. Non-stoichiometric GaAs can result in slightly smaller values of A for As-rich GaAs and slightly greater values for Ga-rich GaAs [22], but these variations in A are even less than the 0.02% mentioned above.

TABLE 2.1

Room Temperature (300 K) Properties of GaAs
(Discussed in the text and in reference [3])

Lattice constant	5.65 Å
Density	5.317 g/cm^3
Atomic density	4.4279 × 10^{22} atoms/cm^3
Molecular weight	144.642 (69.720 + 74.922)
Bulk modulus	7.55 × 10^{11} dyn/cm^2
Sheer modulus	3.26 × 10^{11} dyn/cm^2
Linear expansion coefficient	5.73 × 10^{-6} K^{-1}
Specific heat	0.327 J/g-K
Lattice thermal conductivity	0.55 W/cm-K
Dielectric constant	12.85
Bandgap	1.423 eV
Threshold field	3.3 KV/cm
Peak drift velocity	2.1 × 10^7 cm/s
Electron mobility (undoped)	8500 cm^2/V-s
Hole mobility (undoped)	400 cm^2/V-s
Melting point	1238 °C

Figures 2.4 and 2.5 show truncations of the unit cube by specific planes within the crystal. It is useful to be able to speak of specific directions and planes within the crystal structure. *Miller indices* are used for this purpose and consist of triplets of numbers corresponding to the three spatial directions. For example, the direction [110] may be thought of as a vector having components of $x = 1$, $y = 1$, and $z = 0$, where x, y, and z are oriented with respect to the unit cube (Figure 2.3). A negative direction is indicated by placing a bar over the index. Crystallographic planes are designated by the direction normal to the plane's surface. Various types of brackets are used to enclose the indices; these have the following designations:

[ ] indicates a direction: [111], [112], [100], etc.

( ) indicates a plane: (110), (101), etc.

< > indicates a family of directions; e.g., <110> represents [110], [011], etc.

{ } indicates a family of planes

Sometimes the above conventions are not adhered to rigorously and the reader should be aware of such instances.

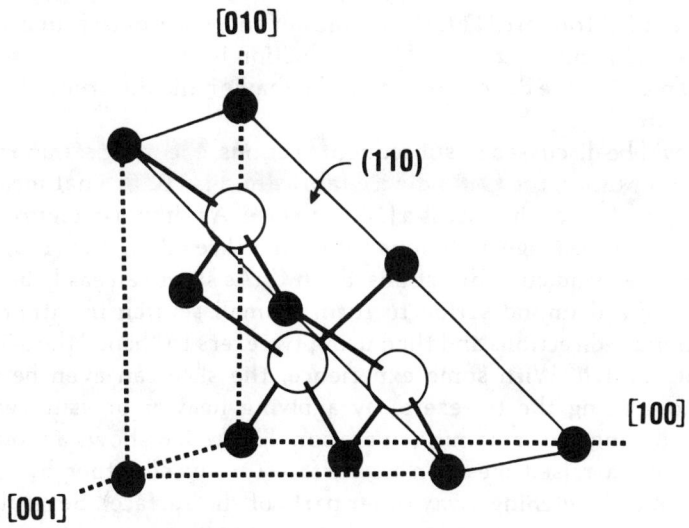

Figure 2.4 Truncation of the GaAs unit cube by the (110) plane (after reference [3]).

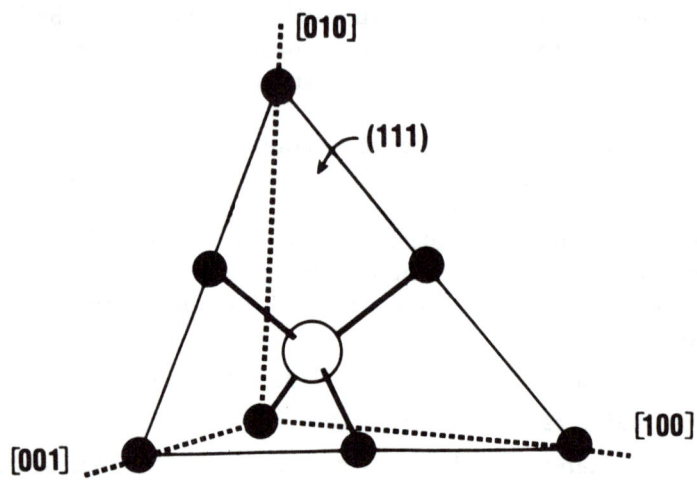

Figure 2.5   Truncation of the GaAs unit cube by the (111) plane (after reference [3]).

Note that the {111} family of planes contain only one type of atom, either Ga or As. The letter A or B is attached to the plane to designate the Ga or As plane respectively. Thus, of the eight planes contained in the {111} family, four are {111A} and contain only Ga atoms; four are {111B} and contain only As atoms. This distinction is important because the A and B planes have different chemical behavior and different growth (or etch) rates.

As will be discussed in subsequent sections, the most commonly used slice orientation for GaAs device fabrication is (100). That means that the top surface of the slice is a {100} surface. As shown in Figure 2.6, the two natural cleavage directions for such a slice (along the (110) planes) are two perpendicular directions. Such GaAs slices can easily be cleaved by using a diamond scribe to form a small scratch in either of the appropriate directions and then using tweezers to "bend" the slice apart at the scratch. With some experience, the slice can even be cleaved (without using the tweezers) by applying heavier pressure with the scribe during the scratching operation. Figure 2.6 shows a (100) GaAs slice with a raised mesa on its surface (produced either by localized growth, or by etching away other parts of the surface). Several crystal planes (facets) are exposed. Note that the {111B} surfaces (As planes) are smaller than the {111A} surfaces (Ga planes), indicating faster growth (or slower etching) of the {111A} (Ga) planes. Also shown in Figure 2.6 are profiles of an etched trough (using a common acid etch). Note that the profile is different for troughs oriented 90° apart. This lack of symmetry results from the interaction of etches with the GaAs crystal lattice. It is an important issue in device fabrication and is considered in detail in Chapter 5 (Wet Etching).

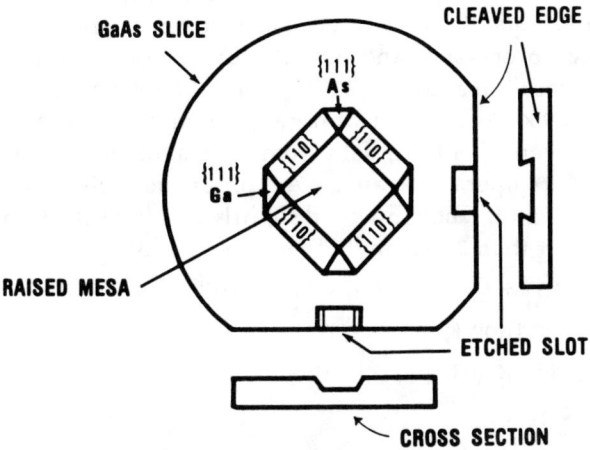

Figure 2.6 (100) oriented GaAs slice illustrating cleavage planes and anisotropy as evidenced by a raised mesa and etched slots.

The crystal structure of GaAs will appear more dense when viewed in some directions than when viewed in others. This is illustrated in Figure 2.7 and has relevance for epitaxy and ion implantation, as will be discussed in those sections.

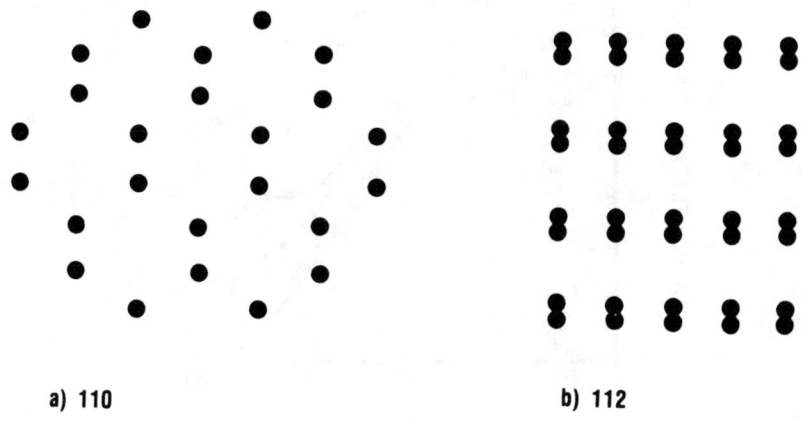

Figure 2.7 View of GaAs lattice from different directions: (a) along the [110] axis; (b) along the [112] axis.

2.2.2 Thermal Characteristics

The behavior of the coefficient of thermal expansion, α, (defined by $dL/dT = \alpha L$, where L is length) of GaAs is a rather complicated matter when considered over a wide temperature range [11], especially less than 200 K or above 1000 K. Fortunately, such extremes are not relevant to general device processing and use. A number of investigators have published values for α at 300 K [22-27], the values ranging between 4.84 $\times 10^{-6}\ T^{-1}$ and $6.9 \times 10^{-6}\ T^{-1}$. A quadratic approximation of the form $\alpha = A + BT - CT^2$ has been used to fit data from several of these investigations over moderate temperature ranges. A "consensus" polynomial was recommended [28] and, when modified slightly to adjust the balance point to 300 K (rather than 293 K in [28]), is given by [3]

$$\alpha_T = 4.24 \times 10^{-6} + 5.82 \times 10^{-9}\ T - 2.82 \times 10^{-12}\ T^2\ K^{-1}$$

for $200 < T < 1000$ K

which gives a 300 K value of

$$\alpha_{300} = 5.73 \times 10^{-6}\ T^{-1}$$

(which is remarkably close to the value of 5.7×10^{-6} quoted by the earliest of the above references, [24]).

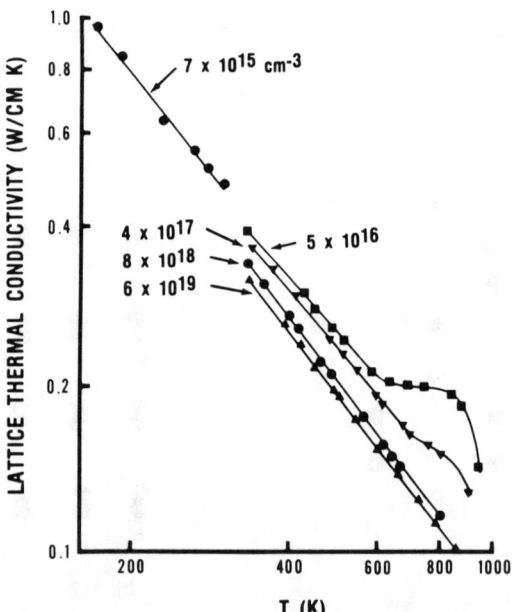

Figure 2.8 Lattice thermal conductivity (after reference [3]; data from [30,32]). Note lower conductivity with increased doping.

The coefficient of thermal conductivity, taken as equal to the lattice thermal conductivity, κ_L, also exhibits complicated behavior over extreme temperature ranges, but it is reasonably well-behaved over more moderate temperature ranges [3, 29-32]. Figure 2.8 [3] illustrates several investigators' measurements over the approximate range of 200 K to 1000 K. This data shows a general trend of decreasing thermal conductivity with increased doping. The 300 K value of κ_L for undoped GaAs is approximately

$\kappa_L = 0.55$ W/cm-K

Note that GaAs is a poor thermal conductor compared to silicon ($\kappa_L = 1.5$ W/cm-K).

2.2.3 Dielectric Constant

Knowledge of the low frequency dielectric constant, K_o, is necessary for calculating parasitic capacitances and, particularly, for designing transmission lines on GaAs integrated circuits. (Low frequency in this case means below optical frequencies and encompasses the frequency range of dc to 10^{11} Hz.) However, ascertaining the dielectric constant has not been as trivial as might be expected. As one reviewer states [3], "One might reasonably expect this to be straightforward to the point of dullness. For GaAs, this has not been the case at all." The point is emphasized by considering the dozen or so room temperature measurements of K_o reported in one review article [33]. Even eliminating two anomalously low values still leaves an unsatisfying range of 11.6 to 13.3 among the remaining measurements. That reviewer [33] recommended a choice of $K_o(300) = 12.9$, very close to the value of 12.91 chosen by another investigator [34]. More recent investigations have claimed $K_o(300) = 12.9 \pm 0.07$ for the range 4 to 18 GHz [35]. These and other measurements [36-38] were used by a recent reviewer [3] to recommend the expression

$K_o(T) = 12.4 (1 + 1.2 \times 10^{-4} T)$

which yields a 300 K value of 12.85. It therefore seems reasonable to take a 300 K value of

$K_o(300) = 12.9$

for three-digit accuracy. Note that this value is significantly different from the value of 12.5 that was published in 1961 [39] and used widely for some time thereafter.

2.2.4 Energy Band Structure

An elementary understanding of the energy band structure of GaAs is necessary to properly understand the transport properties (electron and hole current flow) in doped GaAs crystals. This section reviews the relevant information in a very simplified manner. A complete treatment of these matters may be found in reference [40]. The following discussion is intended for people lacking a background in solid-state physics; some over-simplifications are employed.

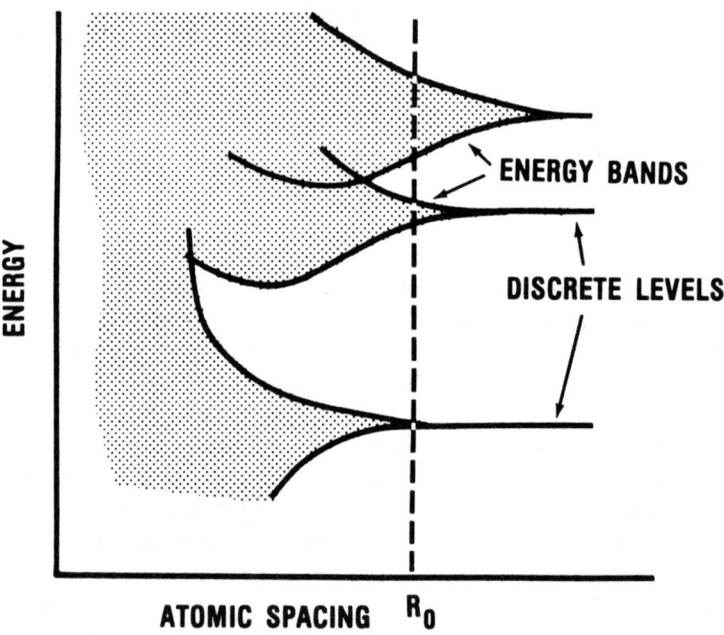

Figure 2.9 Energy band diagram showing creation of energy bands as discrete atoms assemble into a solid.

As is well-known, electrons in isolated atoms can have only certain discrete energy values. This is the result of the laws of quantum mechanics. As individual atoms are brought closer together and assembled into a solid, the same physical laws result in electrons being restricted not to single energy levels, but to broadbands of energies (Figure 2.9). The two bands of interest, called the *valance band* and the *conduction band*, are separated by an energy gap (Figure 2.10 (a) and (b)). At 0 K all electrons are in the valance band and all its electron states are filled. The laws of quantum mechanics do not allow two electrons to occupy the same *energy state*; hence, at 0 K no electron movement is possible and the material is a perfect insulator. Above 0 K some electrons have sufficient thermal energy to make a transition to the conduction band. Here they are free to move and can conduct current through the crystal. Back in the valance band the absence of the electron is called a *hole*, and can be treated as though it were a positive charge. These holes can move in the valance band and also conduct current. In silicon, both types of carriers (electrons and holes) must generally be considered in device physics. In GaAs, however, the hole mobility may be over an order of magnitude less than the electron mobility (see section 2.2.5) and only the electron current need be considered in many GaAs devices. It is in this sense that such GaAs devices are *majority carrier devices* or *unipolar devices*.

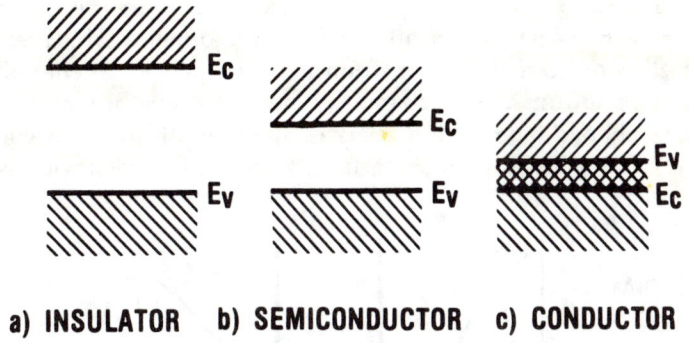

a) INSULATOR   b) SEMICONDUCTOR   c) CONDUCTOR

Figure 2.10 Bandgap of three types of solids: (a) insulator; (b) semiconductor; (c) conductor.

The energy gap between the valance and conduction bands is an extremely important parameter. If it is large, few electrons will have enough thermal energy to reach the conduction band and the solid will be an insulator. Diamond has an energy gap greater than 5 eV and is a very good insulator. If the bandgap is very small (or even nonexistent; i.e., overlapping bands) then many electrons can reach the conduction band and the solid will be a conductor. The intermediate case is the semiconductor. These cases are illustrated in Figure 2.10. It was noted in section 2.1 that pure silicon is not as resistive as pure GaAs. This follows from their respective, room temperature band gaps of 1.11 eV and 1.43 eV. However, actual material is neither absolutely pure nor perfectly crystalline; defects and impurities play a significant role in the resistivity of actual GaAs material (section 2.3).

A more complete band diagram for GaAs and for silicon is shown in Figure 2.11, which shows that the energy bands are functions of the magnitudes of the electron (crystal) momentum. These relationships also depend on the crystalographic direction of the momentum as indicated in the Figure. GaAs is a *direct bandgap* semiconductor because the minima of the conduction band is directly over the maxima of the valance band. Transitions between bands generally involve movement from an energy state near the maximum of the lower band to an energy state near the minima of the upper band. Other transitions require greater energy and are less likely. In direct bandgap semiconductors such a transition only requires a change in energy; no change in momentum is needed. Electrons can make this transition by emitting or absorbing a photon. This phenomena is essential to GaAs lasers and LEDs. Silicon, however, is an *indirect bandgap* semiconductor in that the minima and maxima of the two bands occur at different momenta. This means that for an electron to make a transition between bands, it generally must change momentum as well as energy; hence, it must interact with the lattice to change momentum. This is much less likely to occur; thus, indirect semiconductors are not as suitable for optical devices such as lasers.

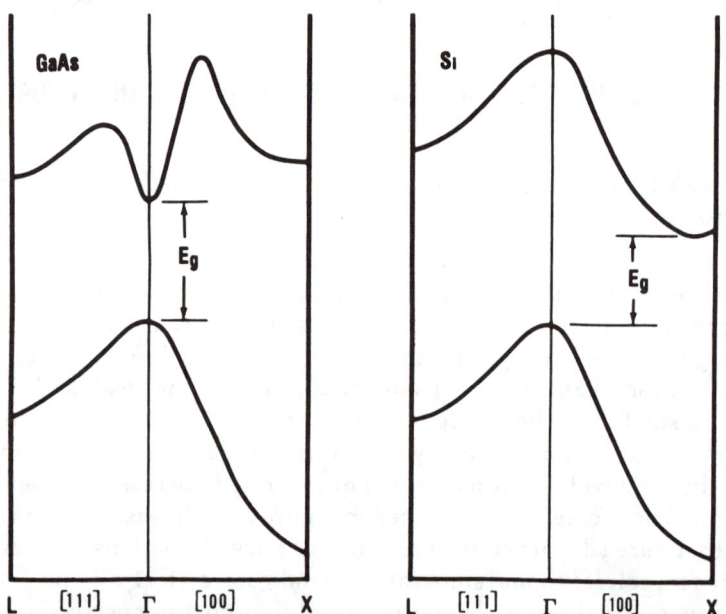

Figure 2.11 Energy band diagram of GaAs and silicon.

In all these cases the term *bandgap* refers to the energy difference between the minima and maxima of the conduction and valance band, even if they do not occur at the same momentum (Figure 2.11). The bandgap is a function of temperature. For temperatures below 1000 K, the GaAs bandgap is accurately given by the expression [41]

$$E(T) = 1.519 - 5.405 \times 10^{-4} \frac{T^2}{T + 204} \; eV$$

which gives a room temperature value of

$E(300) = 1.42$ eV

It should also be noted that there are minima other than the lowest one (Figure 2.11) in the GaAs conduction band. If electrons have sufficient energy and interact (scatter) with the lattice, they can make transitions to these other minima or *valleys*. This phenomena is important in GaAs because it explains the decrease in electron velocity as the electric field in the crystal increases above a certain critical value (Figure 2.1). Electrons in these upper valleys act as though their mass is greater than in the lower valleys [40]. Hence, their velocity is reduced after making such a transition. This phenomena results in an apparent *negative resistivity* (an increase in electric field causing lower electron velocity) and is responsible for the *Gunn effect* in GaAs (see Chapter 3).

The energy band structure described above assumed a pure semiconductor and a crystal without defects or surfaces. Atoms were surrounded only by other atoms in the crystal. Any departure from this homogeneity, such as the presence of crystal defects, surfaces, or dopant atoms, changes the energy band structure near the inhomogeneity. In general, other energy levels then appear within the bandgap of the semiconductor as indicated in Figure 2.12. These levels are called *shallow* if they are near a band edge and *deep* if they are far from a band edge (Figure 2.12).

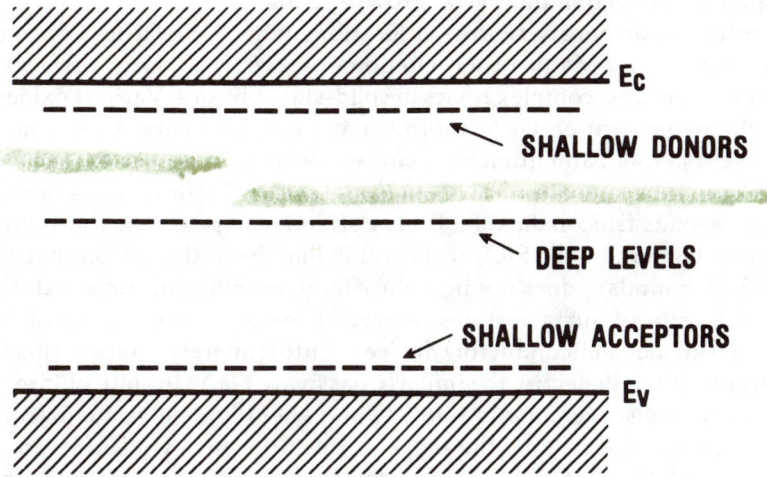

Figure 2.12 Energy states within the GaAs bandgap caused by impurity atoms or crystal defects.

Dopant atoms are intentionally introduced into the GaAs crystal structure to form conductive regions. These atoms are *n*-type dopants if they introduce extra electrons. In this case, the extra energy levels are near the conduction band and the electrons have enough thermal energy to make the small energy jump into the conduction band (Figure 2.12). Such atoms are often introduced either during expitaxial growth of the surface layers of the slice (section 2.4) or by ion implantation and anneal (section 2.5). Of course, atoms are also unintentionally included into the crystal structure. It is virtually impossible to keep them out. The nature of the problem is illustrated by noting that intentionally introduced dopant atoms usually compose only one part in 10^5 to 10^6 of the crystal atoms. Undesired impurities tend to dope the GaAs crystal, decreasing its resistivity. Consequently, chromium or oxygen is often introduced into the crystal to compensate for such doping and render the material semi-insulating (see section 2.3).

Crystal defects can introduce intrabandgap levels. These undesired levels are also called electron *traps* because they can trap electrons if they are far from the band edge. Such traps can increase the resistivity of the material. The general questions of the number and cause of electron traps in GaAs are still a subject of intense research and debate. The principal deep electron trap in GaAs is located 0.76 eV below the conduction band edge and is designated as *EL2*. It generally is believed to be related to the presence of an As atom on a Ga lattice site (the antisite defect As_{Ga}), perhaps complexed with a neighboring vacancy [42,43], although this is not yet certain. These traps will be discussed in greater detail in the section on crystal growth (2.3).

Surfaces obviously represent locations where the lattice of atoms is discontinuous. Many other phenomena occur; in fact, surfaces are among the most complex topics in solid-state physics. Various oxides are certainly present on GaAs (both Ga and As), elemental arsenic may be present [39,44], and numerous surface defects. All cause extra electron states in the bandgap and are called *surface states*. In the case of silicon, the native oxide (silicon dioxide) is very effective in *passivating* these surface states. Generally, the SiO_2 acts enough like the native silicon that the Si to SiO_2 boundary does not introduce large numbers of surface states. A low density of surface states is needed for proper operation of MOS (metal-oxide-semiconductor) devices. Unfortunately, there seems to be no equivalent dielectric to similarly passivate GaAs in spite of intensive efforts to discover one [45-48]. (The generally deleterious nature of surface states is beyond the scope of this brief review.) Unfortunately, use of dielectric coatings on GaAs devices for environmental and scratch protection is also often referred to as *passivation*. It certainly is not passivation in the sense used above. It may also be referred to as *glassivation*, especially if SiO_2 is used.

2.2.5 Electron and Hole Transport

As described above, current flow in many GaAs devices is mainly through electron flow, rather than by holes. The hole mobility is an order of magnitude less than the electron mobility. Each is a function of temperature and carrier (electron or hole) concentration. General reviews may be found in references [33,34,49].

A general clarification is in order before continuing the discussion of electron and hole mobility. There are at least two types of mobility: the *drift mobility*, μ_d, and the *Hall mobility*, μ_H. The drift mobility is the quantity used in section 2.1 and represents the response of the carrier to a driving electric field. The hall mobility represents the response of the carrier to a driving magnetic field perpendicular to the electric field. This is the well-known Hall effect in which carriers moving in a solid under an electric field (in the x direction) are subjected to a perpendicular magnetic field (in the y direction), thus developing a velocity perpendicular to both (in the z direction). There are well-known techniques for measuring each type of mobility (Chapter 18). In general, the two are not equal. The details of this distinction may be found in reference [40]. For the temperatures and doping ranges generally found in GaAs devices, the two mobilities are not greatly different.

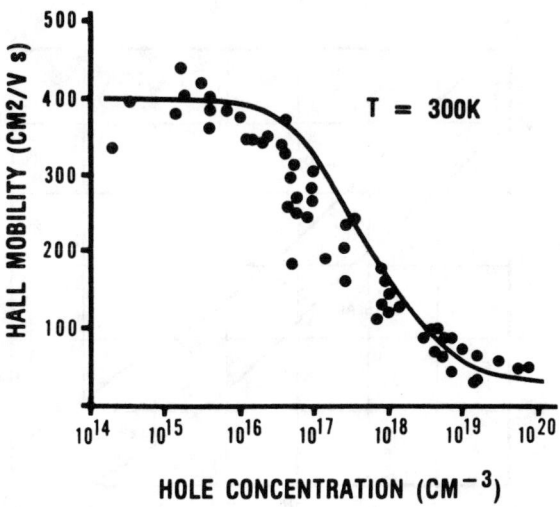

Figure 2.13 Hall mobility of holes in GaAs at 300 K as a function of hole concentration (after reference [49]; data from [50-56]).

Figure 2.13, taken from reference [49], shows data [50-56] of Hall mobility for holes at 300 K as a function of the carrier (hole) concentration. Note that these values are generally below 400 cm^2/V-s, and substantially less than that for doping concentrations above 1×10^{17}.

Figure 2.14 shows similar data for electron mobility [57]. Note that for reasonable FET (field effect transistor) doping ranges (1×10^{17}) the mobility is about 4000 cm^2/V-s. The resistivity of GaAs that follows from these mobilities is shown in Figure 2.15 [40].

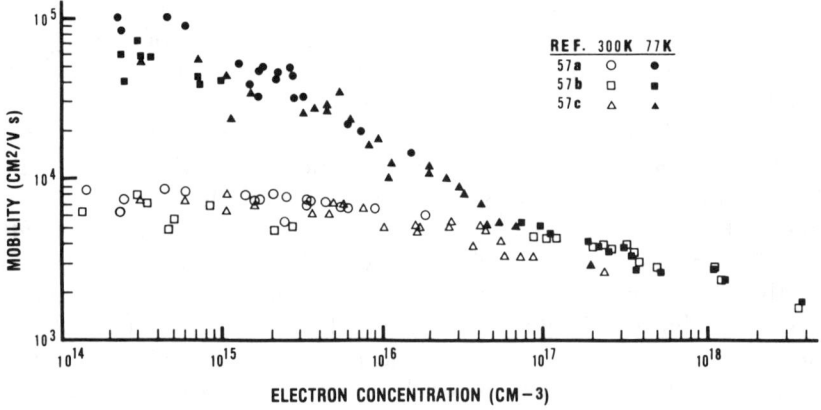

Figure 2.14 Electron mobility of GaAs at 300 K and 77 K as a function of carrier concentration [57].

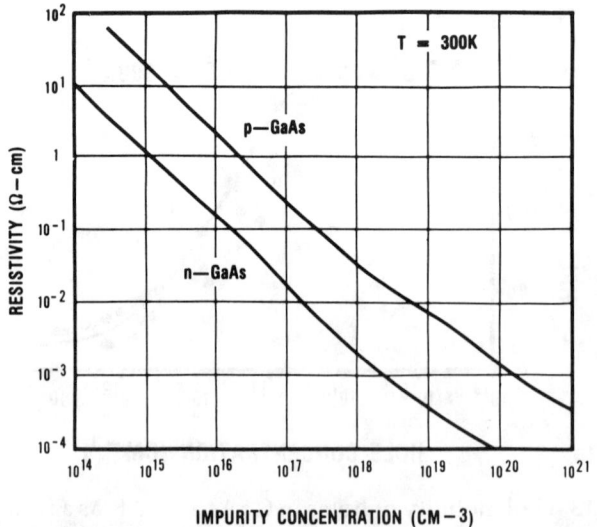

Figure 2.15 Resistivity of GaAs as a function of impurity concentration (after Sze [40]; data from S.M. Sze and J.C. Irvin, *Solid State Electron.*, 11, 1968, p. 599).

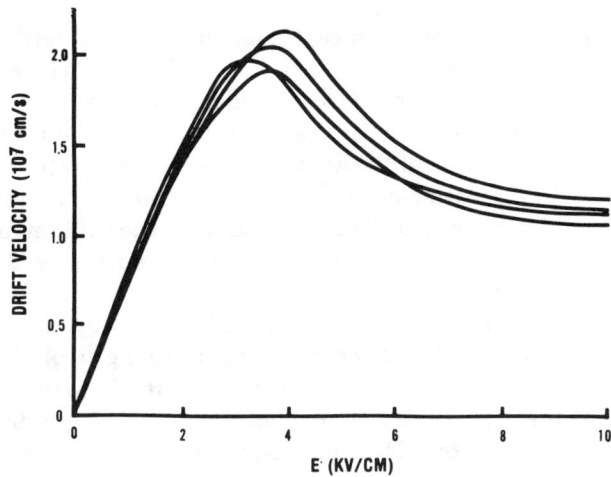

Figure 2.16 Drift velocity of conduction electrons in GaAs at 300 K as a function of electric field, experimental data (after reference [3]).

Figure 2.16 of electron velocity as a function of electric field is similar to Figure 2.1, except that it shows several sets of experimental data. The *threshold electric field*, E_t, is the value at which the velocity reaches the maximum. Doping inhomogeneities and other problems make it difficult to measure E_t and the peak electron velocity with great accuracy. These values may be taken at room temperature (for 10% accuracy) as [3]

$E_{th}(300) = 3.3$ KV/cm

$v_{max} = 2.1 \times 10^7$ cm/sec

2.3 BULK CRYSTAL GROWTH AND SLICE GENERATION

GaAs is grown in crystal ingots, then individual slices are sawed from the ingot, are lapped and polished. There are two principal methods of growing GaAs bulk crystals: *horizontal Bridgman* (HB) and *liquid encapsulated Czochralski* (LEC). The two methods produce crystals with distinctive characteristics. Although LEC material has recently gained popularity as the medium of choice for ion implanted GaAs devices [58], it is far too early to reach a definitive conclusion. Research is active in both types of material and both show promise of near-term improvements. It should be noted that GaAs crystal growth remains as much a technical art as a science, and that the following discussion is only a general overview. Each of the two methods of growth (HB and LEC) have numerous variations; only the major approaches are described.

2.3.1 Horizontal Bridgman (HB)

In the horizontal Bridgman process, starting material (either pure gallium or polycrystalline GaAs) is placed in a long *boat* which is sealed in a quartz ampoule several feet long. The ampoule is backfilled with an inert gas. A GaAs *seed* crystal is placed at the front of the boat and elemental arsenic is placed in the neck of the ampoule. This configuration is illustrated in Figure 2.17. Heating elements are used to generate a temperature profile in which the arsenic becomes gaseous, the gallium (or poly-GaAs) is brought to the melting point of GaAs, and the seed region temperature is just below the GaAs solidification point. After the As has reacted with the Ga, either the ampoule or the heaters are slowly moved so that the temperature front moves along the length of the boat. Crystal growth then follows this temperature front. Growth is usually chosen to be in the <111> direction. This is a slow-growth direction. Horizontal Bridgman material is sometimes called *boat grown* because of the use of the boat.

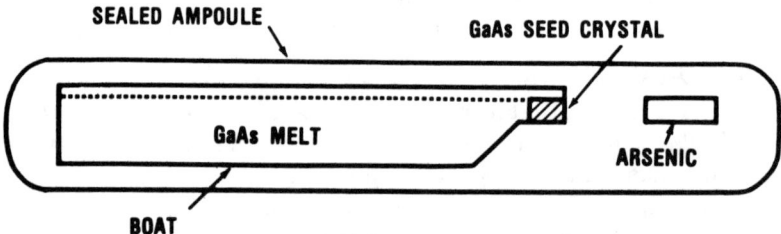

Figure 2.17 Schematic diagram of a horizontal Bridgman (HB) growth system.

The completed crystal has a cross-sectional shape matching the shape of the boat: somewhat circular up to the level of the melt. If the crystal were sawed perpendicular to its axis, the resulting slices would be (111) material. But usually (100) slices are desired, particularly for epitaxial growth reasons (see section 2.4), and also for the desirable perpendicular cleavage planes of (100) material. For this reason, HB crystals are usually sawed at an angle of 54.7° to the ingot axis and the resulting slices are then (100). This angular sawing has one disadvantage: compositional variations along the axis of the crystal are converted into variations across individual slices. The slices generally have the shape shown in Figure 2.18 and are referred to as *D-shaped wafers*. Note that the D-shaped material has a larger area than the cross section of the original ingot because of the angled sawing. Round slices are required for true production operations and this one trait makes D-shaped material undesirable. Such material may be made round by several methods. One is to cut each wafer round. Another is to glue the D-shaped slices together (surface to surface) and grind the resulting composite ingot round. Although the above operations may seem cumbersome, several vendors supply round HB material.

The differences between HB and LEC material will be considered in the following discussion of the LEC technique.

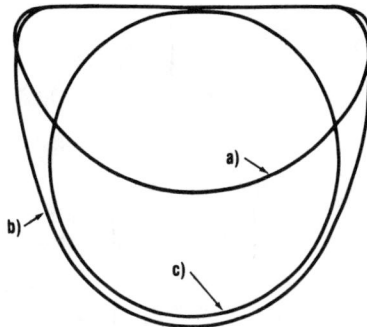

Figure 2.18 Shape of horizontal Bridgman slices: (a) cut normal to crystal axis (111); (b) cut at 54.7° (100); (c) round slice obtained from case (b).

2.3.2 Liquid Encapsulated Czochralski (LEC)

In the liquid encapsulated Czochralski (LEC) process, the crystal is grown in the vertical direction by slowly pulling the ingot from a melt (Figure 2.19). The melt consists of molten GaAs and is confined by a layer of liquid boric oxide (B_2O_3) that floats on the surface of the melt. The pioneering work of using the LEC process to grow semiconductor crystals, including the use of the boric oxide, was reported in 1962 [59]. Use of the technique to grow GaAs single crystals was reported in 1965 [60]. A GaAs seed crystal is used to begin the process when appropriate temperature profiles are present. Growth is initiated by using the seed to penetrate the boric oxide and contact the melt. The melt is contained in a crucible. The seed and the crucible usually are rotated to eliminate azimuthal thermal gradients. The diameter of the crystal is controlled by the pull rate and other operating parameters. The crystal growth machines used in this process are sometimes referred to as *pullers*. The LEC process may be subdivided into low pressure LEC and high pressure LEC. The high pressure LEC process is performed under high ambient pressure (up to 50 atmospheres). Low pressure LEC is performed at about 1 atmosphere. The high pressure procedure was the original method, partly for complex historical reasons. However, there seems to be no reason that the GaAs LEC process cannot be performed as well near ambient pressures; several institutions have had great success [61] with it. The low pressure work, however, has generally required the use of different pullers. Changes in ambient pressure cause changes in the heat flow characteristics. For this reason the high pressure machines cannot easily be used for the low pressure process without substantial modification. Low pressure machines are fundamentally smaller and less expensive. It seems likely that these will dominate in the future.

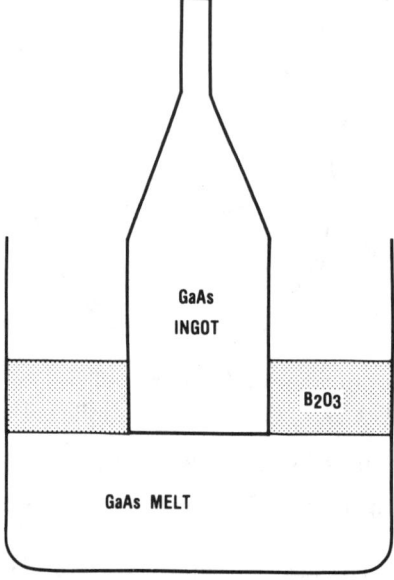

Figure 2.19 Liquid encapsulated Czochralski (LEC) growth of a GaAs crystal.

The GaAs melt must be initially prepared from Ga and As. This process is called compounding. In the high pressure LEC pullers, the Ga and As is placed below the boric oxide and the entire assembly is heated. The high ambient pressure over the boric oxide contains the rather violent reactions that occur as the Ga and As heat and react exothermally. Without such containment, local overheating would cause loss of arsenic and loss of control of the melt composition. The necessity for high pressure containment during compounding was a cogent argument for high pressure pullers; low pressure LEC required a "precompounding" operation in a separate machine. However, a successful technique for in situ compounding in low pressure LEC pullers was developed in which the Ga is first heated to the melting point of GaAs with boric oxide floating on top. Then the neck of a quartz ampoule is inserted through the boric oxide into the gallium, and arsenic is added in a controlled manner [62]. This arsenic reacts immediately with the gallium until the melt is saturated. The low pressure in situ compounding technique gives improved control over stoichiometry (some amount of arsenic always escapes in the high pressure compounding operation). In either type of compounding operation, a desired impurity such as chromium can be placed in the crucible before compounding. Alternatively, if conductive substrates are desired, a dopant such as tellurium can be placed in the crucible.

The crucible which contains the melt is either quartz (SiO_2) or *pyrolytic boron nitride* (PBM). The quartz crucible adds more silicon to the melt than the PBM crucible. It is interesting to note that quartz crucibles are used only once. They crack when the molten GaAs material is cooled at the conclusion of growth.

Bulk GaAs cannot be prepared in truly pure form. Remembering that intentional doping may consist of one atom for every 10^5 to 10^6 crystal atoms, it is easy to appreciate the difficulty of lowering impurities to inconsequential amounts. Silicon is available from the quartz parts used in bulk growth. Carbon is available from graphite heaters and may be transported in gaseous compounds. The ability to grow semi-insulating GaAs depends on complex compensation methods. Generally, shallow donors are compensated by deep acceptors (e.g., Si and Cr), or shallow acceptors are compensated by deep donors (e.g., C and EL2) [42]. If Cr is intentionally added (in the GaAs melt) to compensate for silicon impurities, the material is referred to as "Cr doped." This is generally required for LEC material grown from quartz crucibles, although the boric oxide tends to getter some of the silicon. If PBN crucibles are used, the defect EL2 may serve to compensate for impurities such as carbon. In this case no dopant is added to the GaAs melt and the material is referred to as undoped. Of course, it is undoped only in the sense that a dopant was not intentionally added. Chromium-doped material is sometimes considered undesirable because Cr diffusion occurs during heating (such as following ion implantation). Chromium redistribution has been the subject of hundreds of papers over the years. The undesirable effects are not as severe in "lightly" doped Cr substrates (a Cr concentration of 10^{15}). Such substrates can yield steeper doping profiles upon ion implantation and have been used very successfully to fabricate many types of devices.

Chromium is generally used in quantities of about 0.1 to 0.3 ppma (parts per million, atomic). Higher amounts degrade the mobility of ion implanted material and decrease the activation [58] (section 2.5). Chromium concentrations near the solid solubility limit (about 1 ppma) greatly increase the probability of crystal defects. Yet, care must be taken or this amount of Cr will not be sufficient to compensate for the unintentionally incorporated dopants. The use of chromium has another drawback. Because it has a low distribution coefficient (6×10^{-4}), chromium concentration increases from seed to tail in LEC material. Similar effects occur in HB material [58]. This difference can require different ion implantation conditions for each wafer to maintain constant electrical properties. In fact, it is often desirable to perform trial implants of two slices from each new ingot (one near the seed end, one near the tail end) and adjust the implant dose for the other slices accordingly. For these reasons, undoped material is attractive for ion implantation.

No crystal is perfect, and GaAs crystals have many dislocations in spite of the efforts to improve quality. These dislocations have deleterious effects. They introduce trapping states in the bandgap and they alter the etching properties of the slice, but most importantly, they can alter electrical performance of devices. Exactly how this occurs is still under investigation, but it seems likely that dislocations affect the activation of ion implanted species. Studies have provided rather dramatic evidence showing that source-drain current and threshold voltage of logic FETs are strongly correlated with dislocation density [63]. Other work has shown that the threshold voltage of such logic FETs is affected even up to distances of 50 μm from a dislocation [64]. It has also been demonstrated that sheet resistance and carrier concentration correlate closely with dislocation density [65]. In general, LEC material exhibits a greater dislocation density than HB material, and this is its major disadvantage compared to HB. Dislocation density depends strongly on the size of the ingot being grown. Small crystals can be virtually dislocation-free. There is also substantial scatter in the values obtained from crystal to crystal. LEC dislocation densities can easily be 10^4 to 10^5 cm^{-2} for 2-inch or 3-inch material. Comparable HB material generally exhibits lower dislocation densities, ranging from 8000 to 25,000 cm^{-2}. These are values that can be produced routinely; significantly better values are occasionally obtained. It is also likely that the evolving technology will steadily produce better material. Dislocation density can be determined by etching the slice in hot KOH (about 400° C for 30 minutes), which preferentially etches at dislocations, and then counting the etched pits. Hence the terms *dislocation density* and *etch pit density* are sometimes used interchangeably.

Dislocations generally arise from temperature gradients present during crystal growth. These are difficult to control. It is desirable to reduce both axial and radial temperature gradients to a minimum. One effort, using thick boric oxide layers to reduce thermal gradients, has reported one case of 5000 dislocations per cm^2 for 2-inch material, equaling the best results for HB [58]. There is reason to believe that future work will lower the dislocation densities that can be obtained routinely in 3-inch LEC material.

Because of the radial nonuniformities in temperature and growth kinetics, the dislocation density is not uniform over a slice, but generally exhibits a "W" pattern across any diameter as shown in Figure 2.20 [65]. As indicated above, the sheet resistance tends to follow the same pattern (Figure 2.20).

Round slices with orienting *flats* are desired for production. For LEC material, the general procedure is as follows. The ingot is ground so that it is round and of the correct diameter. X-ray and/or etching methods are used to orient the crystal, and a flat is ground along one side to act as an orientation marker. Normally a smaller *minor flat* is also ground. Then the ingot is sawed into separate slices. These slices are lapped on both sides and polished on one or both sides. The completed slices are generally 0.015 to 0.035 inches thick, appearing as shown in Figure 2.21. These substrates are then ready for ion implantation or epitaxial growth.

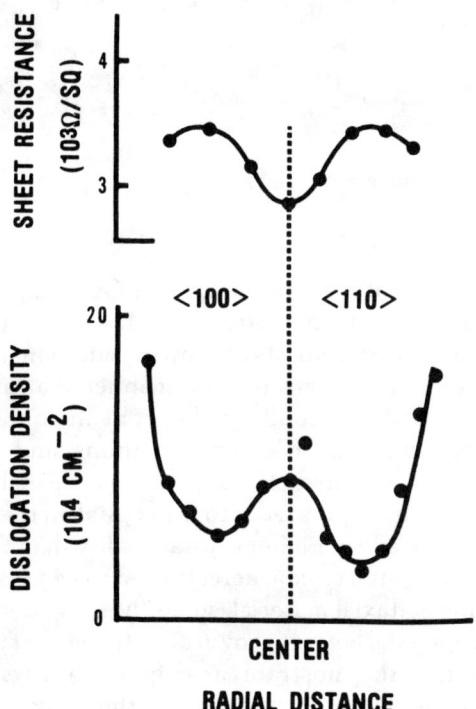

Figure 2.20 Sheet resistance and distribution of dislocations across any diameter of a round LEC slice (after [65]). The W pattern of the dislocations is typical of LEC material.

42 GaAs Processing Techniques

Figure 2.21 GaAs LEC slice with flats.

2.4 EPITAXY

There are two methods of forming conductive GaAs surface layers on GaAs substrates: epitaxy and ion implantation. Ion implantation is considered in section 2.5. Epitaxy consists of growing additional GaAs material on the surface of GaAs substrates in a manner that preserves the crystal structure. Ga and As atoms are brought in contact with the crystal surface under temperatures, concentrations, and other conditions that result in crystalline growth. Dopants may be included among these atoms, and hence, incorporated into the crystal on the lattice sites. Epitaxial layers are generally of higher crystal quality than the substrate they are grown on. Although crystal defects tend to be preserved in the immediate overlaying epitaxial material, some "healing" does take place. If an active (doped) epitaxial layer is grown directly on the substrate, the impurities and defects in the substrate can degrade the crystalline properties, and therefore the electrical properties, of the epitaxial layer (these active layers may be only a few tenths of a micron thick). Therefore, buffer layers of undoped GaAs usually are epitaxially grown on the substrate; the active layer is then grown on the buffer layer before the doping is introduced. In some cases, a topmost layer of highly doped material may be grown to provide a contact layer (Chapter 3). These three layers (buffer, active, contact) are sometimes referred to as n_o, n, and n^+ layers.

Epitaxial growth is generally performed on material that has been cut so that the crystal surface is slightly (~2°) away from a <100> surface. This orientation results in the best morphology. For ion implantation, that misorientation is not necessary.

There are three basic types of epitaxy that have been used for GaAs: *liquid phase epitaxy* (LPE), *vapor phase epitaxy* (VPE), and *molecular beam epitaxy* (MBE). Each will be described separately.

2.4.1 Liquid Phase Epitaxy

Liquid phase epitaxy (LPE) is the oldest technique used to grow epitaxial layers on GaAs crystals. It was used in much of the early laboratory work. However, it is rapidly losing popularity because of the limitations discussed below, and because it is virtually irrelevant for production of GaAs microwave devices. However, it remains an inexpensive method of epitaxy and is capable of growing many material compositions including GaAlAs.

The LPE procedure is illustrated in Figure 2.22. A GaAs substrate is placed in a *slider* that can be moved across the surface of molten material contained in a boat. This material is gallium (or another material) that has been saturated with the desired material (GaAs, etc.). The temperature profiles are such that the melt is supercooled to just below its solidification point and atoms solidify onto the crystal substrate. Dopants may be included in the melt. Different portions of the boat can contain different melts so that different compositions may be grown as the slider pulls the substrate across each zone. The major problem of LPE material is the difficulty of growing uniform layers over large surface areas.

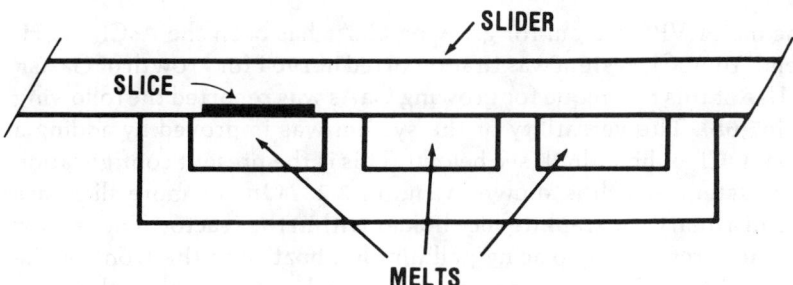

Figure 2.22 Schematic diagram of a liquid phase epitaxy system.

It should be noted that these objections refer to LPE material for microwave device application. LPE remains a successful production technique for light emitting diodes and other such structures that do not require the thin, uniform, high quality epitaxial layers needed for microwave devices.

2.4.2 Vapor Phase Epitaxy

Vapor phase epitaxy (VPE) has been the most popular form of GaAs epitaxial growth. However, its popularity is threatened by the superior results of molecular beam epitaxy (next subsection), the latter being limited only by technological complexity and expense. In VPE growth, the Ga, As, and dopant atoms are brought to the slice in a gaseous phase. Under appropriate temperatures and other conditions, reactions take place on the substrate surface that result in these atoms being deposited on the surface where they replicate the underlying crystal structure.

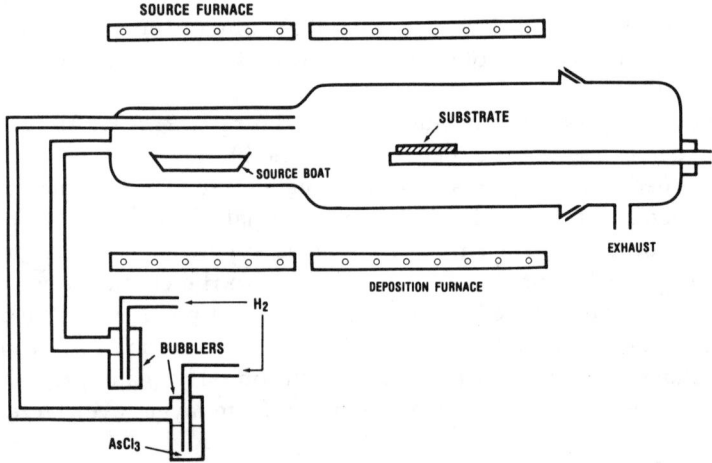

Figure 2.23 Schematic diagram of a vapor phase epitaxy system.

The major VPE system for growing GaAs has been the $AsCl_3/Ga/H_2$ system. An $AsCl_3$ system was first reported in 1964 for growth of GaAsP [66]. Use of this technique for growing GaAs was reported the following year [67,68]. The versatility of the system was improved by adding a second $AsCl_3$ bubbler [69] (see below). This is the present configuration of the system, which is shown in Figure 2.23. One or more slices are placed in a quartz or graphite slice holder within the reactor. The reactor is initially prepared by placing gallium in a boat near the front of the reactor tube. $AsCl_3$, which is a liquid, is placed in the bubblers shown in Figure 2.23. A major advantage of the $AsCl_3/Ga/H_2$ system is that the two starting materials (Ga and $AsCl_3$) may be obtained in very high purity. Before the reactor can be used for growth, a GaAs crust must be formed on the Ga metal in the boat. This is done by bubbling hydrogen gas through the liquid $AsCl_3$ and passing this gaseous flow over the gallium. The reactor is heated by ovens. Temperatures and gas flows are adjusted so that arsenic is made available to saturate the gallium under the reaction

$$4AsCl_3 + 6H_2 \rightarrow 12HCl + As_4 \qquad (2.1)$$

After a GaAs crust has completely formed over the Ga source, the reactor may be used for epitaxial growth. A limited number of growth runs may be accomplished before the source needs to be replenished.

For epitaxial growth, one or more slices are placed in a quartz or graphite slice holder in the reactor tube. The substrate may receive an in situ etch before the growth process begins. Temperatures are adjusted so that the zone containing the gallium boat is hotter that the zone containing the substrate. The substrate zone is usually near 850° C. Hydrogen flows through the main bubbler and across the GaAs crust in the gallium boat. The exact reactions that occur are somewhat complex, but the process may be represented by the reaction

$$GaAs(s) + HCl(g) = GaCl(g) + \tfrac{1}{2}H_2(g) + \tfrac{1}{4}As_4(g) \qquad (2.2)$$

where the letters s and g indicate either a solid or gaseous state. The reaction proceeds from left to right in the higher temperature zone over the source. The reverse reaction then occurs in the lower temperature zone at the substrate. The first bubbler is used for active layer growth. The second bubbler may be used in pregrowth etching and/or buffer layer growth. Dopants are introduced using another gaseous source such as H_2S for n-type doping.

A modification of the above procedure is to use high-purity GaAs for the source instead of a saturated gallium melt. This is known as a *solid source*, and the system is referred to as $AsCl_3/GaAs/H_2$. The purity of the solid GaAs source (taken from bulk-grown GaAs) is generally not as good as can be obtained using the gallium saturation technique. Nevertheless, the purity seems adequate for many needs since solid sources are used in many VPE systems.

Another category of VPE involves transport of the gallium by using organic molecules. This procedure is designated MOCVD for *metal-organic chemical vapor deposition*. It is also sometimes referred to as *organometalic CVD* (OMCVD). Use of this method to grow III-V compounds was first reported in 1968 [70-72]. Trimethylgallium is often used as the source of gallium. Arsine (AsH_3), a highly poisonous gas, is used as the arsenic source. The advantage of MOCVD is its ability to grow AlGaAs layers. Organic molecules such as trimethylaluminum work well in transporting Al, a requirement that has proven impossible to obtain using conventional (non-organic) VPE. AlGaAs is an important constituent in some devices, as described in Chapter 3.

2.4.3 Molecular Beam Epitaxy

Molecular beam epitaxy (MBE) is the most recent major method developed for epitaxial growth. It may be described as a sophisticated evaporation technique performed in ultrahigh vacuum. In this procedure, the substrate is placed in a high vacuum and elemental species are evaporated from ovens, impinging upon the heated substrate. Here they assemble into crystalline order. With proper control of the sources (Ga, As, Al, Si, etc.), almost any material composition and doping can be achieved. Further, the composition may be controlled with a resolution of virtually one atomic layer. Figure 2.24 shows a schematic view of the growth chamber of an MBE machine. The separate ovens, often called effusion cells, each contain an individual element. The heated materials vaporize and exit the effusion cell through one end, go past a shutter, and travel through the growth chamber until reaching the substrate. More complex effusion cells, called cracking furnaces, can reduce molecular species to the dimer form (e.g., As_4 to As_2). The substrate is heated to provide sufficient energy for surface diffusion and incorporation of the species. This temperature is typically 500° to 600° C for GaAs. The slice usually is rotated to aid uniformity. Elements may be switched on and off using the shutter in front of each effusion cell.

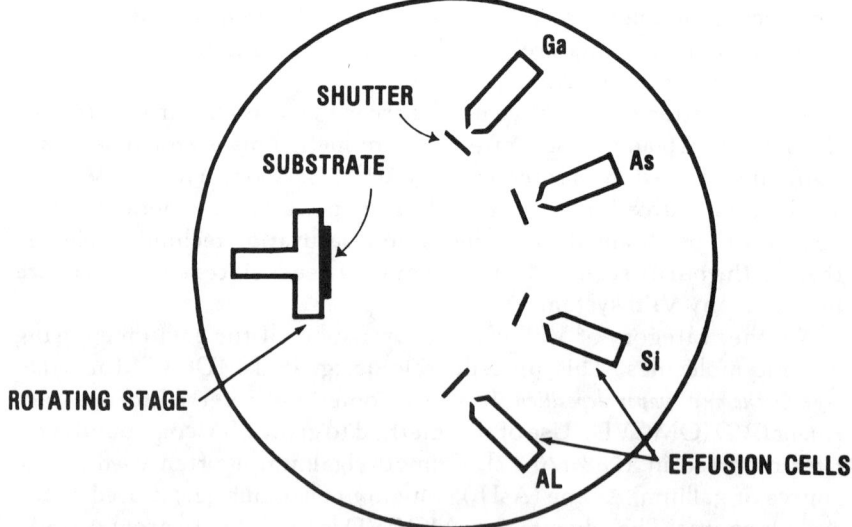

Figure 2.24 Schematic diagram of a molecular beam epitaxy system.

MBE's advantages are that it can produce almost any epitaxial layer composition, layer thickness, and doping; and it can do so with high accuracy and uniformity across a slice. Disadvantages include high vacuum requirements, complex and costly equipment, and slow growth rate. An ultrahigh vacuum is required in the growth chamber, generally in the range 10^{-10} to 10^{-11} Torr. This is an exceedingly difficult requirement, especially in the presence of heated substrates and heated effusion ovens. These require the presence of cooled shrouds and baffles to shield the heat sources. It is hardly an understatement to describe the design and construction of the growth chamber as a technological masterpiece.

Growth rate is typically 1 μm/hour (about one atomic layer per second) although growth rates up to 10 μm/hour may be attainable. As would be expected from difficult technology, cost is high. Yet MBE machines are available from several commercial vendors and are in use in many laboratories. Some have slice loading magazines so that several substrates may be loaded into the vacuum system at once.

MBE machines usually incorporate instruments to analyze the growth process and the resulting crystal structure. The technique of *reflection high energy electron diffraction* (RHEED) is generally employed for this evaluation (Chapter 19). Such instruments may be in the growth chamber or in an adjacent chamber. At this writing, MBE is being used in research and very limited prototype production. The potential of the technology is enormous and its use is certain to expand.

2.5 ION IMPLANTATION

Ion implantation is presently the preferred method to form active layers for production GaAs FET devices. It is likely to remain popular and useful for a long time. In this procedure, dopant atoms are introduced into the substrate surface by ion implantation. Typical energies and doses are 30 KeV to 400 KeV and 10^{12} to 10^{14} atoms/cm^2. Such implantation greatly damages the crystal lattice and the implanted atoms come to rest at random locations within the material. A high temperature annealing step (about 850° C) is performed to anneal out the lattice damage and allow the implanted atoms to move onto lattice sites. This is known as *activating* the implant. Not all the dopant atoms are incorporated at lattice sites and supply carriers. Activation is generally in the range of 75% to 95% and depends on the implant and anneal conditions.

Ion implantation has been an important element in silicon IC manufacture for some time, so commercial ion implanters are readily available and implantation technology is well-established. Ion implantation can be performed with high uniformity over a slice and with good uniformity from slice to slice. It is also more economical than VPE (the capacity of commercial implanters can exceed 100 wafers per hour). These are the major reasons that it is has become the most popular method for general production use.

It has one other significant advantage: doping can be performed locally by selectively masking the slice so that ion implantation occurs only at desired locations. No other technology offers this flexibility. This feature is especially useful when several different doping levels are needed on the same slice, such as for low noise (high doping) and power (medium doping) FETs on GaAs monolithic ICs. Various portions of GaAs logic devices also employ localized ion implantation, sometimes in conjunction with VPE or MBE (see Chapter 3).

For implant energies less than about 1 MeV, there are two mechanisms responsible for energy loss of the ions as they enter the substrate and come to rest. The first is elastic scattering from the nuclei of the substrate atoms. Virtually all the angular deviation is caused by nuclear scattering. The second is inelastic interactions between the ions and the electrons of the substrate. This force acts as viscous damping and contributes little to the angular deviations. Typical implant species are silicon or selenium ions for n-type dopants (also S, Sn, Te) and beryllium for p-type dopants (also C, Mg, Zn, Cd). The substrate may be implanted at room temperature or it may be heated.

A disadvantage of ion implantation is that the doping profile transition (between active layer and substrate) cannot be as sharp as with epitaxial methods. This is a result of the statistics of the implantation and the general diffusion that takes place during high temperature anneal. Implanted atoms scatter off the atoms of the host crystal and come to rest in a near-Gaussian distribution as shown in Figure 2.25. A true Gaussian has the form

$$N(x) = N_o \exp [-(x - R_p)^2 / 2\sigma^2]$$

where R_p is the depth of the distribution peak. The tail of the curve extends deeper than a true Gaussian. A second, shallower implant is sometimes used to fill in the area near the surface and make the doping profile more uniform. Of course, these implanted profiles may be altered by diffusion during the subsequent anneal step.

Ion implantation usually is performed at a slight angle to the surface of the substrate, on the order of 6° to 10°. This is done to avoid an effect called *channeling*. If the crystal lattice is viewed in an appropriate direction, there are open cells which contain no atoms (Figure 2.7); these are called channels. An ion moving parallel to a channeling direction can travel deep into the crystal before scattering occurs. This effect depends critically on the angle of the crystal to the ion beam and this beam angle will vary slightly across the slice. For these reasons, channeling results in implant profiles that are very difficult to predict and reproduce accurately. The general theory of ion implantation into crystals has been worked out and standard tables exist detailing the implant parameters that result when different species are implanted into various materials at a given energy. Reference [73] is the major reference for semiconductor work and data from it is presented in the abbreviated ion implantation table (Table 2.2) at the end if this chapter. In the case of some species, tables do not exist for implanting into GaAs. In these cases, tables for germanium may be regarded as essentially equivalent. This is because the mass of the Ge atom is very close to the average mass of Ga and As atoms; their exterior electronic shells also average out. These are the dominant factors of the substrate material that effect the implant range distribution.

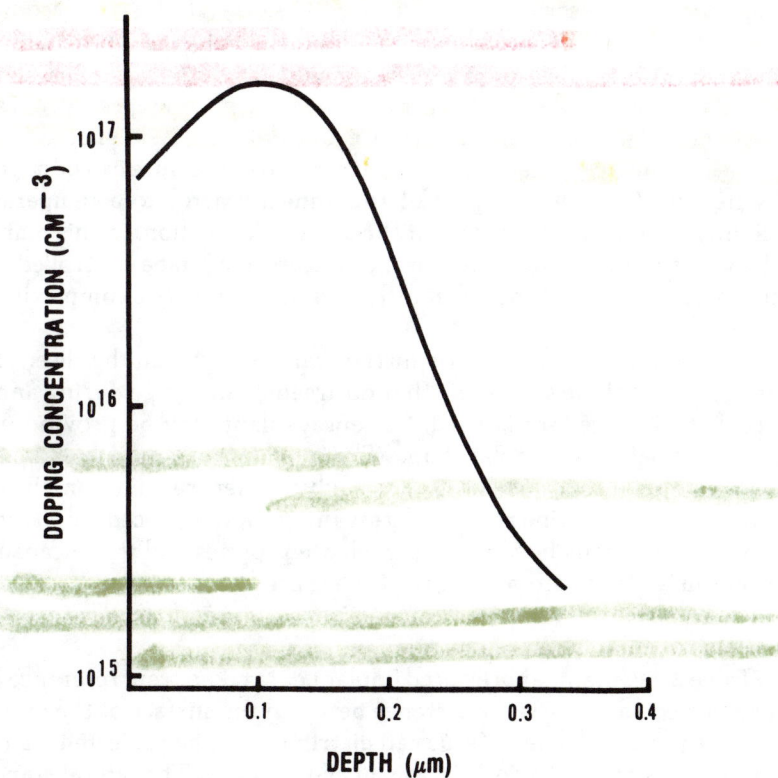

Figure 2.25 Gaussian-like doping concentration profile resulting from ion implantation.

Annealing usually takes place in high temperature ovens although other techniques have achieved some success, including the use of strip heaters [74], infrared halogen lamps [75], arc lamps [76], electron beams [77], and lasers [78]. These latter techniques attempt to activate the implant over short time periods (a few seconds) and so reduce the diffusion that occurs during oven anneals. One note of caution: some of the above techniques, especially the use of lasers, have had far more success on silicon than on GaAs. The volatile nature of GaAs and its tendency to evolve As (ruining stochiometry) make the use of these procedures more troublesome for GaAs material. Ovens remain the most popular anneal method, but flash lamps have demonstrated both good results and high activation [75, 76], and are likely to increase in popularity. Oven annealing is usually performed between 800° C and 900° C. At these temperatures, which are necessary to activate the implant and heal crystal lattice damage, the crystal tends to decompose and emit arsenic. Hence, this arsenic must be kept from leaving the crystal during anneal to preserve stoichiometry and crystal structure. This is typically done in one of three ways: arsenic overpressure, use of an encapsulant on the slice, or use of another GaAs slice over the annealing slice (proximity anneal). Arsenic overpressure can be supplied by heating GaAs in one part of the anneal reactor to a temperature slightly hotter than is present at the annealing location. In this manner, the vapor pressure (overpressure) of the arsenic can be controlled. This technique works well, but obviously requires some extra complexity and care.

Encapsulants, usually silicon nitride, can be grown on the slice. These effectively seal the surface so that no arsenic can escape during annealing. If this procedure is used, the encapsulant may be grown on the substrate before ion implantation. The implant can be performed directly through the encapsulating layer which then remains on the slice during annealing. One possible disadvantage of this procedure is *knock on*, in which collisions between the implanted species and the encapsulant atoms drive these atoms into the GaAs slice. If silicon nitride is used as the encapuslant, the implant calculations are rather easy because Si_3N_4 has approximately the same stopping power as GaAs.

Table 2.2 contains abbreviated implant data taken from reference [73]. The projected range is the distance between the surface of the material and the peak of the near-Gaussian distribution. The projected standard deviation is the half width of this near-Gaussian. The lateral standard deviation represents scattering perpendicular to the incoming direction. This scattering must be considered when selectivity implanting through a mask. Some ions will come to rest under the edges defined by the mask.

TABLE 2.2
Abbreviated Implant Statistics
(After reference [73])
(microns)

E (KEV)	Hydrogen into Ge*			Boron into Ge*			Carbon into GaAs			Oxygen into GaAs			E (KEV)
	PROJ. RANGE	STAND DEV	LAT STD DEV	PROJ. RANGE	STAND DEV	LAT STD DEV	PROJ. RANGE	STAND DEV	LAT STD DEV	PROJ. RANGE	STAND DEV	LAT STD DEV	
10	0.108	0.070	0.106	0.024	0.020	0.028	0.020	0.017	0.024	0.016	0.013	0.018	10
20	0.218	0.099	0.159	0.047	0.035	0.051	0.039	0.030	0.042	0.030	0.022	0.030	20
30	0.307	0.119	0.197	0.073	0.049	0.071	0.059	0.041	0.058	0.044	0.030	0.043	30
40	0.396	0.133	0.227	0.098	0.061	0.090	0.080	0.051	0.074	0.059	0.038	0.054	40
50	0.480	0.144	0.252	0.124	0.071	0.107	0.101	0.060	0.089	0.075	0.045	0.065	50
60	0.561	0.154	0.274	0.151	0.081	0.124	0.123	0.069	0.103	0.090	0.051	0.076	60
70	0.640	0.162	0.293	0.177	0.091	0.139	0.144	0.077	0.116	0.106	0.058	0.086	70
80	0.716	0.169	0.311	0.203	0.099	0.154	0.166	0.084	0.128	0.122	0.065	0.096	80
90	0.791	0.173	0.327	0.229	0.107	0.168	0.187	0.091	0.140	0.138	0.070	0.105	90
100	0.866	0.181	0.342	0.255	0.115	0.181	0.208	0.098	0.152	0.154	0.076	0.114	100
120	1.013	0.192	0.369	0.307	0.128	0.206	0.251	0.110	0.174	0.186	0.086	0.132	120
140	1.159	0.200	0.394	0.357	0.140	0.229	0.293	0.212	0.193	0.218	0.096	0.149	140
160	1.306	0.209	0.418	0.407	0.151	0.250	0.334	0.131	0.212	0.249	0.105	0.164	160
180	1.456	0.217	0.441	0.456	0.161	0.270	0.375	0.140	0.229	0.281	0.113	0.179	180
200	1.607	0.225	0.463	0.504	0.170	0.288	0.415	0.148	0.246	0.316	0.121	0.193	200
220	1.762	0.233	0.484	0.551	0.179	0.305	0.454	0.156	0.261	0.342	0.128	0.206	220
240	1.921	0.240	0.506	0.598	0.187	0.321	0.493	0.163	0.275	0.373	0.134	0.218	240
260	2.084	0.247	0.527	0.643	0.194	0.336	0.531	0.170	0.289	0.403	0.141	0.230	260
280	2.251	0.254	0.548	0.689	0.201	0.351	0.569	0.176	0.302	0.432	0.147	0.242	280
300	2.423	0.262	0.570	0.733	0.207	0.365	0.606	0.182	0.314	0.462	0.152	0.253	300
320	2.600	0.268	0.591	0.777	0.213	0.378	0.643	0.188	0.326	0.491	0.158	0.264	320
340	2.781	0.276	0.613	0.820	0.219	0.390	0.680	0.193	0.338	0.519	0.163	0.274	340
360	2.968	0.284	0.636	0.863	0.224	0.402	0.716	0.198	0.349	0.548	0.167	0.284	360
380	3.161	0.291	0.658	0.905	0.229	0.414	0.751	0.203	0.359	0.576	0.172	0.293	380
400	3.359	0.299	0.681	0.947	0.234	0.425	0.786	0.207	0.369	0.604	0.176	0.302	400

*Ge has the same average mass as GaAs; see text.

TABLE 2.2 [cont.]

E (KEV)	Selenium into GaAs PROJ. RANGE	Selenium into GaAs STAND DEV	Selenium into GaAs LAT STD DEV	Silicon into GaAs PROJ. RANGE	Silicon into GaAs STAND DEV	Silicon into GaAs LAT STD DEV	Sulfur in GaAs PROJ. RANGE	Sulfur in GaAs STAND DEV	Sulfur in GaAs LAT STD DEV	Tellurium into GaAs PROJ. RANGE	Tellurium into GaAs STAND DEV	Tellurium into GaAs LAT STD DEV	E (KEV)
10	0.006	0.003	0.004	0.010	0.007	0.010	0.009	0.007	0.009	0.006	0.002	0.003	10
20	0.011	0.005	0.006	0.018	0.013	0.016	0.017	0.011	0.014	0.009	0.004	0.004	20
30	0.014	0.007	0.009	0.026	0.017	0.023	0.023	0.015	0.020	0.012	0.005	0.006	30
40	0.018	0.009	0.010	0.034	0.021	0.029	0.030	0.019	0.025	0.014	0.006	0.007	40
50	0.021	0.011	0.012	0.042	0.025	0.034	0.038	0.022	0.029	0.017	0.007	0.008	40
60	0.024	0.012	0.014	0.051	0.029	0.040	0.045	0.026	0.034	0.019	0.008	0.009	60
70	0.028	0.014	0.016	0.059	0.033	0.045	0.052	0.029	0.039	0.022	0.009	0.010	70
80	0.031	0.015	0.018	0.068	0.037	0.051	0.059	0.032	0.044	0.024	0.010	0.011	80
90	0.034	0.016	0.019	0.076	0.041	0.056	0.067	0.036	0.048	0.026	0.011	0.012	90
100	0.037	0.018	0.021	0.085	0.044	0.062	0.074	0.039	0.053	0.029	0.012	0.013	100
120	0.044	0.021	0.024	0.103	0.051	0.072	0.089	0.045	0.061	0.033	0.014	0.015	120
140	0.050	0.023	0.027	0.120	0.058	0.082	0.104	0.051	0.070	0.037	0.016	0.017	140
160	0.056	0.026	0.031	0.138	0.064	0.091	0.120	0.056	0.078	0.041	0.017	0.019	160
180	0.063	0.029	0.034	0.156	0.070	0.101	0.135	0.062	0.087	0.046	0.019	0.021	180
200	0.070	0.031	0.037	0.174	0.075	0.110	0.151	0.067	0.095	0.050	0.021	0.023	200
220	0.076	0.034	0.040	0.192	0.081	0.119	0.167	0.072	0.103	0.054	0.022	0.024	220
240	0.083	0.036	0.043	0.210	0.086	0.127	0.182	0.077	0.110	0.058	0.024	0.026	240
260	0.089	0.039	0.046	0.228	0.091	0.136	0.198	0.081	0.118	0.062	0.026	0.028	260
280	0.096	0.041	0.049	0.246	0.096	0.144	0.214	0.086	0.125	0.067	0.027	0.030	280
300	0.103	0.044	0.052	0.263	0.100	0.152	0.229	0.090	0.132	0.071	0.029	0.031	300
320	0.110	0.046	0.055	0.281	0.105	0.159	0.245	0.094	0.139	0.075	0.030	0.033	320
340	0.116	0.049	0.058	0.299	0.109	0.167	0.260	0.098	0.146	0.079	0.032	0.035	340
360	0.123	0.051	0.061	0.316	0.113	0.174	0.276	0.102	0.152	0.083	0.033	0.036	360
380	0.130	0.053	0.064	0.333	0.117	0.081	0.291	0.106	0.159	0.088	0.035	0.038	380
400	0.137	0.056	0.067	0.351	0.121	0.188	0.307	0.110	0.165	0.092	0.036	0.040	400

REFERENCES

[1] V.M. Goldschmidt, *Trans. Faraday Soc.*, 25, 1929, p.253.

[2] H. Welker, *Z. Naturforsch. A*, 7, 1952, p. 744.

[3] J.S. Blakemore, *J. Appl. Phys.*, 53, 1982, p. R123.

[4] C. Hilsum and A.C. Rose-Innes, *Semiconducting III-V Compounds*. Oxford: Pergamon, 1961.

[5] O. Madelung, *Physics of III-V Compounds*. New York: Wiley, 1964.

[6] M. Neuberger, *III-V Semiconducting Compounds*. New York: IFI/Plenum, 1971.

[7] A.N. Goryunova, *The Chemistry of Diamond-Like Semiconductors*, J.C. Anderson, editor. Cambridge, MA: MIT/Press, 1965.

[8] H. Kressel and J.K. Butler, *Semiconductor Lasers and Heterojunction LEDs*. New York: Academic, 1977.

[9] H.C. Casey and M.B. Panish, *Heterostructure Lasers* in two parts: *Part A: Fundamental Principles* and *Part B: Materials and Operating Characteristics*. New York: Academic, 1978.

[10] C. Hilsum, in *Progress in Semiconductors*, Vol. 9, A.F. Gibson and R.E. Burgess, editors London: Heywood, 1965, p. 135.

[11] A.G. Milnes, *Deep Impurities in Semiconductors*. New York: Wiley, 1973.

[12] *Gallium Arsenide and Related Compounds*, Institute of Physics, London (Volumes in Inst. of Physics Conference Series: relevant volumes are No. 3 (1967), No. 7 (1969), No. 9 (1971), No. 17 (1973), No. 24 (1975), No. 33a&b (1977), No. 45 (1979), No. 56 (1981), and (1983)

[13] R.K. Willardson and A.C. Beer, editors, *Semiconductors and Semimetals*. New York: Academic, from Vol. 1 (1965) onward.

[14] H. Gesch, W. Kellner, and H. Kniepkamp, *IEEE Trans. Electron Devices*, ED-30, 1983, p. 1640.

[15] P.M. Solomon, *Proc. IEEE*, 70, 1982, p. 489.

[16] J.A. Cooper, *Proc. IEEE*, 69, 1981, p. 226.

[17] R. Keyes, in *Digital Technology, Status and Trends*, H. Painke, editor. Munich: Oldenburg-Verlag, 1981, p. 253.

[18] E.T. Watkins and J.M. Schellenberg, *Technical Digest*, 1981 IEEE MTT-S International Microwave Symposium, p. 145.

[19] C. Kittel, *Introduction to Solid State Physics*, 5th Edition. New York: Wiley, 1976.

[20] M. Simmons, *Technical Digest*, 1983 GaAs IC Symposium (Phoenix, Az), IEEE, Piscataway, N.J., p. 124; and references therein.

[21] C.M.H. Driscoll, A.F.W. Willoughby, J.B. Mullin, and B.W. Straughan, in reference [12], No. 24 (1975), p. 275.

[22] M.E. Straumanis and C.D. Kim, *Acta Crystallogr.*, 19, 1965, p. 256.

[23] H. Welker and H. Weiss, in *Solid State Physics*, Vol. 1, ed. by F. Seitz and D. Turnbull. New York: Academic, 1956, p. 1.

[24] S.I. Novikova, *Sov. Phys. Solid. State*, 3, 1961, p. 129.

[25] L. Bernstein and R.J. Beals, *J. Appl. Phys.*, 32, 1961, p. 122.

[26] E.D. Pierron, D.L. Parker, and J.B. McNeely, *Acta Crystallogr.*, 21, 1966, p. 290.

[27] R. Feder and T. Light, *J. Appl. Phys.*, 39, 1968, p. 4870.

[28] Y.S. Touloukian and E.H. Buyco, editors, *Thermophysical Properties of Matter*, Vol. 13 ("Thermal Expansion"). New York: Plenum, 1976, p. 747.

[29] M.G. Holland, *Phys. Rev.*, 134, 1964, p. A471.

[30] M.G. Holland, in *Proceedings of the 7th International Conference on Physics of Semiconductors, Paris 1964*, ed. by M. Hulin. Paris: Dunod, 1964, p. 713.

[31] R.O. Carlson, G.A. Slack, and S.J. Silverman, *J. Appl. Physics*, 36, 1965, p. 505.

[32] A. Amith, I. Kudman, and E.F. Steigmeier, *Phys. Rev.* 138, 1965, p. A1270.

[33] G.E. Stillman, C.M. Wolfe, and J.O. Dimmock, in reference [13], Vol. 12 (1977), p. 169.

[34] D.L. Rode, in reference [13], Vol. 10 (1975), p. 1.

[35] R.E. Neidert, *Electron. Lett.*, 16, 1980, p. 244.

[36] K.S. Champlin and G.H. Glover, *Appl. Phys. Lett.*, 12, 1968, p. 231.

[37] T. Lu, G.H. Glover, and K.S. Champlin, *Appl. Phys. Lett.*, 13, 1968, p. 404.

[38] I. Strzalkowski, S. Joshi, and C.R. Crowell, *Appl. Phys. Lett.*, 28, 1976, p. 350.

[39] K.G. Hambleton, C. Hilsum, and B.R. Holeman, *Proc. Phys. Soc. London*, 7, 1961, p. 1147.

[40] S.M. Sze, *Physics of Semiconductor Devices*, 2nd Edition. New York: Wiley, 1981.

[41] C.D. Thurmond, *J. Electrochem. Soc.*, 122, 1975, p. 1133.

[42] D.E. Holmes, R.T. Chen, K.R. Elliott, C.G. Kirkpartrick, and P.W. Yu, *IEEE Trans. Electron Devices*, 29, 1982, p. 1045.

[43] J. Lagowski, H.C. Gatos, J.M. Parsey, K. Wada, M. Kaminska, and W. Wolukiewiez, *Appl. Phys. Lett.*, 40, 1982, p. 342.

[44] D.E. Aspnes, J.B. Theeten, and R.P.H. Chang, *J. Vac. Sci. Technol.*, 16, 1979, p. 1374.

[45] R.L. Farray, R.K. Chang, S. Mroczkawski, and F.H. Pollak, *Appl. Phys. Lett.*, 31, 1977, p. 768.

[46] S.P. Murarka, *Appl. Phys. Lett.*, 26, 1975, p. 180.

[47] H. Hasegawa, K.E. Forward, and H.L. Hartnagel, *Appl. Phys. Lett.*, 26, 1975, p. 567.

[48] R.P.H. Chang and J.J. Coleman, *Appl. Phys. Lett.*, 32, 1978, p. 332.

[49] J.D. Wiley, in reference [13], Vol. 10 (1975), p. 91.

[50] F.D. Rosi, D. Meyerhofer, and R.V. Jensen, *J. Appl. Phys.*, 31, 1960, p. 1105.

[51] D.E. Hill, *Phys. Rev.*, 133, 1964, p. A866.

[52] D.E. Hill, *J. Appl. Phys.*, 41, 1970, p. 1815.

[53] J. Vilms, and J.P. Garrett, *Solid State Electron.*, 15, 1972, p. 443.

[54] F.E. Rosztoczy, F. Ermanis, I. Hayashi, and B. Schwartz, *J. Appl. Phys.*, 41, 1970, p. 264.

[55] O.V. Emel'yanenko, T.S. Lagunova, and D.N. Nasledov, *Sov. Phys. Solid State*, 2, 1960, p. 176.

[56] S.M. Gasanli, O.V. Emel'yanendo, V.K. Ergakov, F.P. Kesamanly, T.S. Lagunova, and D.N. Nasledov, *Sov. Phys. Semicond.*, 5, 1972, p. 1641.

[57] W.R. Frensley (unpublished). Data from the following:
(a) J.V. Dilorenzo, *J. Cryst. Growth*, 17, 1972, p. 189.
(b) D.J. Ashen, et al., *Proc. 1974 GaAs Symposium, Inst. Phys. Conf. Ser. No. 24*, p. 229.
(c) J.V. Dilorenzo and G.E. Moore, Jr., *J. Electrochem. Soc.*, 118, 1971, p. 1823.

[58] E.M. Swiggard, Digest of papers: *1983 GaAs IC Symp. Tech. Digest*, IEEE, Piscataway, NJ, p. 26.

[59] E.P.A. Metz, R.C. Miller, R. Mazelsky, *J. Appl. Phys.*, 33, 1962, p. 2016.

[60] J.B. Mullin, B.W. Straugham, and W.S. Briekell, *J. Phys. Chem. Solids*, 26, 1965, p. 782.

[61] W.M. Duncan, G.H. Westphal, J.B. Sherer, *Electron Dev. Lett.*, 4, 1983, p. 199.

[62] L. Pekavek, *Czech. J. Phys. B*, 20, 1970, p. 857.

[63] Y. Nanishi, S. Ishida, and S. Miyazawa, *Jpn. J. Appl. Phys.*, 22, 1983, p. L54.

[64] S. Miyazawa, T. Honda, Y. Ishii, and S. Ishida, *Appl. Phys. Lett.*, 44, 1984, p. 410.

[65] T. Honda, Y. Ishii, S. Miyazawa, H. Yamasakai, and Y. Nanishi, *Jpn. J. Appl. Phys.*, 22, 1983, p. L270.

[66] W.F. Finch and E.W. Mehal, *J. Electrochem. Soc.*, 111, 1964, p. 814.

[67] D. Effer, *J. Electrochem. Soc.*, 112, 1965, p. 1020.

[68] J.R. Knight, D. Effer, and P.R. Evans, *Solid State Electronics*, 8, 1965, p. 178.

[69] T. Nozaki, M. Ogawa, H. Terao, and H. Watanabe, *Inst. Phys. Conf. Ser.*, 24, 1975, p. 46.

[70] H.M. Manasevit, *Appl. Phys. Lett.*, 12, 1968, p. 156.

[71] H.M. Manasevit and W.I. Simpson, *J. Electrochem. Soc.*, 116, 1969, p. 1725.

[72] The entire issue of *J. Cryst. Growth*, 55 (1), October 1981.

[73] J.F. Gibbons, W.S. Johnson, S.W. Mylroie, *Projected Range Statistics, Semiconductors and Related Materials*, 2nd Edition. Stroudsburg, PA: Dowden, Hutchinson, & Ross, 1975.

[74] R.L. Chapman, J.C.C. Fan, J.P. Donnelly, and B.Y. Tsaur, *Appl. Phys. Lett.*, 40, 1982, p. 805.

[75] M. Kuzuhara, H. Kohzu, and Y. Takayama, *Appl. Phys. Lett.*, 41, 1982, p. 755.

[76] K. Tabatabaie-Alavi, A.N.M. Masum Choudhurn, and C.G. Fonstad, *Appl. Phys. Lett.*, 43, 1983, p. 505.

[77] M. Bujatti, A. Cetronio, R. Nipoti, and E. Olzi, *Appl. Phys. Lett.*, 40, 1982, p. 334.

[78] N. Tsukada, S. Sugata, and Y. Mita, *Appl. Phys. Lett.*, 42, 1983, p. 424.

CHAPTER 3

GaAs DEVICES: A TUTORIAL

This chapter briefly reviews the geometry and operation of the major GaAs devices, both analog and digital. A general understanding of GaAs devices is a necessary foundation for the remainder of the book; this chapter serves as an overview for people unfamiliar with these devices. Reference [1] contains a more complete description of the diode devices, but with a minimum of mathematics. A rigorous treatment of the physics of major semiconductor devices may be found in reference [2].

Section 3.1 describes the basic Schottky diode and introduces the voltage-controlled depletion region that exists at metal-semiconductor interfaces. Section 3.2 describes the major GaAs device, the field effect transistor (FET). Emphasis is placed on enhancement mode and depletion mode MESFETs (Metal Semiconductor FETs) for analog and digital use, although mention is also made of the JFET (Junction FET) and novel geometry FETs such as the VFET (vertical FET) and the permeable base transistor (PBT). Section 3.3 describes the major microwave diodes — the IMPATT diode and the Gunn diode. Section 3.4 describes GaAs devices that incorporate other material compositions (GaAlAs) such as the high electron mobility transistor (HEMT). Such devices are not yet approaching production status, but show tremendous promise. Virtually all topics in this book apply to their fabrication, and so they are included in this review.

3.1 SCHOTTKY DIODES AND VARACTORs

The Schottky diode is shown in two configurations in Figure 3.1. One is a planar style fabricated on a semi-insulating substrate (Figure 3.1(a)); the other is fabricated on a conductive substrate which forms one of the electrical contacts (Figure 3.1 (b)). This device incorporates the two major elements of many GaAs devices: an ohmic contact and a Schottky barrier. Each is a complex topic and a full chapter, encompassing theory, fabrication, and characterization, is subsequently devoted to each one. For now, the ohmic contact may be regarded simply as a metal contact that allows electrical current to flow into or out of the semiconductor in a linear, ohmic manner.

A *Schottky barrier* results when a metal is placed in intimate contact with a semiconductor such as GaAs (Chapter 12). It has the electrical characteristics of a diode, as shown in Figure 3.2. As described in detail in Chapter 12, there exists a region under the Schottky metal contact that is depleted of carriers (electrons for *n*-type GaAs). This zone is called the *depletion region*. The distance it extends into the slice is a function of the voltage applied between the metal and the semiconductor as indicated in Figure 3.3. Because this region is depleted of carriers, it acts as an insulator between the metal and the surrounding (undepleted) semiconductor material. Therefore, it functions as a capacitor, as shown in Figure 3.3. The boundary between the depletion region and the undepleted region is not perfectly sharp, but it is useful to assume that it is. In this case, the depth of the depletion region is given by (see Chapter 12)

$$w^2 = \frac{2\epsilon(V + V_{bi})}{qN}$$

where w is the depth of the region, q is the electron charge, ϵ is the permittivity of the semiconductor, N is the doping concentration, and V is the applied voltage. V_{bi} is the "built-in" voltage and represents the fact that a depletion region exists even in the absence of external volltage. (Confusion sometimes arises regarding the signs of the voltages in the above expression. Hence $V < 0$ wil expand the depletion region.) If end effects are ignored and the depletion region is assumed to have a sharp boundary, the capacitance associated with the Schottky barrier is simply that of a parallel plate capacitor (Figure 3.3):

$$C = \frac{\epsilon A}{w}$$

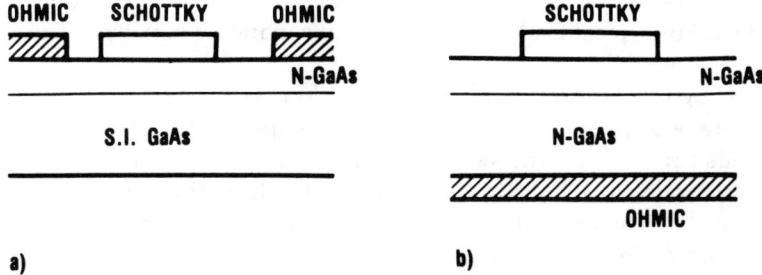

Figure 3.1 Two configurations of a Schottky diode: a) planar diode fabricated on conductive material over a semi-insulating substrate; b) diode fabricated on a conductive substrate which serves as the ohmic contact.

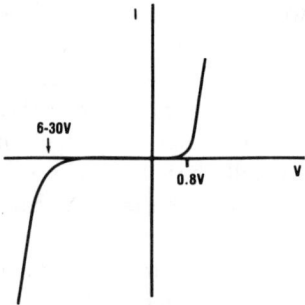

Figure 3.2 IV characteristic of Schottky diode.

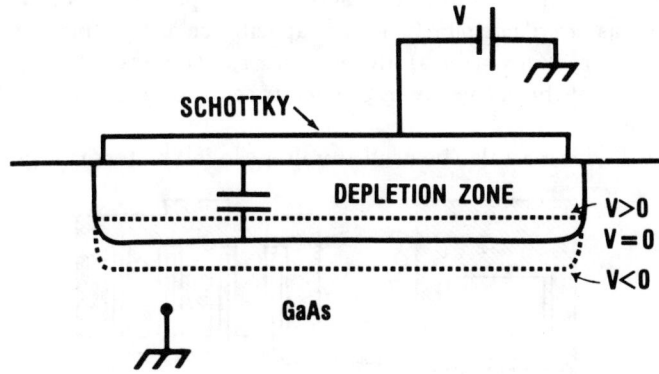

Figure 3.3 Depletion region under a Schottky diode formed on n-GaAs. Free electrons are excluded from the region, which acts as the dielectric of a capacitor. The depth that the depletion zone extends into the GaAs is a function of the doping profile and the applied voltage.

where C is the capacitance and A is the area of the Schottky metal contact. For typical GaAs devices this capacitance is usually in the picofarad range.

As shown in Figure 3.3, a depletion region exists even when no external voltage is applied. If a negative voltage is applied, the depletion region extends further into the semiconductor. If a positive voltage is applied (counteracting the built-in voltage) the depth of the depletion region decreases. If the applied voltage becomes too large in the positive (forward) direction or in the negative (reverse) direction, breakdown phenomena occur and a substantial current flows, as shown in Figure 3.2. The forward breakdown voltage is approximately 0.8 V. The reverse breakdown voltage depends on the doping concentration and other factors (see Chapter 12), but generally is in the range of 6 to 30 V. For voltages between the forward and reverse breakdown, the width of the depletion region (and its capacitance) may be controlled by the applied voltage.

The Schottky diode may be used as a simple diode or as a voltage-controlled capacitance (a variable reactance or *VARACTOR*). When used as a VARACTOR, the amplitude of the rf signal is generally small. In this case, the dc bias voltage determines the capacitance of the diode and the rf voltage swing is so small that it does not substantially alter that capacitance. VARACTORs may be used as tuning elements in *voltage-controlled oscillators* (VCOs).

Figure 3.4 shows two typical geometries used for Schottky diodes. The type shown in Figure 3.4(a) is often used for C-V profiling (Chapter 18). The interdigitated style (Figure 3.4(b)) is used to decrease the parasitic resistance (usually undesirable) that exists between the bottom of the depletion region and the ohmic contacts. Various parameters of VARACTORs such as *capacitance ratio* (ratio of capacitance at zero bias to capacitance at the breakdown bias) and *breakdown voltage* are affected by the doping profile of the active layer. Capacitance ratios greater than 10 to 1 are possible.

Shottky diodes may also be used as voltage level shifters in digital logic circuits.

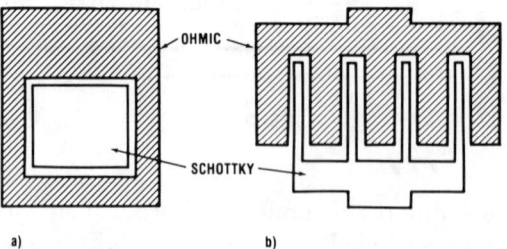

Figure 3.4 Two typical geometries used for Schottky diodes: a) single electrode style, often used to obtain CV profiles (Chapter 18); b) interdigitated style, used to minimize series resistance.

3.2 FIELD EFFECT TRANSISTOR

The *field effect transistor* (FET) has been the dominant GaAs device for analog purposes such as amplifiers (both power and low noise), oscillators, and switches. It has also figured prominently in the development of GaAs digital logic. Two types of FETs exist — the *junction FET* (JFET) and the *metal-semiconductor* FET (MESFET). The JFET uses a *p-n* junction instead of a Schottky barrier for the gate. A *p-n* junction is similar to a Schottky barrier in that it has a depletion region and a built-in voltage associated with it. The JFET acts exactly like the MESFET. The MESFET, which is by far the more popular, and is easier to fabricate, uses a metal Schottky barrier gate (Chapter 12). It is the type of FET that will be emphasized here.

The JFET was first analyzed in 1952 [3]; the first operating JFET was reported the next year [4, 5]. The MESFET was proposed in 1966 [6] and fabrication reported the following year [7].

3.2.1 Basic Geometry and Operation

Cross-sectional diagrams of both types of FETs are shown in Figure 3.5. The FET consists of two ohmic contacts called the *source* and the *drain* that allow current to flow into or out of the GaAs. Between these two contacts is a metal strip called the *gate*. For historical reasons the dimension of the gate parallel to the source-drain space is called the *gate length* (Figure 3.5) and the other direction is called the *gate width*. Because gate lengths are often less than one micron, and gate widths are many microns, this terminology is the reverse of common English usage. Only the surface layer (usually < 0.3 μm) of the GaAs is electrically conductive because the underlying material is insulating. The FET is isolated (by etching or other means) so that the only path for current flow is directly between the source and the drain.

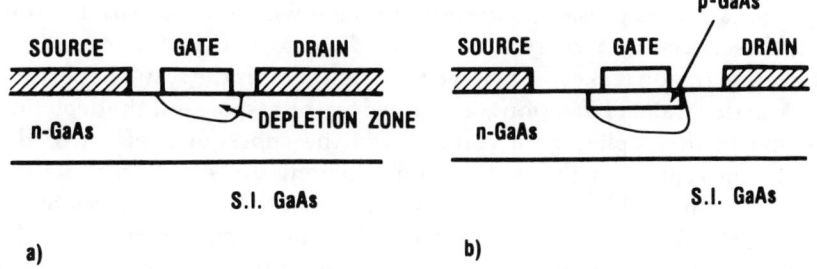

Figure 3.5 The two major types of GaAs field effect transistors (FETs): a) metal-semiconductor FET (MESFET); b) junction FET (JFET).

Usually the source of the FET is made the common or ground terminal. A positive dc voltage (usually 4 to 10 V) is applied to the drain terminal. In the absence of the gate, the IV characteristic (a plot of source-drain current as a function of source-drain voltage) appears as shown in Figure 3.6. The source-drain current, I_{sd}, becomes saturated for reasons described in section 2.2. It may appear on a measuring instrument (such as a curve tracer) as indicated in either Figure 3.6(a) or (b). The latter appearance often occurs due to negative differential resistance in the device (section 2.2) and/or oscillations in the measuring circuitry. The voltage at which the current changes from linear to saturated is called the saturation voltage, V_{sat}. The quantity I_{sat} may be defined as the maximum current in the case of Figure 3.6(b). In the case shown in Figure 3.6(a), I_{sat} is defined to be the saturation current at some arbitrarily selected voltage. (Note: as viewed on an instrument such as a curve tracer, the IV characteristic may appear double valued. That is, the trace formed by the upward voltage sweep will not exactly overlay the trace formed by the downward voltage sweep. This phenomenon is known as *looping*.)

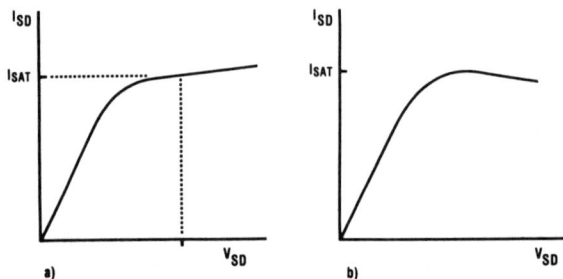

Figure 3.6 The IV characteristic of FET before gate fabrication (source-drain current as a function of source-drain voltage): a) without oscillation effects; b) with oscillations or negative differential resistance (see text).

Once a gate is present, a depletion region will exist beneath it, from which source-drain current is excluded. Note that in Figure 3.5 the depletion region is skewed toward the drain. This is the result of positive bias on the drain. FET action is explained by the response of the depletion region to the applied gate voltage, and the subsequent effect of the depletion region on the source-drain current. An exact treatment is rather complex [2], especially for long gate length (long compared to the channel thickness). For short gate length (as most modern FETs are) the following discussion qualitatively represents the situation. Applying a voltage to the gate (with respect to the source) changes the depletion zone depth (described in section 3.1) and hence changes the source-drain current. If the gate voltage is negative enough, the depletion region will extend to the bottom of the active channel and (almost) no current will flow. This condition is known as *pinch-off*. The required gate voltage, called the *pinch-off voltage*, is usually −1 to −6 V for most FETs. (It should be noted that gate voltage is usually negative; colloquial usage often assumes this and therefore "3 volts" may really mean −3 volts.) A small change in gate voltage can cause a large change in source-drain current. This is the fundamental principle that makes the FET work as an amplifier. Figure 3.7(a) contains a typical IV characteristic of an FET. Note that in this case (gate present) the quantity that was referred to as I_{sat} above (without a gate) is now designated I_{dss}, the saturated source-drain current. These two terms, I_{sat} and I_{dss}, may sometimes (improperly) be used interchangeably in colloquial conversation.

If forward bias (positive voltage) is applied to the gate, the depth of the depletion region will decrease from its zero bias value; therefore, current greater that I_{dss} will flow as shown in Figure 3.7(b). The maximum current that will flow under forward gate bias, designated I_{max}, is an important parameter in power FETs.

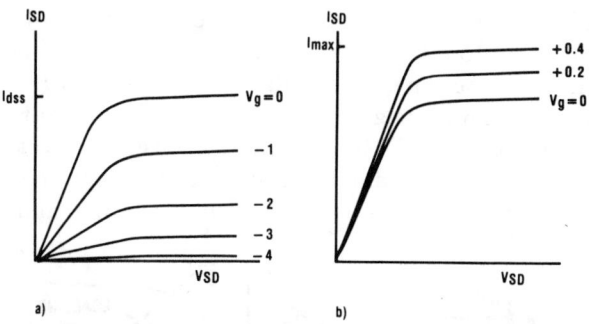

Figure 3.7 The IV characteristic of a FET: a) as usually displayed on a curve-tracer, using only negative gate bias; b) with positive gate bias applied.

The change in source-drain current as a function of gate voltage is an important FET parameter. It is called the *transconductance*, g_m, and is measured in inverse ohms, or mhos. In practice, it is usually given in millimhos. Although g_m is fundamentally defined as dI_{sd}/dV_g, common practice is to measure this quantity using a fixed gate voltage step, such as 1 volt. It is important to know the conditions under which the transconductance was measured because it is a function of both the gate voltage and the gate voltage step. This is apparent in Figure 3.7. The change in current that occurs between $V_g = 0$ V and −1 V is greater than that which occurs between −3 V and −4 V. Hence, $g_m (0\ V) > g_m (-3\ V)$. Transconductance is often specified at 0 V, but because of this effect, using a gate voltage step of 0.5 V will yield a larger g_m than using a step of 1.0 V.

Transconductance is obviously also a function of the total gate width of the device. Therefore, the gate width must be known or, alternatively, the transconductance is specified for some unit gate width, usually 1 mm. The transconductance of GaAs Fets are typically 100 to 200 mmho/mm at $V_g = 0$ V.

The basic operation of the FET as an amplifier is shown in Figure 3.8. In the common-source configuration the rf input is applied between the source and the gate, and the rf output signal is taken between the source and the drain. Dc bias voltage must also be supplied to the gate and to the drain. Other circuit elements are generally necessary to prevent the rf signal from entering the dc bias networks and to prevent the dc bias from reaching other portions of the circuit; these elements are not included in the figure. As indicated in Figure 3.8(b), during ac operation the voltage and current of the FET follow an ac *load line*, the slope of which is related to the impedance matching. For high output power the FET must operate over as much of the load line as possible. It is limited on one end by I_{max} and the saturation voltage (forward breakdown of the gate occurs near I_{max}). It is limited on the other end by the reverse breakdown of the gate. For small signal operation, the FET will operate over a small portion of the load line. In this case the gate circuit will draw almost no current. For high power operation, the gate circuit may draw substantial current as the rf cycle swings the FET into forward and reverse breakdown. This description has been exceedingly basic. Circuit operation of microwave FETs is a formidable subject in electrical engineering, and many books treat this topic [8,9].

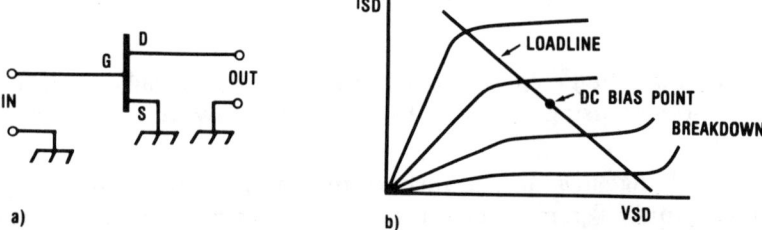

Figure 3.8 The FET used to amplify an ac signal: a) simplified circuit diagram for use of a FET in the common source configuration; b) the load line representing the operating voltages and current of the FET. The slope of the load line is related to the impedance matching.

The type of FET described above is a *depletion mode* FET because source-drain current flows at zero gate bias and negative gate bias must be applied to expand the depletion region and reduce the current to zero. In almost all cases, analog FETs (amplifiers, oscillators, etc.) are depletion mode FETs. However, it obviously is possible to construct the FET so that the zero bias depletion region fills the channel and pinches it off, preventing current flow at zero gate bias. Such a device is an *enhancement mode* FET and positive gate bias must be applied to shrink the depletion region and allow current flow. The useful gate voltage range is limited, however, because the gate will undergo forward breakdown near V_g = +0.8 V. Enhancement mode FETs have high g_m, draw little power, and are *normally off*, all of which make them very attractive for use in digital logic circuits.

3.2.2 Small-Signal Equivalent Circuit

For computer aided circuit design and other reasons, it has been found enormously useful to model FETs using an equivalent circuit model of the type shown in Figure 3.9. The circuit employs a voltage-controlled current source (available in computer circuit analysis programs) to model the effect of gate voltage on source-drain current. The voltage across the gate capacitance (C_{gs} in Figure 3.9) controls the current source, often with a phase delay included. Thus, the current generator produces a current given by

$$I = g_m V_{cgs} e^{-j\omega t}$$

where this g_m is the *intrinsic transconductance*, which is greater than the transconductance measured externally because of parasitic source resistance (see Chapter 18), V_{cgs} is the rf voltage amplitude across the depletion zone capacitance C_{gs}, ω is the angular frequency of the applied rf voltage, and t is the phase delay (usually in the low picosecond range). The current appears across the drain resistance, R_{ds}. Sometimes other circuit elements are included in the model to represent other parameters such as pad capacitances or bond wire inductances. The elements in the box labeled "intrinsic" represent the fundamental elements responsible for FET action. Other resistors and capacitors generally represent undesirable parasitic elements which can be reduced by good design, but never eliminated. Many of the circuit element values are functions of the applied gate and drain dc biases. Therefore, this type of model is valid only for small-signal rf operation in which the amplitude of the rf signal does not appreciably change the voltages caused by the dc bias. Large signal analysis is possible but is beyond the scope of this review.

Because such circuit models can accurately predict the small-signal microwave rf performance of FETs, these circuits can be used to represent FETs in computer-aided circuit design programs. Also, because the equivalent circuit values are closely related to the physical parameters of the FET, they are useful for analysis of FETs.

66 GaAs Processing Techniques

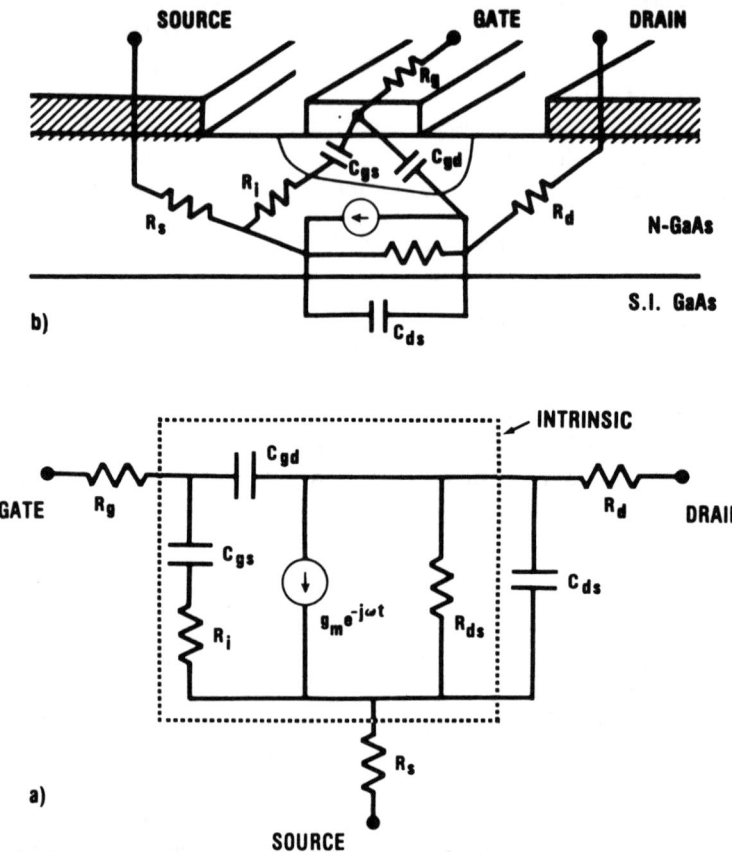

Figure 3.9 Small signal equivalent circuit of a FET: a) equivalent circuit; b) physical relationship to the FET structure.

Whether the equivalent circuit values are desired for analysis or for use in circuit design, they must be deduced from measurements on a specific FET. Some of these values may be obtained by direct, dc measurements. The source resistance is one example (see Chapter 18). But other parameters, such as C_{gd}, must be derived from rf (usually microwave) measurements. This is a two-step process. First, the rf characteristics of the FET are determined by measuring its *scattering parameters* (discussed below). Second, computer modeling programs are used to determine the circuit parameters that will generate those scattering parameters (the circuit elements that were derived from dc measurements are entered into this optimizing program as fixed values). This procedure yields the remainder of the desired circuit values.

The scattering parameters (S-parameters) are four complex numbers, $S_{11}, S_{12}, S_{21},$ and S_{22}, that represent the relationship between incident and reflected voltage waves (having amplitude and phase) at the input and output of the FET. If a_i and b_i represent normalized complex voltage waves incident and reflected at the ith port (1 = input; 2 = output) then

$$b_1 = S_{11}a_1 + S_{12}a_2$$

$$b_2 = S_{21}a_1 + S_{22}a_2$$

where all variables are functions of frequency. The S-parameters are simply related to important quantities. For example,

$$|S_{11}|^2 = \frac{\text{Power reflected from the input}}{\text{Power incident on the input}}$$

$$|S_{21}|^2 = \text{Power gain with load and source impdeance, } Z$$

where Z is the reference impedance for all measurements, usually 50 ohms. Similarly, $|S_{12}|$ is related to reverse isolation.

Knowledge of the scattering parameters is essential in the design of circuits that include microwave FETs. Commercial equipment exists to measure these scattering parameters up to 26 GHz. Even so, great care must be taken in designing appropriate test fixtures to hold the FET, and in eliminating the effects of the test fixture from the measurements. A very good and succinct description of S-parameters and their use in circuit design may be found in a Hewlett-Packard application note [10].

The equivalent circuit of Figure 3.9 may be used to illustrate the definition of two generally used terms regarding FETs. The cutoff frequency, f_t, is defined to be the frequency at which the current gain of the intrinsic FET is zero (the current through C_{gs} is equal to the current of the current generator) and is given by the expression

$$f_t = \frac{g_m}{2\pi C_{gs}}$$

The maximum frequency, f_{max} is the maximum frequency of operation when the remaining, parasitic elements are included.

Several basic configurations of FETs are shown in Figure 3.10. There are two basic types. One is called *interdigitated* and is represented by Figure 3.10(a) and (b). In this case the gate fingers are fed from the ends and total gate width is obtained by paralleling as many fingers as desired. A

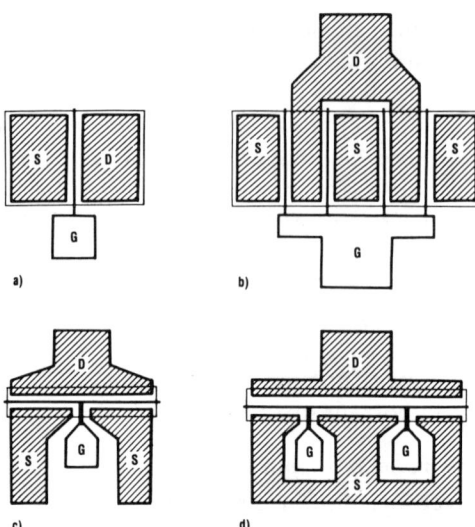

Figure 3.10 Several FET configurations: a) single finger, end-fed FET; b) multiple finger, interdigitated gate FET; c) T-gate FET; d) Pi-gate FET.

large amount of total gate width can be fabricated within a small area. The second type of FET configuration is represented in Figure 3.10(c) and (d). In this case the gate is one single, continuous strip that is fed at one or more central points. This configuration is often called a *T* gate or a *Pi* gate because of the appearance from above. This configuration is generally limited to smaller gate widths because of size problems; there are difficulties in properly feeding a microwave signal into a physically large device. A fuller explanation of this point, and the various issues involved in choosing the width (the long direction!) of individual gate fingers is beyond the scope of this review. However, measurement of various FET parameters must take the configuration of the FET into account. This will be explored in Chapter 18.

The above has been a generic description of the MESFET. Modern FETs have various modifications from the above structure depending on the function of the FET. These will be considered briefly in the following subsections.

3.2.3 Power FETs

Figure 3.11 shows a more realistic cross section of a power (or low noise) FET. There are two grooves or slots etched into the GaAs between the source and drain contacts. Most analog FETs employ the smaller etched slot into which the gate is placed. Some also employ the larger slot. Use of the narrow slot results in what is known as a *recessed gate structure*. That slot is usually etched into the GaAs using the gate photoresist as a mask. The etching action in the GaAs undercuts the photoresist mask. Thus, upon gate metal evaporation, the gate metal is automatically placed in the middle of the etched slot. Such fabrication processes are called *self-aligned*. This gate recess etching is usually done while monitoring the source-drain current and the slot is etched until the current is reduced to a predetermined target value (Chapter 12). The gate recess has several advantages. The extra channel thickness on either side of the gate slot results in decreased source-to-gate and gate-to-drain parasitic resistances. As shown in Figure 3.11, the recessed gate geometry also places the bottom of the gate below the surrounding GaAs surface and its associated surface depletion region (section 2.2.4). Therefore there is no impediment to current flow when the gate voltage swings strongly forward (shrinking the depletion zone under the gate) as it does in power FET operation. If the geometry were planar (if there were no recessed slot) the depletion region existing under the surrounding GaAs surfaces would restrict the ability of the gate's depletion region to modulate the current under positive gate bias.

The recessed gate geometry does have one disadvantage. It increases gate-to-drain feedback, represented by C_{gd} in the equivalent circuit of Figure 3.9. Increased feedback capacitance reduces the gain and can make the FET less stable by decreasing the isolation between input and output. The exact reasons for this increase in feedback capacitance are not firmly documented, but almost certainly involve the extension of the depletion region toward the drain side of the slot (Figure 3.11). Planar FETs exhibit the least feedback capacitance. Hence the second, wider recess slot of Figure 3.11 is sometimes employed so that the narrow slot need not be as deep. The relative dimensions of the two slots represent tradeoffs among parasitic resistance, feedback capacitance, breakdown voltage, and other parameters. The dual recess geometry also results in improved reliability [11].

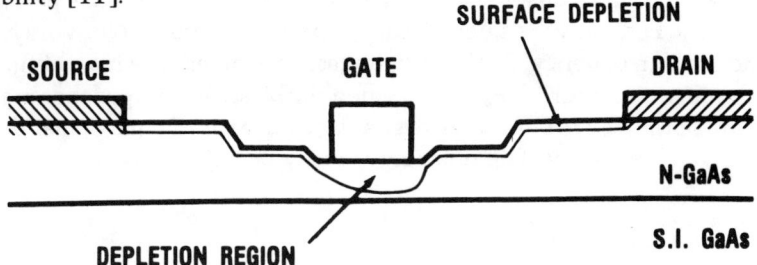

Figure 3.11 Generic cross-sectional configuration of a GaAs FET. The depletion region is skewed toward the drain because of the positive drain bias.

Power FETs generate a great amount of heat, which can degrade FET performance and affect reliability. Because GaAs is a poor thermal conductor, power FETs are usually 50 μm to 150 μm thick and are soldered onto thermally conductive material.

The performance of a FET is limited by various parasitic elements. One of the most important is source inductance: the amount of inductance that exists between source (in the GaAs, next to the gate) and true ground — usually the metal on which the FET is mounted. In most applications, most of this inductance arises from bond wires connecting the heat sink (ground) to the bonding pads on the FET, and also to any air bridge interconnection between source pads. Use of *VIAs*, plated-through holes from the front to the back of the FET, can reduce this inductance. Figure 3.12 shows a cross section of a power FET incorporating a combination of VIAs and air bridges. VIAs are also extremely important for monolithic ICs because FETs on such an IC may not be near the edge of the chip, but will require a low inductance ground.

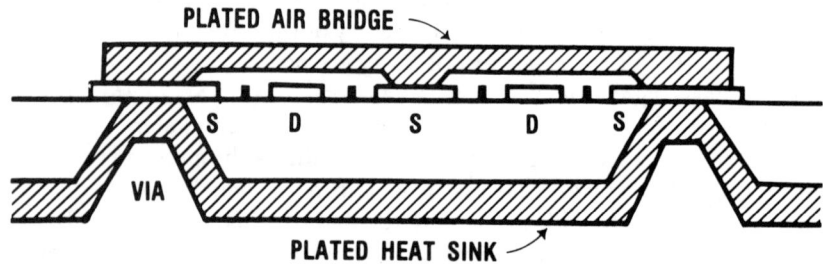

Figure 3.12 Cross section of a power FET incorporating air bridges, via holes, and a plated heat sink.

3.2.4 Low Noise FETs

Low noise FETs differ slightly from power FETs. The noise performance of a FET may be quantified by the noise figure, NF, which is a function of frequency, FET bias voltages, and impedance matching. The optimum noise figure, NF_o, occurs when the biases and impedance matching are optimized to obtain the best noise figure for that particular device at a given frequency. The noise figure may be defined by

$$NF = \frac{(signal\ /\ noise)_{in}}{(signal\ /\ noise)_{out}}$$

That is, it is the excess noise that is added to the signal by the FET. A semi-empirical equation for low-noise FET performance is [12]

$$NF_o = 1 + 2\pi K_f f C_{gs} \sqrt{(R_s + R_g)/g_m}\ 10^{-3}$$

where K_f is a frequency-dependent constant having to do with material parameters, f is the frequency, C_{gs} is the gate-source capacitance, R_s is the source resistance, R_g is the gate resistance, and g_m is the transconductance. This means that the most important elements of a low noise FET are a short gate (usually 0.5 μm or less) to minimize C_g, high channel doping to increase g_m and decrease R_s, and a short source-to-gate spacing to decrease R_s. Another semi-empirical equation has been developed that relates optimum noise figure to physical device parameters [12]:

$$NF_o = 1 + K_f f \frac{(NL^5)^{1/6}}{a} \left[\frac{17z^2}{hL_g} + \frac{2.1}{a_1^{.5} N_1^{.66}} + \frac{1.1L_2}{a_2 N_2^{.82}} + \frac{1.1L_3}{a_3 N_3^{.82}}\right]^{1/2}$$

where z is the gate finger width, L_g is the physical gate length, L is the effective gate length, and a and N represent the thicknesses and doping concentrations in various zones of the FET. All these are defined in Figure 3.13. Both of the above equations have proven accurate once the constant K_f is experimentally determined for the material. K_f is approximately 0.040. Hence, these equations serve as excellent guides for low noise FET design.

In general, high source-drain current contributes to noise by electron scattering and this noise is reduced as the current is reduced. However, reducing the current too close to pinch-off reduces the transconductance which causes increased noise figure because of decreased gain. There will exist an optimum gate bias that represents the best compromise. This is a somewhat oversimplified analysis, but it illustrates the basic considerations. Low noise FETs are often biased near 15% of I_{dss}. The general cross section of a low noise FET will be similar to that of the power FET in Figure 3.11, except the gate may not be placed in the middle of the source-drain spacing, but offset toward the source side. Generally, the source-drain spacing will be smaller also. Usually the gate recess slot is used, but some low noise FETs have been built in a planar fashion.

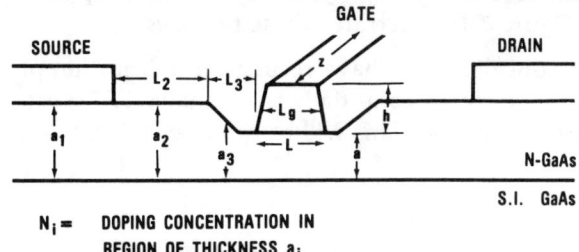

Figure 3.13 Diagram of a low noise FET defining the quantities used in expressions to predict optimum noise figure (see text).

3.2.5 FETs For Digital Logic

FETs for use in digital logic circuits are generally enhancement mode FETs because of low power consumption and high transconductance. Enhancement mode FETs have very thin channels beneath the gate because the zero bias depletion zone must extend completely through the active channel and pinch off the FET. Above the saturation voltage, the dependence of source-drain current of FETs on the gate voltage is usually approximately quadratic (for both enhancement and depletion mode FETs). That is, the dependence of source-drain current on gate voltage may be represented by the equation

$$I_{sd} = K(V_g - V_t)^2$$

over most of the operating range, as indicated in Figure 3.14. V_t is defined as the *threshold voltage*; this terminology and symbol are used for enhancement mode FETs. For depletion mode FETs the pinch-off voltage, V_p, is often used in place of V_t in the above equation.

FETs used in digital logic circuits are intended to be binary devices: the gate voltage is either high or low and current is either high or low. Further, the difference between the high and low gate voltage should be small so that the transition between states can be rapid. This requires a tight tolerance on the threshold voltage, V_t. An important distinction between digital and analog FETs is that analog FETs can generally tolerate far larger variations in the physical parameters. But logic circuits may require thousands of FETs that have nearly identical characteristics, especially threshold voltage. It is very difficult to achieve the desired V_t uniformity if a recessed gate geometry is used. The recess etching process tends to be insufficiently uniform over a slice, particularly for the thin-channel enhancement mode FETs. For this reason, enhancement mode FETs for digital logic purposes are usually planar devices. Yet this planarity raises another difficulty. The planar, enhancement mode FET has a very thin channel; therefore, it would tend to have high source-to-gate and gate-to-drain parasitic resistances. The surface depletion that exists even under free GaAs surfaces compounds the problem. It is, therefore, useful to extend highly doped n^+ material up to the very edge of the gate. Figure 3.15 illustrates these features.

The above considerations have resulted in fabrication processes for digital FETs that differ substantially from analog FETs, especially with regard to gate fabrication. This will be considered in Chapter 12.

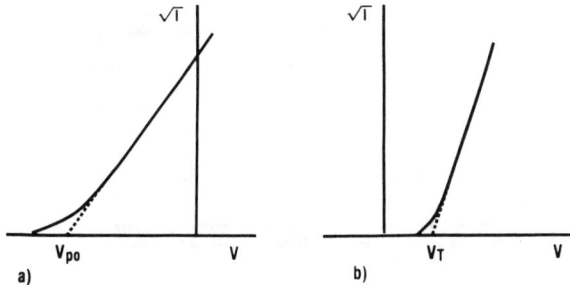

Figure 3.14 Transfer functions of: (a) a depletion mode FET and, (b) an enhancement mode FET. The square root of the source-drain current is plotted as a function of gate-source voltage (drain voltage is fixed).

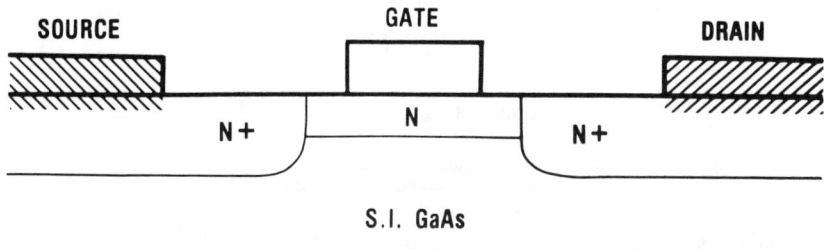

Figure 3.15 Cross section of an enhancement FET used for digital logic. Highly doped n^+ GaAs material that extends nearly to the gate edge serves to lower series resistance.

3.2.6 Novel-Geometry FETs

Two other interesting types of GaAs FETs are being investigated in the laboratory. These are the *permeable base transistor* (PBT) and the *vertical FET* (VFET), shown in Figures 3.16 and 3.17, respectively. The operation of both is essentially identical. Current flows vertically between the metal gate fingers. These fingers are formed by standard evaporation techniques, and so result in exceedingly small gate lengths. This small, effective gate length promises high frequency operation. The only major difference between the two devices is that the PBT uses epitaxial overgrowth to completely fill the layer over the gates with GaAs material. This requires use of a refractory metal such as tungsten. Achieving complete overgrowth has proven difficult. In fact, there are significant processing problems associated with both devices. They are not likely to mature in the near future.

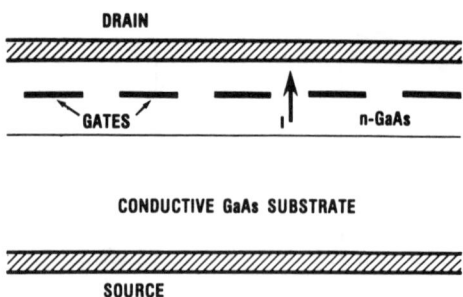

Figure 3.16 Cross section of permeable base transistor (PBT).

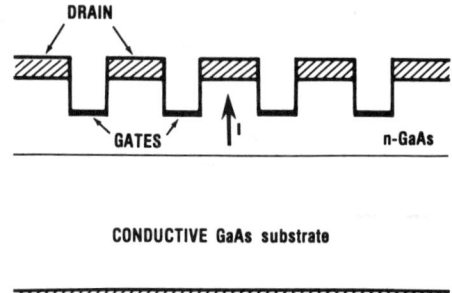

Figure 3.17 Cross section of vertical FET (VFET).

3.3 MICROWAVE DIODES (IMPATT AND GUNN)

This section reviews two-terminal GaAs devices used to generate or amplify microwave signals. (Although the Schottky diode described in section 3.1 is a diode used at microwave frequencies, it is used only as a diode and not as a signal generator or amplifier.) These devices have the advantage that they can produce higher power and operate at higher frequencies than FETs. Production FETs usually operate below 18 GHz although laboratory devices have functioned at 60 GHz [13]. Microwave diodes can operate above 100 GHz. The disadvantages of these devices include noise and lack of isolation. They are noisier than FETs, although Gunn diodes can have relatively low noise. The two-terminal nature of these devices results in little isolation between input and output; therefore, impedance matching is crucial.

The diodes considered here fall into one of two classes. The first class consists of diodes, containing a *p-n* junction, that are operated in the avalanche mode and that incorporate carrier drift. These are the IMPATT diodes and their variations. The second class consists of diodes that make use of the transfer of electrons from the central valley of the conduction band to an upper valley resulting in reduced velocity (see section 2.2.4). These are designated *transferred-electron devices* (TEDs) or Gunn diodes.

All of these diodes make use of negative resistance effects. The Gunn diodes do not contain a *p-n* junction and the negative resistance effects derive from the bulk characteristics of the GaAs material. They are called diodes in the sense that they are two-terminal devices. The negative resistance effects in the IMPATT class result from the time delays involved in the avalanche and drift processes. Both classes represent complex devices that can operate in several modes. A more detailed, but still narrative, treatment may be found in reference [1]. A detailed description of the device physics of these diodes may be found in reference [2].

3.3.1 IMPATT Devices

IMPATT stands for *impact ionization avalanche transit time*. The IMPATT diode exists in a number of variations and operating modes. Only the major ones will be considered here. Negative resistance at microwave frequencies is produced by using the time delay of the avalanche process (*impact ionization avalanche*) and the delay of the drift process (*transit time*) to cause the rf current to be approximately a half cycle out of phase with the rf voltage. Achieving negative resistance by transit time effects was first suggested in 1954 [14]. W.T. Read proposed a diode structure in 1958 composed of p^+-n-i-n^+ semiconductor layers (the i stands for intrinsic) [15]. This particular type of IMPATT is known as a *Read diode*. In this, and all cases below, the semiconductor layers can be reversed in type, and the role of electrons and holes interchanged. Thus, a n^+-p-i-p^+ structure is also a Read diode. The first oscillations of an actual Read diode were not reported until 1965 [16] because of fabrication difficulties. The first IMPATT oscillations (in any diode structure), however, were observed in a silicon diode in 1965 [17]. Review information on IMPATT diodes through 1976 may be found in reference [18]. Reference [19], a 1973 book, is a good general reference on IMPATTs.

The basic operation of IMPATT diodes may be illustrated using the Read structure shown in Figure 3.18. The diode is biased so that the p^+-n junction is reverse, biased into avalanche. This applied bias also depletes the intrinsic (i) region. Hence, electrons generated by impact ionization (avalanche) at the p^+-n junction drift across the intrinsic region at the saturated drift velocity and are collected at the anode. The applied rf voltage results in more or less avalanche current being generated (Figure 3.19). The diode is constructed and operated so that the two time delays (avalanche and drift) result in the rf current being half a cycle out of

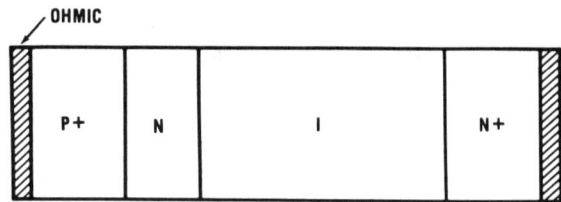

Figure 3.18 Basic structure of the Read (IMPATT) diode.

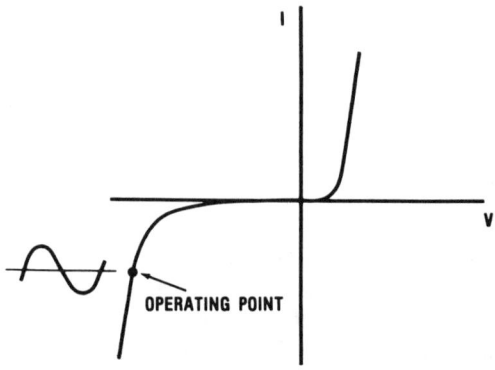

Figure 3.19 IV characteristic of a Read diode, showing the operating point.

phase with the rf voltage. The diode supplies energy to an outside resonant circuit instead of absorbing it. The Read diode illustrates the basic IMPATT principle. But many other geometries, even a simple *p-n* junction, can result in IMPATT operation [20].

Avalanche breakdown generates both holes and electrons. In the Read geometry, however, only the electrons are given a drift region. Such an IMPATT device is known as *single drift*. Generally, it is limited in efficiency to less than 15% (the theoretical limit is 30%). A *double drift* geometry, illustrated in Figure 3.20, consists of a p^+-p-n-n^+ structure. Avalanche occurs at the center *p-n* junction and both the electrons and the holes may drift through the *n* and *p* regions, respectively. This type of geometry has less capacitance per unit area, which means that larger devices can be used for a given circuit impedance. This results in high efficiency and high power [21].

Two further modifications of the Read structure are designated *hi-lo* and *lo-hi-lo*. The respective doping profiles are shown in Figure 3.21(a) and (b). The hi-lo structure replaces the intrinsic region of the Read structure with an *n* layer. The lo-hi-lo structure results in an almost uniform electric field over the first lo portion, so the maximum electric field can be lower than in the Read structure. At this writing, these device structures have only been realized in GaAs and InP.

There is a vast richness of device structures and operating modes that exist in IMPATT-like devices that is beyond the scope of this review. Reference [2] is an excellent source to pursue these topics.

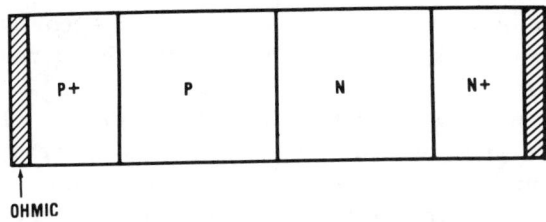

Figure 3.20 Double drift configuration of the IMPATT diode.

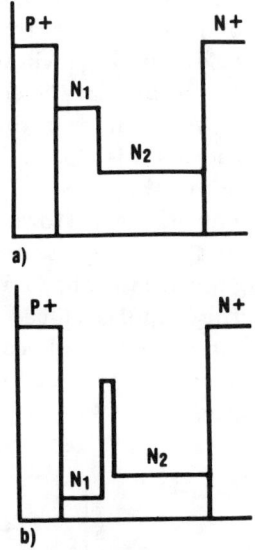

Figure 3.21 Other doping profiles used in IMPATT diodes: (a) hi-lo; (b) lo-hi-lo.

3.3.2 The Gunn Diode

The first device exhibiting microwave behavior caused by the transferred-electron process was reported by Gunn in 1963 [22, 23]. The next year it was pointed out [24] that Gunn's results could be explained by the *negative differential resistance* (NDR) effect [25,26]. Although Ridley, Watkins, and Hilsum initially described the NDR effects [25-28], Gunn's device results have resulted in his name usually being used to describe such devices. Hence, *transferred-electron devices* (TEDs) are generally called *Gunn diodes* (in the USA) and the underlying effect is generally called the *Gunn effect*. Comprehensive reviews may be found in references [29-33].

The Gunn or TED process was described in section 2.2.2. That section may be summarized by stating that electrons in GaAs continue to gain energy in an electric field until they have sufficient energy to transfer out of the central valley of the conduction band and into the satellite valley. In this valley the electrons behave as though they have a mass approximately twenty times greater than they had in the central valley. Thus, they exhibit a velocity reduction in response to an increase in electric field. Figure 3.22 illustrates the fundamental Gunn effect. If the field in the semiconductor is above the threshold field, E_t, electron velocity will begin to decrease and charge will accumulate. This results in a dipole forming between the accumulated electrons and nearby (depleted) regions (Figure 3.22(b)). When much of the voltage across the semiconductor can be dropped across this dipole, the electric field outside of the dipole drops below the threshold level. Therefore, once one dipole has formed, it is likely that no others will form while it still exists. Such a dipole is called a *domain* and will respond to the electric field in the semiconductor by drifting toward the anode. When it reaches the anode a current spike appears in the output circuit and the process begins again. Thus, the fundamental Gunn effect would consist of an output of current spikes separated by the time it takes a domain to drift through the length of the semiconductor. This is known as the *transit time mode* of a Gunn diode.

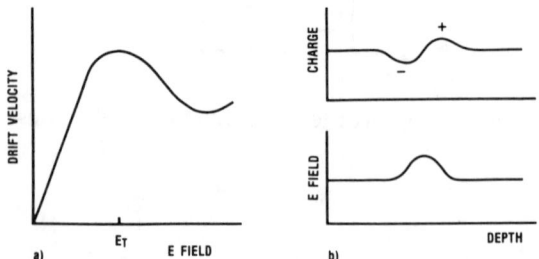

Figure 3.22 The negative differential resistance, or Gunn, effect: (a) velocity-field diagram; (b) formation of dipole.

The transit time mode, which is not the usual mode Gunn diodes are operated in, has a number of disadvantages. The frequency is determined only by the diode parameters and not by any external circuit. Efficiency (rf power out divided by dc power in) is also low, usually well under 10%. A more efficient and useful mode of Gunn diode operation is the delayed-domain mode. In this mode the Gunn diode is connected to an external tuned circuit and the operating parameters are adjusted so that at the operating frequency (determined by the external circuit) the negative excursions of the rf signal bias the diode below the threshold field so that no further domains can be generated, even after the present domain has reached the anode. As the rf signal swings more positive, the threshold field is surpassed and another domain can form. Thus, output current pulses are forced to occur at a frequency determined by the external circuit. Efficiency is higher in this mode.

Still another operating mode, the simplest one in principle, is the *limited space charge accummulation* (LSA) mode. In this mode the frequency (determined by an external circuit) is much greater than the frequency in the transit time mode. The excursions of the rf voltage above the threshold field are so brief that stable domains do not have time to form. Therefore, the electric field in the device remains above the threshold value for most of the rf cycle (because no domain is present to cause the field to fall outside of it). Hence, the electrons see a negative resistance as they flow across the device.

There are other operating modes, and many other subtleties of the Gunn diode. References [1] and [2] contain more complete information. Without going into detail, it should be noted that most operating modes require that the doping-length product of the device be greater than 10^{12} cm^{-2}:

$$NL > 10^{12} \text{ cm}^{-2}$$

3.4 HETEROJUNCTION DEVICES

A semiconductor junction between like materials that are doped differently is called a *homojunction*. The typical example is a *p-n* junction in any semiconductor. A heterojunction is a junction between materials of different composition, such as between GaAs and AlGaAs. Aluminum may be included the GaAs lattice in place of Ga atoms. The new material, AlGaAs, has a different bandgap and different properties from GaAs. These properties are functions of the amount of Al used. For device use, a ratio of 3 atoms of Al to 7 atoms of Ga is common. The significant feature of the GaAs/AlGaAs system is that the bandgap of the AlGaAs is larger than that of the GaAs. As will be seen below, this feature is useful

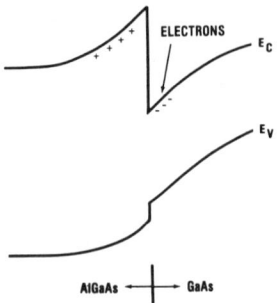

Figure 3.23 Energy band diagram of AlGaAs in contact with GaAs.

in constructing devices. Generally, the semiconductor layers for these devices are grown using molecular beam epitaxy, although metal-organic VPE has also been used successfully.

3.4.1 High Electron Mobility Transistor (HEMT)

The mobility of electrons in GaAs is limited by the impurity atoms introduced to dope the layer (*impurity scattering*). It would, therefore, be desirable to have electron flow occur in an undoped GaAs layer, where no impurity scattering limits mobility. This would be especially useful for digital applications because high mobility implies a fast turn-on characteristic. The *high electron mobility transistor* (HEMT) employs a doped AlGaAs layer adjacent to an undoped GaAs layer. Electrons are supplied from the doped AlGaAs layer, but electron flow takes place in the undoped, adjacent GaAs layer. This is possible because the bandgap discontinuity between the two materials causes electrons (supplied by donors in the AlGaAs) to stay in the GaAs as indicated in Figure 3.23. The electrons, however, are electrostatically constrained to stay close to the donor atoms. Therefore the electrons stay very close to the boundary between the two materials and are sometimes referred to as a two-dimensional electron gas. This gives rise to several acronyms, including 2DEG and TEG, the latter sometime being used in the acronym TEGFET. The HEMT has the potential of rivaling the superconducting Josephson junction technology for logic speed.

A typical cross-sectional structure of a HEMT is shown in Figure 3.24. A very thin (usually about 70 Å) layer of undoped AlGaAs is included between the doped AlGaAs and the undoped GaAs. This is because in the absence of such a layer some interface scattering of electrons from dopant atoms very near the heterojunction will exist. It is difficult to make excellent ohmic contacts to AlGaAs. Therefore, a layer of highly doped GaAs is usually grown over the AlGaAs to aid ohmic contact fabrication. These features are shown in Figure 3.24.

3.4.2 Heterojunction Bipolar Transistor (HJBT)

This device is different from the others considered because it is a bipolar device. It has great promise for digital applications. Electron flow is vertical through a thin base region instead of under a metal gate; this makes it fast. Further, the *threshold voltage* is set by the bandgaps and so is highly uniform over a slice. A cross-sectional diagram is shown in Figure 3.25. The device has the disadvantage that fabrication involves a complex materials processing sequence. It usually involves both molecular beam epitaxy and several implant steps.

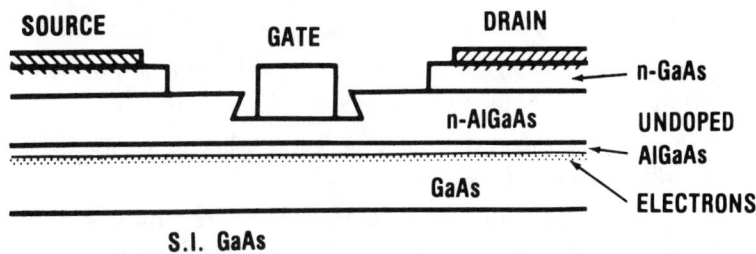

Figure 3.24 High electron mobility transistor (HEMT).

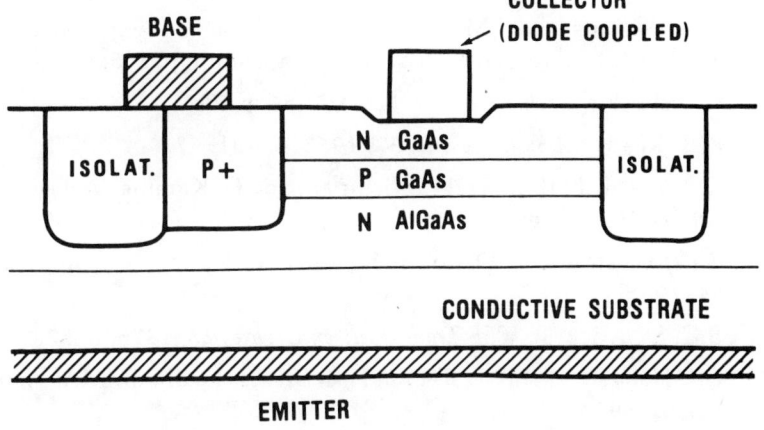

Figure 3.25 Heterojunction bipolar transistor (HJBT).

REFERENCES

[1] Ben G. Streetman, *Solid State Electronic Devices*, 2nd edition. New Jersey: Prentice-Hall, 1980.

[2] S.M. Sze, *Physics of Semiconducting Devices*, 2nd edition. New York: Wiley, 1981.

[3] W. Shockley, *Proc. IRE*, 40, 1952, p. 1365.

[4] G.C. Dacey and I.M. Ross, *Proc. IRE*, 41, 1953, p. 970.

[5] G.C. Dacey and I.M. Ross, *Bell Syst. Tech. J.*, 34, 1955, p. 1149.

[6] C.A. Mead, *Proc. IEEE*, 54, 1966, p. 307.

[7] W.W. Hooper and W.I. Lehrer, *Proc. IEEE*, 55, 1967, p. 1237.

[8] J.V. DiLorenzo, D.D. Khandelwal, *GaAs FET Principles and Technology*. Dedham, MA: Artech House, 1982.

[9] R. Soares, J. Graffeuil, and J. Oberegon, editors, *Applications of GaAs MESFETs*. Dedham, MA: Artech House, 1983.

[10] HP Application Note 95-1: "S-Parameter Techniques for Faster, More Accurate, Network Design," available through any Hewlett-Packard sales office.

[11] S.H. Wemple, et al., *IEEE Trans. Electron Devices*, 28, 1981, p. 834.

[12] H. Fukui, *IEEE Trans. Electron Devices*, 26, 1979, p. 1032.

[13] E.T. Watkins and J.M. Schellenberg, *Technical Digest*, 1981 IEEE MTT-S International Microwave Symposium, p. 145.

[14] W. Schockley, *Bell Syst. Tech. J.*, 33, 1954, p. 799.

[15] W.T. Read, *Bell Syst. Tech. J.*, 37, 1958, p. 401.

[16] C.A. Lee, R.L. Batdorf, W. Wiegman, and G. Kaminsky, *Appl. Phys. Lett.*, 6, 1965, p. 89.

[17] R.L. Johnston, B.C. DeLoach, Jr., and B. G. Cohen, *Bell Syst. Tech. J.*, 44, 1965, p. 369.

[18] B.C. DeLoach, Jr. *IEEE Trans. Electron Devices*, 23, 1976, p. 57.

[19] G.I. Haddad, *Avalanche Transit-Time Devices*. Dedham, MA: Artech House, 1973.

[20] T. Misawa, *IEEE Trans. Electron Devices*, 13, 1966, p. 137.

[21] R.E. Davis and D.E. Iglesias, *Proc. IEEE*, 59, 1971, p. 1222.

[22] J.B. Gunn, *Solid State Commun.*, 1, 1963, p. 88.

[23] J.B. Gunn, *IBM J. Res. Dev.*, 8, 1964, p. 141.

[24] H. Kroemer, *Proc. IEEE*, 52, 1964, p. 1736.

[25] B.K. Ridley and T.B. Watkins, *Proc. Physc Soc. Lond.*, 78, 1961, p. 293.

[26] C. Hilsum, *Proc. IRE*, 50, 1962, p. 185.

[27] B.K. Ridley, *J. Appl. Phys.*, 48, 1977, p. 754.

[28] C. Hilsum, *Solid State Electron.*, 21, 1978, p. 5.

[29] J.E. Carroll, *Hot Electron Microwave Generators*. London: Edward Arnold, 1970.

[30] P.J. Bulman, G.S. Hobson, and B.S. Taylor. *Transferred Electron Devices*. New York: Academic, 1972.

[31] L. Eastman, *Gallium Arsenide Microwave Bulk and Transit Time Devices*. Dedham, MA: Artech House, 1972.

[32] B. G. Bosch and R.W. H. Engelmann, *Gunn-Effect Electronics*. New York: Wiley, 1975.

[33] H.W. Thim, in *Handbook on Semiconductors*, Vol. 4 *Device Physics*, ed. by C. Hilsum. Amsterdam: North-Holland, 1980.

PART II.
GENERAL PROCESS TECHNIQUES

CHAPTER 4

CLEANING AND CLEANLINESS

4.1 INTRODUCTION

As used in this chapter, cleaning refers to removing undesired material from the slice before subsequent process steps (some of this material may be remnants from previous steps). Cleanliness refers to preventing contamination, to maintaining the level of cleanliness already present. Environmental and handling considerations (section 4.2) address cleanliness.

Cleanliness is crucial to achieving high yields and reproducible processes in the production of any semiconductor device. The effect of particles in causing defects is obvious. Less apparent, but even more insidious, are the effects of "minor" chemical contamination and poor environmental control. These can degrade or destroy virtually any aspect of the fabrication process. Metal adhesion, resist application and patterning, wet etching, dry etching, and plating are only a few examples. The semiconductor industry is full of horror stories of process lots ruined by some subtle departure from purity. In such cases, availability of the diagnostic techniques described in Chapter 19 can be enormously useful in aiding a rapid diagnosis of the problem.

Cleanliness is a relative term. No surface or chemical in the real world is completely pure. We shall adopt the common sense approach of defining *clean* as clean enough to result in a reliable, reproducible procedure for the particular process step in question. This standard can be rather stringent. A few monolayers of a contaminant can radically alter metal adhesion. Fortunately, most chemicals and gases can be reliably obtained in a high purity form because of the historical need of the silicon industry for them. Many chemicals are intended precisely for such use and are designated *semiconductor grade*. Nevertheless, it is still possible to obtain a bad lot. Initial purity is only part of the battle. Good etiquette must be rigorously observed to prevent subsequent contamination. Minor traces of contaminants on beakers, stirring rods, thermometers, tweezers, slice holders, or other items can degrade the purity of the chemicals.

Early research work in GaAs did not demand high cleanliness standards. Device area was small and only a few "good" devices were needed for research projects. Such fabrication could probably have been successful even in a parking lot. But those days are over. Production of semiconductor devices places a high premium on uniformity (both across-the-slice and slice-to-slice) and yield. Uniformity, of course, is one requirement for high yield. The near-term uses for GaAs products (Chapter 1), especially *monolithic microwave integrated circuits* (MMICs), require that thousands to millions of MMICs be produced with nearly identical operating characteristics. Uniformity in performance follows directly from uniformity in material and process.

Yield is tied to cost because affordable products demand good yield. GaAs analog microwave devices and ICs are relatively small by silicon standards, but GaAs digital logic circuits employ thousands of gates. Hence, they begin to face the same challenge faced by silicon ICs: one bad gate out of thousands can ruin the entire circuit. Elimination of particles becomes more important as the number of gates increases. Particles on the slice can cause pattern defects resulting in electrical failures such as shorts in capacitors or opens in gate fingers. Particles that cause openings in dielectric films can result in reliability problems. Particles that become included in resist films interfere with proper thickness uniformity and exposure. Particles can also damage masks used in contact photolithography. Resist operations in general must be free from particle contamination.

If defects are randomly distributed over the slice, as would be the case for particles falling on the surface, the yield is given by a Poisson distribution. The usual formula is

$$Y = e^{-DA_c}$$

where Y is the yield, D is the defect density, and A_c is the critical or active area of the slice. The critical area is the portion of the slice in which defects cause device failures. The source-drain space of FETs is one example. Note that the critical area will generally be much smaller than the total area of the slice. This formula pessimistically predicts yield to fall exponentially with increased active area. But total defects (crystal defects, for example) may not be uniformly distributed over a slice, and this consideration led to a more optimistic formula widely used in the silicon industry [1]:

$$Y = \int_0^\infty f(D) \, e^{-DA_c} \, dD$$

where $f(D)$ is a probability density function giving the defect density distribution. A number of distribution functions have been used in this model, including an exponential distribution which gives [2]

$$Y = 1/(1 + DA_c)$$

There are arguments that these modifications are too optimistic [3]. But these distinctions are mostly relevant for VLSI (*very large scale integration*) geometries employing tens of thousands of gates. GaAs has not yet reached that scale. For discrete devices and MMICs especially, the basic Poisson approach is a good rule of thumb.

A great many of the cleaning and cleanliness techniques developed in the silicon semiconductor industry can be applied directly to GaAs fabrication. Of course, the differences in the two types of devices (silicon based, and GaAs based) lead to different sensitivities to various contaminations.

4.2 ENVIRONMENT AND HANDLING (CLEANLINESS)

Cleanliness is preserved by good environmental and slice handling techniques. These include control of the number and size of particles present in the fabrication area, the temperature, and the humidity. Slice handling techniques are particularly important given the propensity of GaAs to cleave or shatter under mild stress. The cleanliness standards require that most fabrication occur in environmentally controlled areas called clean rooms. These rooms generally use only yellow light so that resist may be handled freely (photoresist is not sensitive to yellow light; ordinary room light would expose it before it could be used). Appropriate yellow filters are placed over light fixtures and windows. Because of this feature, clean rooms are sometimes referred to as *yellow rooms*.

4.2.1 Particles

Clean rooms are catagorized by the number of particles contained in the air. This is given by the *class*. *Class-X* means that a cubic foot of air contains less than X particles that are 0.5 μm or greater in diameter. Thus, a class-100 clean room has one hundred or less such particles in a cubic foot of air. The general environment in an average building usually exceeds class-100,000. (The amount of debris present in ordinary rooms was impressed on the author when he once unpacked and uncovered a clean, six-inch telescopic mirror. It was impressive to hold the mirror and see particles begin to visibly cover its surface in a period of only tens of seconds.) Class-10,000 is relatively easy to obtain and can be used for assembly operations or even some device fabrication. But serious fabrication efforts usually occur in environments near class-100. Class-100 requires good filtering techniques and protective clothing on personnel. Advanced silicon work is already using class-10 in some areas. The present nomenclature is based on 0.5 μm particles. At this writing, there is considerable discussion of the merits of adopting 0.125 μm diameter particles as the standard. Commercial machines exist to measure the particle count, which may vary greatly as a function of location within the clean room and with surrounding activity. Assuming that there is decent filtering of incoming air, the major source of particles is personnel. Table 4.1 gives typical examples of the increase in the particle contamination caused by various personnel activities [4]. Of course, the exact results are complex and depend on many factors, and so the table should be taken as indicative only. The major procedures used to reduce particle contamination within the general environment are to filter all incoming air and to cloth personnel in protective garb.

Filtering may be done in several ways. The most effective (and most expensive) is to force air through filters placed continuously on the ceiling, and extract the air through perforated flooring. Adjustable dampers may be placed in the floor segments so that the room can be *balanced* to provide vertical laminar flow at equal flow rates across the entire area. This type of installation is expensive because of the large filter area and the special ceiling and raised floor that must be constructed. A less expensive technique is to place a bank of filters in one wall and extract air at the other end of the room. The area nearest the filters will be the cleanest. However, good air flow may be impeded by equipment or personnel within the room. All such filtering techniques require rather substantial *blowers* to force air through the filters. Air flow should be sufficient to make the room *positive pressure* so that when doors are opened air flows out instead of in. This prevents particles and moisture from being brought in from the outside. The air that is extracted (through the floors or walls) is recirculated through the filters. However, a significant fraction of the room air may be removed as *exhaust air* through such items as chemical fume hoods. These may draw air at approximately 100 SCFM (*standard cubic feet per minute*) for every linear foot of length. Such exhaust is clearly unsuitable for recirculation and must be replaced by *make-up air* from the outside.

TABLE 4.1
Typical Particle Increases Caused by Personnel
(0.2 to 50 μm particles; reference [4])

Activity	Times Increase Over Ambient
Normal walking	1.2 to 2
Sitting quietly	1 to 1.2
Gather 4-5 people at one location	1.5 to 3
Normal breathing	1
Breathing of smoker up to 20 min. after smoking	2 to 5
Sneezing	5 to 20
Rubbing skin on hands & face	1 to 2
Laminar-flow work station with hands inside	1.01
Brushing sleeve (protective clothing)	1.5 to 3
Removing handkerchief from pocket	3 to 10

Protective clothing for personnel may include smocks, hoods, face covers, gloves, safety glasses, and booties (to slip over shoes). Booties may not be required in vertical laminar flow rooms — the downward flow of air takes out particles brought in on shoes or stirred up by walking. In all other cases, shoes and walking can be a considerable source of particles. Pads of sticky sheets may be placed near entrance areas to remove loose dirt from shoe bottoms. They are located so that personnel naturally walk across them when entering the clean area. When the top sheet becomes too dirty, it may be pulled up revealing another one underneath. Gloves not only restrict particles, skin oils, and salts from being spread from hands, but also protect hands from dangerous chemicals. Safety glasses protect eyes from chemical splashes, breakage, explosions, or other accidents. (Contact lenses are regarded as a safety hazard in environments that require safety glasses. Dangerous fluids can be trapped beneath them, and they may be difficult to remove quickly, especially if it must be done by another person.) It may be necessary to restrict the use of cosmetics on exposed skin surfaces for the highest cleanliness standards. Plastic shields are available to place in front of microscopes to prevent the breath of the viewer from contaminating the slice. Some or all operations can be performed in laminar flow hoods. These can be substantially cleaner than the adjacent environment, assuming that gloves are used by operators.

Paper is another source of particles. Lint-free paper is available for clean room use. Lead pencils and/or erasures within clean rooms can generate large amounts of particles and should be prohibited. Back side processing operations such as lapping, sawing, or scribing tend to be innately "dirty." These are usually performed in locations separate from the main process area.

4.2.2 Temperature and Humidity

Temperature and humidity need to be controlled within the clean room, principally for resist work. If *relative humidity* (RH) is too low (below about 20% RH) static becomes more of a problem (see below). Extreme dryness can harm photoresist, which needs a certain amount of moisture. But there is greater danger to resist operations from high humidity (over about 50% RH). This applies to application, exposure, and developing. The resist can exhibit adhesion problems and/or cracking. Poor control of humidity in resist areas can cause enormous problems; it should be controlled to within 5% RH. Because of its importance, relative humidity should be continuously monitored. Atmospheric humidity loads are usually described in terms of the pounds of water vapor generated per hour. The units are *grains* (gr) of water vapor, where one grain is 1/7000 pound. Table 4.2 indicates typical moisture loads. In addition to the sources indicated in the table, moisture can enter through walls, cracks, and ductwork if these are not properly sealed. Airlocks are generally used at entrances to reduce moisture diffusion from outside air when doors are opened, and also to reduce the entrance of particles stirred up in the area adjacent to the door. Nevertheless, air flow should be outward to assure that moisture and particles tend to leave the clean room, rather than enter.

Humidity control usually requires removal of moisture from the air; it is relatively easy to add moisture if needed. There are two general methods used to lower relative humidity. The first is the use of desiccants. These are quite effective and cost efficient if the initial humidity level is low. Higher humidity levels require dehumidification by condensation. In this case the incoming air is cooled to remove the moisture by condensation and then heated if necessary. Some systems use combinations of both approaches [5]. In typical operating conditions, most of the water vapor load is from incoming air. Hence, the system must be able to accommodate the wide ranges of moisture content in the incoming air. Even though most of the moisture load is from external, make-up air, some is generated within the clean room (Table 4.2) and the air must be dry enough to act as a "sponge" and absorb this internal load. A rough

TABLE 4.2
Typical Moisture Contributions of Various Sources
(one grain (gr) is 1/7000 pound; reference [5])

Source	Water Vapor Added (gr/hour)
One person doing light work	1500 to 3500
Door opening	1000 to 3000
Exposed wet surfaces	140 to 700 (per ft^2 area)
Outside air (1000 CFM)	130,000 to 350,000 (depending on climate)

TABLE 4.3
Relative Humidity and Water Content of Air
(reference [6])

Relative Humidity %	Specific Humidity (gr/lb of air), for T in °F					
	40°	50°	60°	70°	80°	90°
10	4	5	8	10.5	15	20
20	7	10	15	22	30	42
30	10.5	16	23	33	46	63
40	14	21	30	44	60	84
50	17.5	26	38	55	76	105
60	21	31.5	46	65	92	125
70	25	37	54	76	107	149
80	28.5	42.5	61	87	123	171
90	32	48	69	99	140	193
100 (dew point)	36	53	77	110	156	217

calculation can indicate if this is possible. Let L be the internal moisture load (in gr/hr). Table 4.2 can be used to approximate L using the number of people, number of door openings, exposed wet surface area, etc. Let W_r and W_a be the water content of the room and of the incoming air, respectively (each expressed in gr/lb). W_r can be determined from Table 4.3 knowing the temperature and relative humidity of the room air. W_a can be determined if the incoming air is dehumidified by the condensation technique: knowing the temperature to which the air was cooled gives the water content (the *dew point*; Table 4.3). If the incoming air flow rate is given in standard cubic feet per minute (SCFM), and the density of air is approximately 0.075 lb/ft^3, then the maximum load, L_m, that the incoming air can accommodate is given by

$$L_m = \text{SCFM} \times 60 \text{ min/hr} \times 0.075 \text{ lb/ft}^3 \times (W_r - W_a)$$
$$= 4.5 \text{ SCFM} (W_r - W_a) \text{ gr/hr} \qquad (4.1)$$

If $L_m > L$, the incoming air can handle the moisture load generated in the room. Note that the condensation method limits the minimum value of W_a that can be obtained. That is, air is generally cooled to no less than about 40° F to prevent frost problems. At this temperature, $W_a = 36$ gr/lb. If this value, as used in equation (4.1), results in $L_m < L$, then the humidity in the room will rise. The amount of change can be determined by solving

$$L = 4.5 \text{ SCFM} (W_r - W_a)$$

for W_r and then using Table 4.3 to determine what relative humidity that value corresponds to. If the condensation method is insufficient to accommodate the internal moisture load, the system must be augmented by a desiccant method. The amount of difficulty in using condensation alone will depend on the amount of room air that is exhausted and replaced by make-up air. The exhaust air may contain dangerous fumes and so cannot be recycled.

4.2.3 Static Electricity

Static is not as severe a problem in GaAs fabrication as it is for many silicon devices which often employ thin oxides (MOS devices) that can easily be damaged by minor static charge. Although completed GaAs devices can be damaged by static charge, especially during bonding operations where there is a path through a Schottky gate to an ohmic contact, GaAs devices seem rather insensitive to static damage during fabrication. Unfortunately, this does not mean that static can be ignored. Static charge attracts and binds particles to slices or other surfaces. The electrostatic force between the surface and the particle can be substantial, making removal extremely difficult. Static can be reduced by several means, not the least of which is good humidity control (especially in winter). Conductive floor material or waxes are available to alleviate charge buildup on personnel. De-ionizing nuclear cartridges can be used on air guns, where rapid air flow tends to generate static charge. It is even possible to install ion "spray" grids for the entire room, although these are not a standard item in most clean rooms. Many items intended for use within clean rooms can be obtained constructed of *static free* (i.e., somewhat conductive) materials. These include slice containers, storage cabinets, and even anti-static clothing and booties. Commercial instruments exist to measure static charge.

4.2.4 Slice Handling

GaAs slices are notoriously susceptable to breakage. Dropping a slice from a height of only a few inches can shatter it. Even if the slice remains whole, strains may be introduced that result in breakage at a subsequent step. Minor scratches can initiate cleaves. Handling by tweezers should be avoided as much as possible although vacuum tweezers work well. Most importantly, slices should never be touched by ungloved hands. Oils and salts from human skin are easily transferred to objects by the slightest touch, and are then very difficult to remove. In fact, a severe test of any cleaning process is attempting to remove a finger print. This type of contamination can be insidious because the contaminant can transfer from object to object. Therefore, one must not touch any object that itself will touch a slice. Obviously this caution could be carried to ridiculous extremes, but it is intended to apply to actions such a touching the end of vacuum tweezers with a bare finger to see if the vacuum is on.

In-process storage of slices is usually in closed cabinets continuously purged with dry nitrogen. This keeps oxygen, atmospheric vapors, and moisture away from the slices. This seems especially helpful if the slices have a resist film spun on them or are being stored for extended durations, such as overnight or over weekends. (Normally, resist films are not left on slices any extended time before exposure and development.)

4.3 CLEANING TECHNIQUES

Cleaning operations are performed before all major steps during device processing. These steps may employ organic solvents, vapor degreasing, acids, bases, and/or plasma etches. Chemical purity is obviously important, and many of the commonly-used solvents, acids, and bases may be obtained in *semiconductor grade* (SC grade), which is one step above reagent grade. The major improvement in purity is the reduction of metal contaminants. There are also a number of commercially available, proprietary compositions useful for cleaning, resist stripping, or oxide removal. Although these may be largely mixtures of common solvents or acids, they may also include buffering agents or other constituents that aid in the chemical action. As a general rule, they are worth using.

4.3.1 Solvent Cleaning

Organic solvents are effective in removing oils, greases, waxes, and organic material such as photoresist. Organic solvents are also innocuous to almost all materials intended to be permanently present on GaAs devices. Such materials include GaAs, metals, and dielectrics.

Some properties of solvents used in GaAs processing are listed in Table 4.4. Solvent cleaning is usually done at elevated temperatures, even at the boiling point of the solvent. It may also take place in a vapor degreaser using solvent vapors. This is a particularly clean method in that the solvent is constantly being distilled; only vapors condense on the slice. However, even this seemingly clean procedure must be used with caution. Contaminants on slice holders may be moved onto slices. In fact, vapor degreasing is not used much in semiconductor processing. The more common practice is to immerse the slice in the heated solvent. Agitation during these steps can also aid in removal of inorganic (or "stubborn" organic) particles. However, solvent cleaning can also remove organics from around inorganic particles, leaving them dried out and more firmly adherent to the surface than before.

The debris that is removed during solvent cleaning accumulates in the solvent. Except in the case of vapor degreasing, these contaminants can be deposited on subsequent slices. Therefore, solvents must be discarded and replaced at regular intervals. It is also possible to filter the solvent in a continuous manner to prevent accumulation of debris. Solvents may be toxic, flammable, and explosive. There is also evidence that some may be carcinogens, neoplastic agents, and/or teratogens (a *neoplastic agent* is a substance that causes nonmalignant tumors; a *teratogen* is a substance that causes physical defects in a developing embryo). The evidence for such claims is based on animal studies using large quantities of the substance, and the validity of such studies continues to be debated. Nevertheless, clearly it is wise to be prudent. Table 4.4 notes which solvents are suspected of these traits. Because solvents tend to be used when hot, or even boiling, good ventilation and safety practices should be observed to prevent personnel from being exposed to solvent vapors. Use of solvents in well-exhausted fume hoods is recommended. Disposal of used solvents may require special arrangements to prevent contamination of the environment.

There are several proprietary compositions popular in semiconductor processing. One is a composition called J-100 (a trademark of Indust-Ri-Chem Laboratory, Inc.), used as a resist stripper in the silicon industry. It is a combination of solvents and phenol (carbolic acid) and is a very powerful cleaner. Used in full strength, it will slowly attack many metals, such as AuGe/Ni ohmic contacts, and so should be used cautiously after metal patterns are present on the slice. It can be used to remove stubborn material (usually resist remnants). J-100 is very effective in the initial cleanup of slices before any metal is present on them.

TABLE 4.4
Solvents and Properties
(Most data from reference [7])

Solvents	Boiling Point (°C)	Flash Point (°C)[1]	Water Sol.	Density (g/ml)[2]	Safety[3]
Acetone	56.2	-16	100%	.784	F
Benzene	80.1	-11	<1%	.874	F [C,N]
n-Butyl acetate	126.	22	<1%	.876	F
Carbon tetrachloride (tetrachloroethylene)	76.8	none	<1%	1.58	[C,N,T]
Chlorobenzene	132.	29	<1%	1.10	F
ortho-Dichlorobenzene	180.	74	<1%	1.3	
Ethanol	78.5	60	100%	.789	
Ethylene dichloride	83.5	13	<1%	1.25	F [C]
Isobutyl alcohol	108.	35	8.5%	.798	F [C]
Methanol	64.7	12	100%	.787	F
Methyl ethyl ketone	79.6	-1	24%	.800	F [T]
Methylene chloride	39.8	none	1.6%	1.32	
Propanol-1	97.2	25	100%	.800	F [C]
Propanol-2 (isopropyl alcohol)	82.3	22	100%	.781	F
Tetrachloroethylene (perchloroethane)	121.	none	<1%	1.62	
Trichloroethylene	87.2	none	<1%	1.45	

NOTES:
[1] Flash point temperatures depend on the conditions of measurement; different references may give slightly different values.
[2] At 25°C
[3] F = Flammable
 [C] = Possible carcinogen
 [N] = Possible neoplastic agent (cause non-malignant tumor)
 [T] = Possible teratogen (cause physical defects in developing embryo)

96   GaAs Processing Techniques

Waxes are sometimes used in GaAs processing to mount slices on stronger substrates (for lapping, handling after lapping, etc.). They are somewhat soluble in solvents — usually more so in one kind than another. For example, polyglycol is soluble in acetone, paraffin is soluble in tetracholoroethylene (perchloroethylene).

After solvents are used, they themselves must be removed from the slice. Some are more difficult to remove than others. As can be seen in Table 4.4, many are not soluble in water. Most solvents, however, are soluble to some extent in alcohols, which themselves are completely miscible with water. This is one reason that solvent cleaning operations may use a sequence of solvents, the last one usually being an alcohol. A solvent such as xylene or perchloroethylene is useful to remove J-100. As with many other processes, the exact cleaning schedules (solvent types, order, times, and temperatures) of device manufacturers are often considered proprietary. But there is very little that is special or subtle. A minor amount of experimentation should yield quite adequate solvent cleaning procedures. As noted above, most solvents are relatively innocuous to the materials used in GaAs device fabrication. They are also good cleaners. Therefore, solvent cleanups form a major part of a process flow, occurring at almost every step that a cleanup is needed, including prior to application of resist or dielectric films.

4.3.2 Acids and Bases

Acids are used in cleaning steps to remove metal contaminants, to remove oxides, or as part of wet etchants to remove GaAs material and provide a fresh surface. The latter two uses are the principal ones. GaAs etchants (Chapter 5) generally employ an acid as one component; these etchants may be used in cleaning to remove a small amount of surface material and expose fresh, "clean" GaAs. Such etchants usually employ oxidizing agents and hence, leave an oxide layer on the etched surface. Ashering in oxygen plasmas (see below, and Chapter 9) oxidizes the surface also. Ordinary exposure to air will result in oxide on GaAs surfaces. This oxide, depending on previous steps, is generally between 10 and 50 Å thick. An acid can remove this oxide, but the acid alone usually will not etch the GaAs. It may be helpful to remove such oxides before metalizations or application of dielectrics, although some special cases may benefit from the presence of an oxide. Acids used for oxide removal or cleaning are usually used in highly diluted concentrations (using de-ionized water). The choice of acid depends on the materials present on the slice. Dilute HCl is a common choice. Another compound that is excellent for removing oxides is the commercially available compound Bell #2, which consists mainly of buffered HF.

The danger of acids to personnel is well known, but it is useful to emphasize the dangers associated with hydrofluoric acid (HF). This acid is especially dangerous because it can be absorbed through the skin with little or no pain. It can then cause serious tissue and bone damage before pain begins.

Bases or alkaline solutions can act as cleaners for some types of soils. Also, because oxides formed on GaAs tend to be amphoteric, bases as well as acids can be used to dissolve the oxide. Again, these are usually used in dilute concentrations. Bases tend to cause photoresist adhesion problems and caution should be observed when considering their use in the presence of resist patterns.

4.3.3 Plasma Etching

Plasma etching techniques, described in Chapter 9, can be used for cleaning. Oxygen plasmas are used to remove organics from the slice (ashering). The plasma causes O_2 to dissociate into highly reactive oxygen molecules that react with carbon and hydrogen in organics to produce volatile waste products (CO, CO_2, H_2O). Ashering may be used to remove resist, as a *descum* to remove thin films of resist (in developed areas), or as a general technique to remove organics. It is sometimes more successful than liquid solvents in removing stubborn material. Oxygen plasmas will not etch exposed GaAs surfaces, but the process will leave an oxide on such surfaces. These oxides may require removal before subsequent process steps.

4.3.4 Other Issues

Small particles can be very difficult to remove from a slice. Electrostatic forces can bind such low-mass particles to the surface with extraordinary force. Further, forceful blowing of gases may not be effective because the particles may be small enough to be inside the near-stagnant boundary layer present at the slice surface under laminar flow. Mechanical means may be necessary to remove these particles. On a small scale, this may be done by swabbing. On a larger scale, commercial machines exist to perform such operations. These machines may also be used to clean photomasks.

Water is part of many processing operations, including rinsing. For use in semiconductor processing, *de-ionized water* (DI water) is used. The quality of the water is assessed by its resistivity. DI water should be over 10 Megohm-cm. Stagnant water will tend to decrease in resistivity, and it may be necessary to have the water flow for some time following the first time it is turned on each morning. The water should also be filtered in the submicron region to remove particles. The general DI water system should consist of continuous loops — dead ends can support bacterial growth. There are commercial companies that specialize in engineering DI water systems.

Drying is an important consideration after rinsing slices. If drying occurs in a haphazard manner, solvent stains or water stains will remain on the slice. An example is shown in Figure 4.1. These stains may result from trace contaminants in the solutions. Drying proceeds best if the film of liquid (water, alcohol, etc.) is removed in one continuous sheet, rather than breaking up into individual drops before drying. Such drops are good sources of residue that would have been carried off the slice by the above technique. Compressed air is generally not a good choice for slice drying. Compressed air tends to contain moisture and, worse still, trace amounts of oil from the compressor. Usually dry nitrogen is used. It is pure and economical. It can be derived as a by-product of liquid nitrogen evaporation. Liquid nitrogen is often available in processing facilities due to its use in cold traps. Filters and/or nuclear de-ionizing cartridges are used in "air" guns to remove particles and alleviate static.

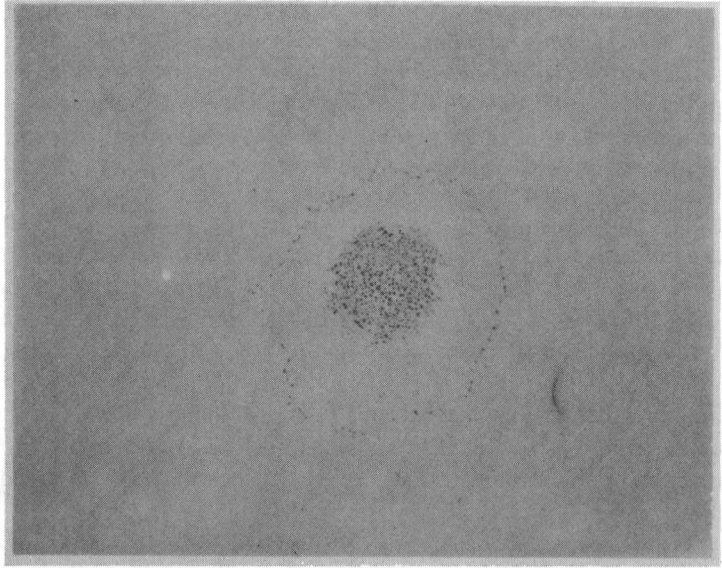

Figure 4.1 Solvent stains on GaAs slice.

REFERENCES

[1] B.T. Murphy, *Proc. IEEE*, 52, 1964, p. 1537.

[2] R.B. Seeds, *IEEE Int. Conv. Rec.*, Part 6, 1967, p. 60.

[3] S.M. Hu, *Solid State Electron.*, 22, 1979, p. 205.

[4] *Semiconductor International*, January 1984, p. 19.

[5] L.G. Harriman III, *Solid State Technol.*, 26 (6), 1983, p. 155.

[6] T. Baumeister and L.S. Marks, editors, *Standard Handbook for Mechanical Engineers*, 7th edition, New York: McGraw-Hill, 1967.

[7] *Solvent Guide*, Burdick & Jackson Laboratories, Inc., (Muskegon, Michigan), 1980.

CHAPTER 5

WET ETCHING

5.1 INTRODUCTION

Wet etching is the use of liquid etchants to remove material. Such etching procedures are important parts of various processing steps. Gallium arsenide is etched to remove damaged material; to form mesa structures for device isolation (Chapter 10); to recess gates in field effect transistors (Chapter 12); to aid in polishing, as part of cleaning procedures, or as an integral part of diagnostic techniques. Other materials, such as metals and dielectrics, may be etched as part of device fabrication procedures. There are important etching techniques other than wet etching. These are described in Chapter 9 (Dry Etching). Etching of gallium arsenide is the main topic of the chapter and will be considered in detail in sections 5.2 and 5.3. Discussion of electrically-aided wet etching (anodic or cathodic etching) is included. Section 5.4 discusses wet etching of other materials likely to be used in GaAs processing.

Wet etching proceeds through chemical reactions that take place at the surface of the material. In order for these to occur, the etchant species must reach the surface, the appropriate reactions must take place, and the resulting products must be removed from the surface. One of two basic mechanisms limits the dissolution rate. First, the etch rate may be controlled by the rate at which reactant species can reach the surface (and/or the rate at which reactant products can be removed). This situation is referred to as *diffusion-limited* or *mass-transport-limited* etching. This is a common occurrence since the diffusion coefficient of species in liquids is very small, on the order of 10^{-9} m^2/sec [1]. Second, the etch rate may be limited by the rate of chemical reactions taking place on the surface. This situation is referred to as *reaction-rate-limited, surface-limited,* or *kinetically-limited etching*. Which regime is limiting is a function of etch composition and temperature.

Diffusion- or mass-transport-limited etchants tend to be more isotropic with respect to crystal orientation. They tend to give a more polishing effect. This occurs because surface protrusions are more available to incoming species and tend to be etched more rapidly than a smoother surface. Because agitation can greatly affect mass transport, etch rates may be highly sensitive to the form and degree of agitation.

Reaction-rate-limited etches tend to preserve surface morphology, although this tendency can be completely dominated by a second tendency: to be highly anisotropic in selectively etching crystalline structures through masking patterns. These etchants are much less sensitive to agitation.

As with all chemical reactions, etching is sensitive to temperature. A 10° C increase in temperature can increase etch rate by a factor of two. Freshly mixed etchants may be hot. Etching which occurs in a small volume of solution can result in a temperature increase of the solution during the etching process. Compound etchants may change composition over time, depending on temperature and storage technique. All these effects must be considered in establishing reproducible and controllable etching processes.

Some etching conditions can produce bubbles, usually of hydrogen, which can adhere to the surface and cause nonuniform etching. This can be alleviated by agitation and/or use of surface-active agents (wetting agents).

5.2 BASIC CONSIDERATIONS IN GaAs ETCHING

The previous section described several fundamental considerations that apply to wet etching in general. Further generalizations can be made about wet etching GaAs. The zincblende crystal structure of GaAs (Chapter 2) leads to etching characteristics which are very different from those of silicon. General reviews of GaAs etching may be found in references [2] through [6]. Papers treating wet etching of GaAs and of other semiconductors may be found in the *Journal of the Electrochemical Society*.

Almost all GaAs etchants operate by first oxidizing the surface and then dissolving that oxide, thereby removing some of the gallium and arsenic atoms. Generally, the etchant will contain one component that acts as the oxidizer and another component which is the dissolving agent. The popular etchant $H_2SO_4/H_2O_2/H_2O$ is an excellent example. The hydrogen peroxide is the oxidizing agent; sulfuric acid dissolves the resulting oxide. In appropriate proportions, this etchant dissolves GaAs rapidly. It etches to some degree in almost all proportions. Yet GaAs will not etch in H_2O_2 alone, nor in H_2SO_4 alone.

The oxidation is essentially an electrochemical process in which localized anodic and cathodic sites are believed to exist at the semiconductor surface. Oxidation takes place at anodic sites and the oxidant is reduced at cathodic sites [3,4,7]. The electrochemical nature of the process makes it sensitive to any mechanism that can supply (or restrict) electrons and/or holes at the surface [3,4]. Examples of such mechanisms are illumination [8] and electrical currents [9-11]. Both anodic and cathodic etching will be considered in section 5.3.

The crystalline nature of the GaAs lattice leads to anisotropic etching in almost all cases in which masking materials are used to pattern the slice for etching. This generality may be illustrated by considering a specific example based on (100) slice orientation (see section 2.2 for a discussion of the nomeclature used to designate crystallographic directions and planes). This is by far the most commonly used slice orientation for device fabrication and, as explained in Chapter 2, has two natural cleavage directions perpendicular to each other. These are the {110} planes. Devices are generally oriented parallel to these planes to allow scribing or to minimize problems in sawing when the slice is cut into separate chips. Figure 5.1 illustrates the anisotropic etching behavior that is generally obtained on such material. The squares in the figure represent mesas that are formed by etching away material outside of the square. A masking material, such as photoresist, can be used to protect the mesa area from the action of the etchant. The figure indicates the general nature of edge profiles that will result using most etchants. In Figure 5.1 (a) the mesa has been oriented so that its edges are parallel to a <011> direction. In this case two of the edges will yield profiles that have outward slopes; the other two edges will have inward sloping or undercut profiles. Obviously, one direction is completely inappropriate for metal crossovers. Such considerations must be anticipated in device design and fabrication (Chapter 10). In Figure 5.1 (b), the mesa has been oriented so that its edges are 45° to a <011> direction. In this case the resulting edge profiles will be midway between the two cases above. They will yield essentially vertical walls. Note, however, that in all cases, there is considerable undercutting of the masking material. This is a major limitation in wet etching GaAs — lateral etch rate is a significant fraction of the vertical etch rate. Achievement of low undercut etch rates usually requires the use of dry etching procedures (Chapter 9). The profiles shown in Figure 5.1 are intended only as general representations. Specific etches (see below) will have slightly different profiles. A few will be more isotropic.

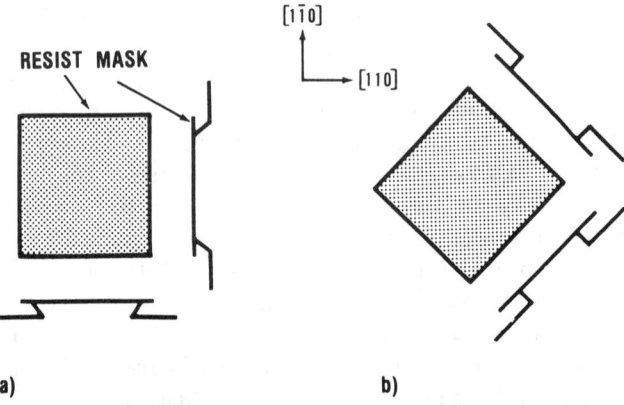

Figure 5.1 Anisotropic etching behavior typically exhibited by most wet etchants on a (100) oriented GaAs slice. (a) Mask edges are parallel to the {011} cleavage planes. (b) Mask edges are 45° to the cleavage planes.

The anisotropic nature of etching behavior follows from the lack of symmetry in the GaAs lattice and the dependence of etch rate on crystal orientation. The etching characteristics of GaAs tend to be dominated by the {111}A planes. These planes often appear in edge profiles resulting from etching through masking patterns. As discussed in section 2.2, the {111}planes consist either of all gallium atoms or all arsenic atoms and are referred to as {111}A or {111}B planes, respectively. The A and B planes have very different chemical properties. The differences can be understood by considering the bonding between atoms as indicated schematically in Figure 5.2. Note that the planes are bonded alternately by sets of three bonds and sets of one bond. The material will cleave between planes joined by the single bonds, rather than between the planes joined by the triple bonds. Thus, one surface will be type A (gallium) and the other will be type B (arsenic). The surface type may be identified using either x-ray or chemical means [2,3]. The two planes joined by triple bonds may be thought of as a double sheet of atoms. A qualitative understanding of the chemical etching properties of various atomic planes may be aided by the following analysis. Assuming electrical neutrality of the surface atoms, the surface gallium atoms (plane A) each have three bonding electrons and the surface arsenic atoms (plane B) each have five bonding electrons. Each atom is connected to the next plane by three bonds. Thus the surface gallium atoms have no free electrons. The surface arsenic atoms have two free electrons and these are available to take part in any reaction. For rate-limited reactions, the {111}B surfaces would be expected to etch faster than the {111}A surfaces. By this argument, a {100} surface, having both Ga and As atoms (with double bonds), would be expected to etch at an intermediate rate. That is, the etch rate would be ordered {111}B > {100} > {111}A. This ordering has been confirmed for several materials having zincblende crystal structure [2], including GaAs when etched using bromine-methanol [12].

The above description is somewhat simplistic. In practice, removal of an atom from the surface not only exposes the layer beneath, but also allows lateral attack of the etchant on the surface layer atoms immediately adjacent to the removed atom. This situation can lead to small-scale roughness (*micro-faceting*) that is superimposed on the large scale behavior described above [13]. Such micro-faceting can make an etched surface appear cloudy to the naked eye. The smoothness of the original surface can also affect the degree of such micro-faceting; a rough surface generates more than a smooth surface generates. Diffusion-

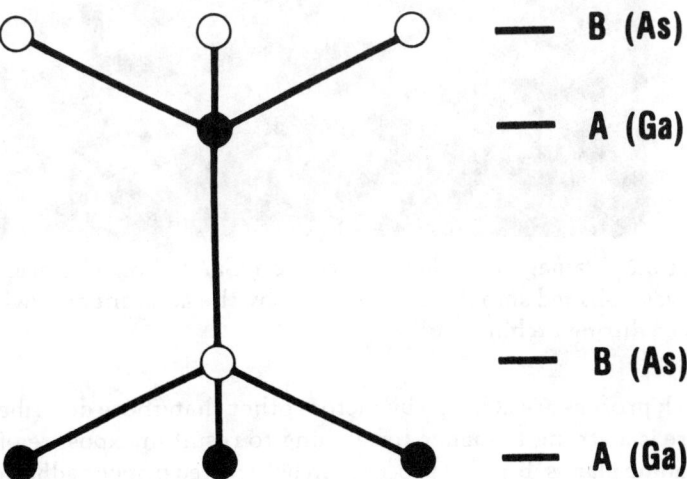

Figure 5.2 Planes of atoms alternately joined by single and triple bonds in the GaAs lattice.

limited polishing etches are used during slice generation and polishing to prevent this effect.

Crystal defects will also affect etch results. In fact, etching in hot KOH is a standard technique used to ascertain dislocation density. Use of nonpolishing etchants will tend to reveal defects or damage. Some of that damage may have been generated in the sawing operation when slices are cut from the boule. Inadequate removal of damaged material and/or poor polishing can result in slices which appear perfectly smooth, but reveal damage patterns when etched. Figure 5.3 shows such an example. The general relationship between etching and crystal defects is complex and will not be pursued here. Further details may be found in references [2] and [3].

Figure 5.3 Damage revealed by etching a GaAs slice. The original surface was polished smooth, as evidenced by the adjacent area which was masked during etching.

Etch profiles are affected by factors other than those described above. There is a strong tendency for etching to result in exposure of various crytalline planes. But this process can be affected by poor adhesion of the masking substance or by diffusion effects. If the masking substance does not adhere perfectly to the slice surface, the resulting profile will be affected. Unbaked photoresist is particularly susceptible to this problem. Diffusion effects near the mask edge can also affect the exact profile that is obtained. This is because the most relevant criterion for faceting is not the absolute etching rate of a given plane, but the etching rate relative to adjacent orientations [14,15] (diffusion-limited etchants are particularly affected). The shape of the edge profile produced by wet etching can be understood on the basis of the etch rate of specific crystalline planes. If the relative etch rate of such planes can be determined, the edge profile can be predicted. The procedure will be summarized here; a detailed description of this technique may be found in reference [16]. The approach is to construct normals to radii vectors which represent the dissolution rate in a given direction. These dissolution rate vectors can be represented by a polar diagram and the normals are then tangents to the polar figure. A facet will appear in the etch profile any place there is a cusp in the polar diagram. The resulting edge profile is represented by the envelope formed by the assemblage of normals. Such a construction is called a *Wulff plot*. This graphical procedure is particularly suitable for computer-generated plots. The success of this approach in predicting etched edge profiles is indicated by the example shown in Figure 5.4 [16]. It shows predicted and experimental profiles obtained by etching (100) GaAs in a sulfuric acid etch using patterns oriented in the two cleavage directions. It should be emphasized that the success of the technique depends on being able to ascertain the rate vectors that are valid for the actual process conditions. Because of effects described above, these may differ from rates obtained using unpatterned surfaces. The practical usefulness of the technique is increased by the fact that sufficiently accurate polar diagrams can often be constructed on the basis of data derived from etching simple test structures [16]. Such polar diagrams can then be used to predict edge profiles when using different slice orientations or initial masking geometries.

As noted above, agitation can increase etch rate, but it also can hinder uniformity over a slice unless the technique provides equivalent agitation at all points across the slice surface. One technique that has been used when high throughput is not required is placement of the slice in the bottom of a beaker, slanted at 30° to 45° and rotated about its axis, of etchant. Ultrasonic agitation is also commonly employed. A novel approach to modifying etch behavior by affecting diffusion is etching in high acceleration fields using a centrifuge [17].

Two aspects of GaAs etching will not be discussed in this book. The first is chemical/mechanical polishing used after cutting slices from the crystal ingot. Such polishing operations are performed using diffusion-limited polishing etchants such as bromine/methanol in combination with mechanical polishing. References [3] and [5] may be consulted for further information. The second topic that will not be discussed further is spray etching. Some of the disadvantages of lateral etching (mask undercutting) typical of GaAs may be alleviated by using spray etching to increase directionality. This process has been used to separate chips instead of sawing or scribing (Chapter 16), but few details have been published.

5.3 GaAs ETCHANTS

Many etchants have been reported for etching GaAs. As indicated above, almost any combination of oxidizer and oxide-dissolver will operate as an etchant. But only a few of these systems have gained substantial popularity in GaAs device processing. These major etching systems will be considered in the following subsections. A final subsection will describe anodic and cathodic etching techniques.

TABLE 5.1

Initial Concentration of Reagents Used in Wet Etching

Reagent	Wt %
HCl	37
HF	49
H_2SO_4	98
H_3PO_4	85
HNO_3	70
CH_3COOH	99
H_2O_2	30
NH_4OH	29 (as NH_3)
NaOH	97

The concentration of initial reagents used in these etching systems in shown in Table 5.1. Etchant mixtures of these reagents are indicated by the notation $n_1:n_2:...$, where the numbers represent volume (not weight!) ratios of the components and are ordered in the same order given in the system name. For example, in the $H_2SO_4/H_2O_2/H_2O$ system, the mixture 1:8:40 means one part sulfuric acid, eight parts hydrogen peroxide, and forty parts water (by volume).

5.3.1 The $H_2SO_4/H_2O_2/H_2O$ System

The most popular etching system for GaAs consists of various proportions of sulfuric acid, hydrogen peroxide, and water. This etching system is often used for etch polishing substrates, mesa etching, and/or gate recess etching. This etchant system has been described in detail in several places [6,13-15].

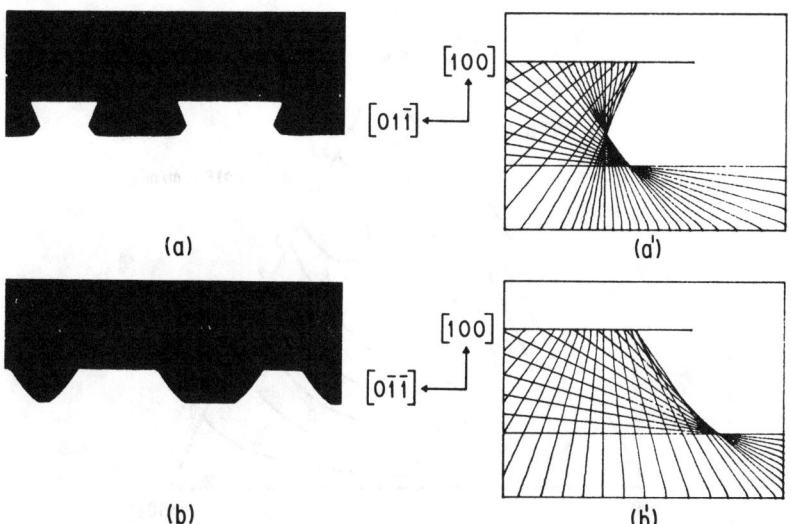

Figure 5.4 Comparison of theoretical and experimental edge profiles produced using a $H_2SO_4/H_2O_2/H_2O$ etchant (reference [16]; reproduced with permission).

Sulfuric acid is a rather viscous liquid. Etchants having a high proportion of sulfuric acid tend to be diffusion-limited in character and are useful for polishing. Combinations of 4:1:1 or 3:1:1 are commonly used to polish substrates and remove surface damage. When masking patterns are present, high proportions of sulfuric acid generally result in holes having nonplanar bottoms [14,18]. This effect is consistent with the diffusion-limited character of high proportion sulfuric acid mixtures and the related mask edge effects referred to in the previous section. Etch rate is rapid; too rapid for good control of such process steps as mesa etching or gate recess. More dilute mixtures are generally used for these purposes. Low sulfuric acid proportions produce flat-bottom holes, consistent with reaction-rate-limited behavior. The transition between "high" and "low" sulfuric acid concentration occurs at approximately 1/3 sulfuric acid [14].

The relative etch rate as a function of the proportions of the components may be inferred from Figure 5.5 [14], which shows absolute etching rates obtained etching (100) material at 0° C. Note that the etch rate approaches zero as either the sulfuric acid or the hydrogen peroxide component approaches zero. It was noted that in regions C and D of the diagram (high H_2O_2 and high H_2SO_4), mirror-like surfaces were obtained over a wide range of etching temperatures; cloudy surfaces resulted

110   GaAs Processing Techniques

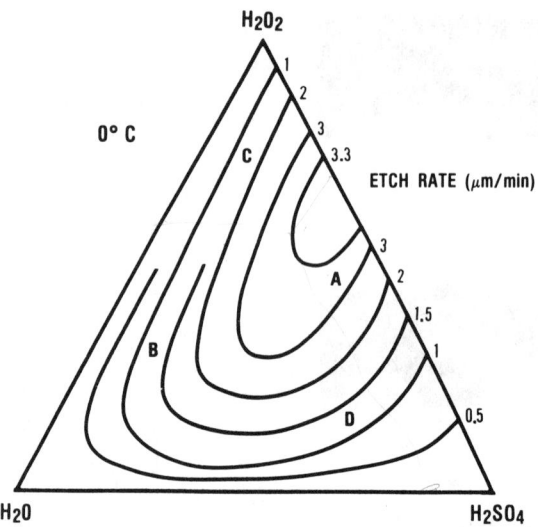

Figure 5.5 Constant etching rate contours for the $H_2SO_4/H_2O_2/H_2O$ system (after reference [14]).

under some conditions in regions A and B [14]. The dependence of etch rate on temperature and agitation (using a high concentration of sulfuric acid) is indicated by the data shown in Figure 5.6, taken from reference [14]. In this case the agitation was ultrasonic.

Various orientations of slices have been etched using the $H_2SO_4/H_2O_2/H_2O$ system. Table 5.2, taken from reference [14], presents the results obtained using low concentration sulfuric acid and indicates the profile obtained. Most processing, however, takes place on (100) oriented slices. Table 5.3, taken from reference [15], presents data for etching (100) GaAs slices using various proportions of the etch components. The slices were etched at room temperature and under gentle agitation (in rotating, angled beakers). Note the flat bottoms (near the mask) formed using low concentration sulfuric acid mixtures.

Photoresist masks exhibit suitable adhesion for most proportions of the components, but only if the resist is baked. Care must also be taken as to concentration of the hydrogen peroxide. The initial reagent is usually 30% H_2O_2. However, it is often stored in bottles with a small hole to prevent dangerous pressure buildup. This situation can result in a decrease in concentration with time, especially when the bottle is near empty. Similarly, the prepared mixture of acid, peroxide, and water will tend to lose peroxide over time. This effect is aided by the high tempera-

TABLE 5.2
Slot Geometries Resulting from Wet Etching of GaAs Substrates of Various Orientations [14]

Substrate	Window Direction	I Plane	I Angle	II Plane	II Angle
(001)	[110]	(111)	55	($\bar{1}\bar{1}$1)	55
	[1$\bar{1}$0]	(1$\bar{1}$1)	-55	($\bar{1}$11)	-55
	[100]	(100)	90	($\bar{1}$00)	90
	[0$\bar{1}$0]	(0$\bar{1}$0)	90	(010)	90
(111)	[1$\bar{1}$0]	(00$\bar{1}$)	-55	($\bar{1}$11)	-70.5
	[11$\bar{2}$]	($\bar{3}$11)	-58.5	(13$\bar{1}$)	-58.5
($\bar{1}\bar{1}\bar{1}$)	[1$\bar{1}$0]	($\bar{1}$11)	70.5	(00$\bar{1}$)	55
	[11$\bar{2}$]	($\bar{1}$31)	58.5	(13$\bar{1}$)	58.5
(110)	[1$\bar{1}$0]	(111)	35	(00$\bar{1}$)	90
	[001]	(100)	45	(010)	45
(112)	[1$\bar{1}$0]	(111)	19.5	($\bar{1}$11)	90
	[11$\bar{1}$]	(1$\bar{2}$1)	-60	($\bar{2}$11)	-60
($\bar{1}\bar{1}$2)	[1$\bar{1}$0]	(00$\bar{1}$)	35	($\bar{1}$11)	90
	[11$\bar{1}$]	(1$\bar{2}$1)	60	($\bar{2}$11)	60
(113)	[1$\bar{1}$0]	(111)	29.5	($\bar{1}$11)	80
	[33$\bar{2}$]	(3$\bar{5}$3)	-55	($\bar{5}$33)	-55
($\bar{1}\bar{1}$3)	[1$\bar{1}$0]	(00$\bar{1}$)	25	($\bar{1}$11)	-80
	[33$\bar{2}$]	(3$\bar{5}$3)	55	($\bar{5}$35)	55

ture that results during mixture of the acid with the other components. For this reason, it may be useful to initially add the acid to the water, wait until the mixture is cool, and then add the hydrogen peroxide. Because of evolution of peroxide, the etchant will exhibit an exponential decrease in etch rate as a function of storage time. Most of the change takes place within 24 to 48 hours. Hence, reproducibility in etch characteristics requires that the etchant be used under restrictions. These might be using only an "aged" mixture, using a mixture within certain time spans after preparation, or storing the mixture in sealed, collapsible containers (such as those used for storing photographic developer).

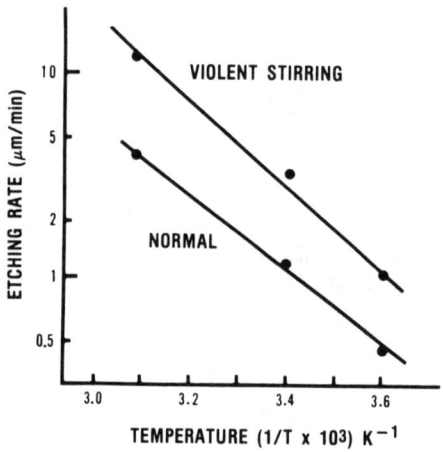

Figure 5.6 Effect of agitation on etch rate using a high sulfuric acid concentration etchant (after reference [14]).

Etchants which include hydrogen peroxide have a tendency to form bubbles while etching. These can cause nonuniform etching and should be removed by agitation or prevented by use of wetting agents.

5.3.2 Bromine-Methanol

The bromine-methanol (Br/CH_3OH) etchant can etch GaAs rapidly and is useful as a polishing or smoothing etchant [3,5]. It is generally used in concentrations near 1% Br, but higher concentrations are also used with different etch characteristics (see below). The etchant has been described in several places [4,6,12,19]. The system has two disadvantages. One is not a technical limitation: bromine is extremely pungent

TABLE 5.3
Characteristics of Acidic Hydrogen Perioxide Etchants for GaAs
(From reference [15], reprinted by permission of the publisher, The Electrochemical Society, Inc.)

ACID	VOLUME RATIOS *	CONCENTRATION (mol/l) ACID	CONCENTRATION (mol/l) H_2O_2	RATIO OF UNDERCUT TO ETCHED DEPTH $<01\bar{1}>$	RATIO OF UNDERCUT TO ETCHED DEPTH $<011>$	RATIO OF UNDERCUT TO ETCHED DEPTH $<100>$	RELATIVE ANISOTROPY	ETCH RATE (100) ($\mu m\ min^{-1}$)	CROSS-SECTIONAL PROFILES (011) SECTION	CROSS-SECTIONAL PROFILES (01$\bar{1}$) SECTION
H_2SO_4	1:8:1	1.8	8.0	0.30	0.30	0.90	1.0	14.6		
H_2SO_4	1:8:40	0.36	1.6	0.89	0.68	1.2	0.55	1.2		
H_2SO_4	1:8:80	0.20	0.90	0.62	0.62	0.86	0.32	0.54		
H_2SO_4	1:8:160	0.10	0.47	0.71	0.71	0.93	0.27	0.26		
H_2SO_4	1:8:1000	0.018	0.079	0.82	0.76	0.95	0.22	0.038		
H_2SO_4	1:1:8	1.8	1.0	0.77	0.53	1.0	0.61	1.3		
H_2SO_4	4:1:5	7.1	1.0	0.49	0.29	0.70	0.83	5.0		
H_2SO_4	8:1:1	14.0	1.0	0.52	0.43	0.61	0.35	1.2		
H_2SO_4	3:1:1	11.0	2.0	0.44	0.44	0.53	0.19	5.9		
HCl	1:4:40	0.27	0.87	0.51	0.28	0.97	1.1	0.22		
HCl	1:1:9	1.1	0.89	0.22	0.18	0.37	0.69	0.20		
HCl	40:4:1	10.6	0.87	0.54	0.54	0.54	~0	>5.0		
HCl	80:4:1	11.2	0.46	0.7	0.7	0.7	~0	1.1		

* ACID (CONCENTRATED): H_2O_2(30%): H_2O

and this etchant must be used in well-ventilated hoods. The other is more serious: bromine is highly caustic and resist is not suitable as a masking substance. Dielectric materials must generally be used.

Bromine-methanol tends to form etched grooves that have nonflat bottoms as shown in Figure 5.7 [6,12]. Figures 5.8 and 5.9 illustrate the relative etch rate of various planes as a function of bromine concentration [11]. (Note that this data is presented using weight percent of bromine, in contrast to the usual practice of quoting volume percent.) The etch rate order of various planes is $\{110\} > \{111\}B > \{100\} > \{111\}A$, but (as usual) the $\{111\}A$ etch rate is significantly less than the others. The above results are for volume concentrations of Br at about 1.3% or less. At higher concentrations, different behavior is seen. Above about 5% Br, the *stop plane* (or very slow etching plane) becomes the $\{332\}$ plane instead of the $\{111\}A$ plane [19]. This affects the resulting geometries and illustrates the complexities that can occur in wet etching. For example, consider a slot oriented in the $[01\bar{1}]$ direction (the direction that yields V-shaped grooves) and slots oriented at 10° intervals from that direction. Low concentration bromine solutions (1%) will yield the V shape only for the on-axis slot. The others will exhibit the reverse mesa shape [19]. But high concentration bromine solutions (>5%) will give

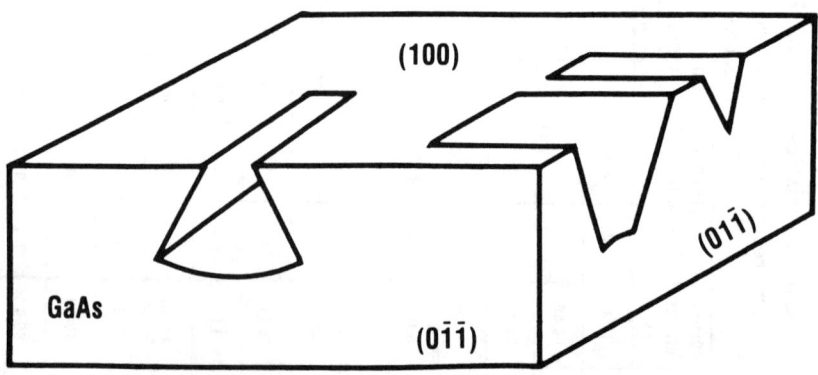

Figure 5.7 Shapes produced using a bromine-methanol etchant on (100) oriented GaAs. The masking slots are oriented parallel to the {011} cleavage planes.

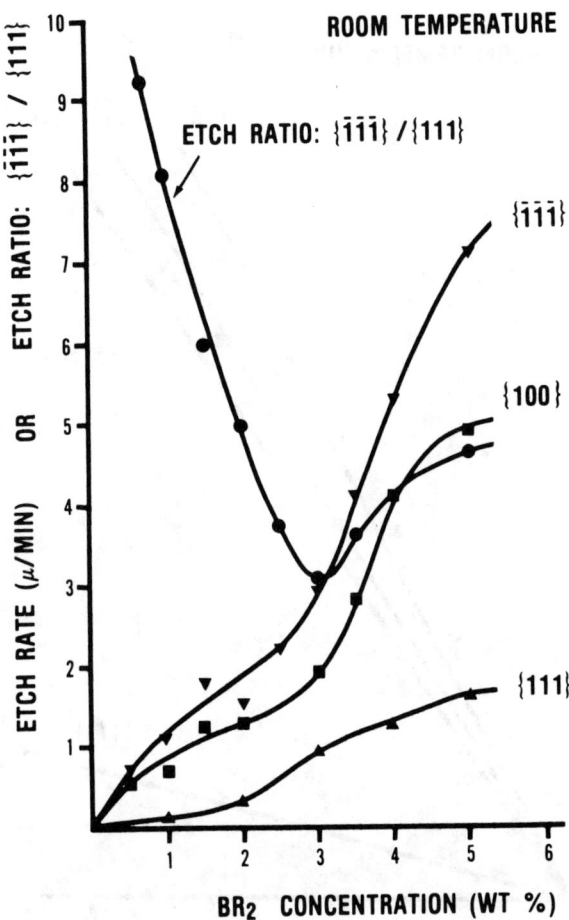

Figure 5.8 Etch rate of bromine-methanol on various crystalline planes of GaAs as a function of bromine concentration. Note concentration is expressed as weight percent (after reference [11]).

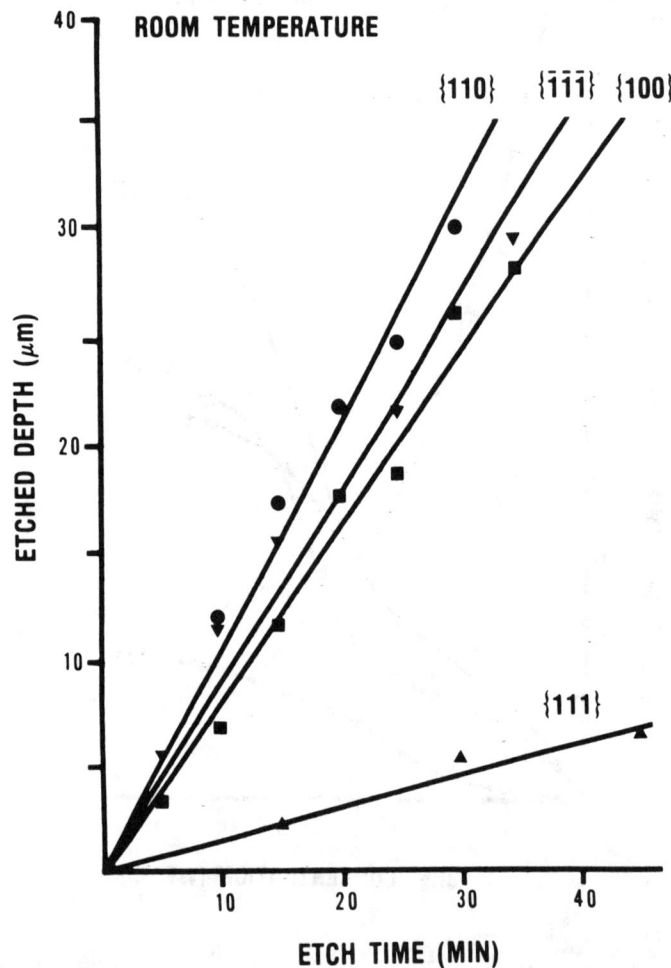

Figure 5.9 Etch depth as a function of time for various crystalline planes of GaAs using a 1% (by weight) mixture of bromine in methanol (after reference [11]).

V-shaped slots for the on-axis and the two slots oriented 10° away; the other slots (at 20°, 30°, etc.) will still exhibit the reverse mesa shape [19]. The center V slot will have serrulated walls. These cases are illustrated in Figure 5.10. The explanation is that the high concentration bromine etchant stops at {332} planes, and these planes form the walls of a V slot located 11.3° (close to 10°) from the [01$\bar{1}$] direction [19]. The preference to reveal {332} planes explains both the serrulated walls on the on-axis groove and the V walls on the grooves 10° away.

5.3.3 HCl-Based Systems

Hydrochloric-based systems have not been as popular in device processing as the above two systems, but they have been studied [6,15,20]. HCl has been used in systems such as HCl/H_2O_2/H_2O [14], HCl/CH_3COOH/$K_2Cr_2O_7$ [19], and in combination with HNO_3 [6]. HCl-based systems are characterized by a tendency to form etch profiles that have curved surfaces rather than straight planes as indicated in Table 5.3 [15]. HCl-based systems are interesting because, unlike almost all other GaAs etchants, they can provide nearly isotropic etching or minimal mask undercutting (but not both simultaneously). Consider the HCl/H_2O_2/H_2O system. Examination of Table 5.3 shows that low concentration HCl etchants tend to be isotropic. In fact, the 80:4:1 mixture appears completely isotropic, even when assessed by its action on circular patterns [15]. High concentration HCl mixtures are not isotropic, but show minimal mask undercutting. This is evident by comparing the profile shown in Table 5.3 for the 1:1:9 HCl mixture to the other profiles contained in the table.

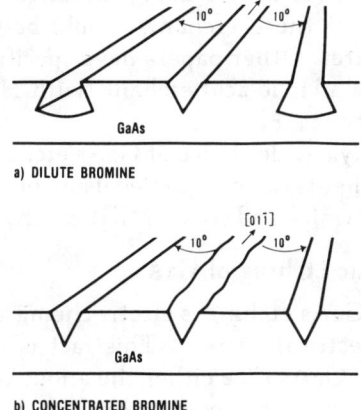

Figure 5.10 Etched slots in GaAs produced using a bromine-methanol etchant: (a) Dilute bromine (~ 1%); (b) concentrated bromine (~ 5%).

5.3.4 Alkaline-Based Systems

Oxides formed on GaAs tend to be amphoteric, and so can be dissolved in either acidic or basic solutions. Hence, alkaline-based systems' operation is similar to the acidic systems' described above. They result in the same straight-walled obtuse or acute (undercut) angles typical of the sulfuric etches. Alkaline systems generally use NaOH [18,21-24] or NH_4OH [25,26] in combination with H_2O_2 and H_2O. They have also been used for chemical/mechanical polishing in appropriate proportions (see the references above).

Although KOH is not a liquid etchant, in the general sense used in this chapter, it should be included because of its use in ascertaining dislocation density in GaAs slices (Chapter 2). Dislocation density is assessed by etching the slice in KOH heated to about 350° C in nickel containers. This etchant forms etch pits at locations on the slice where there is a dislocation in the crystal lattice.

5.3.5 Other Etching Systems

As indicated in section 5.2, almost any oxidizing agent in combination with a solvent will etch GaAs. Therefore, it is no major accomplishment to create still another "new" GaAs etching system. Characteristics usually will not differ greatly from those described above. Table 5.4, taken from reference [6], lists many etching systems that have been investigated. The etch rate is indicated (room temperature, unstirred). Photomicrographs of edge profiles obtained from most of these may be found in that reference. The profiles do not differ greatly from those described above. Those studies were performed using rather concentrated mixtures. In practice, many of these etchants would be more useful in greater dilutions with water. Other papers have specifically addressed unusual etchants such as a citric acid etchant (with H_2O_2) [27] and a $H_3PO_4/H_2O_2/CH_3OH$ etchant [28].

Although there is clearly a wide choice of GaAs etchants available, all GaAs processing requirements can be satisfied using only a few of these. That makes process uniformity and control that much easier.

5.3.6 Anodic and Cathodic Etching of GaAs

Section 5.1 noted that GaAs etching is electrochemical in nature, and therefore sensitive to electrical current. This fact is used in etching procedures that make the GaAs slice either the anode or cathode of an electrolytic cell. Although etching proceeds through the process of oxidation and dissolution, oxide growth requires the presence of holes. If holes can be supplied or denied, oxidation (and therefore etching) will be enhanced or suppressed.

In anodic etching procedures, the slice is made the anode. Such etching has been used for several purposes. One is to selectively remove p-GaAs. In this procedure [29], an alkaline electrolyte is used without a chemical oxidizing agent. Oxidation results from the flow of current. The current flows readily when the surface is p-GaAs, but n-GaAs results in a depletion layer which acts as a diode and restricts current flow. Therefore, the p-GaAs is etched away but the n-GaAs material remains. Another major use of anodic etching is the self-limited etching of n-GaAs on semi-insulating substrates. As just noted, n-GaAs forms a reverse-biased depletion layer which restricts current flow. But the voltage can be increased to the point that reverse breakdown occurs. Then the avalanche process supplies the needed holes [10]. If this procedure occurs in an electrolyte without an oxide-dissolving agent, an oxide will be grown on the slice (anodization). The slice can then be placed in an acid to strip the oxide. Each anodization-dissolution cycle removes a well-defined amount of GaAs [30]. The procedure is referred to as *anodic etching*. Its major advantage is that it is self-limiting. When the n-GaAs layer becomes sufficiently thin, the depletion region will reach the semi-insulating substrate before breakdown occurs. Here, it widens rapidly into the substrate. No breakdown occurs; no holes are generated; no oxide is formed. Another, thicker portion of the slice will continue to be anodized and etched. Hence, an initially nonuniformly thick n-layer can be etched to a more uniform, self-limited thickness. Fortunately, the limiting thickness (which is a function of applied voltage and doping profile) is appropriate for the fabrication of FET devices.

Another use for anodic etching is to fabricate submicron thick GaAs membranes [31]. In this process the surface layer of a conductive GaAs substrate is damaged by ion implantation. This damage apparently reduces the hole lifetime. Electric current supplies holes and the substrate is etched away in the usual anodic etching process. When damaged material is reached, the reduction in hole lifetime causes a corresponding reduction in etch rate. Such experiments indicate that ion-implant induced damage extends deeper than the projected range of the implanted ion [31] by approximately a factor of two.

Cathodic etching uses the GaAs slice as the cathode of the electrolytic cell. This process can be used to selectively remove semi-insulating GaAs, but leave conductive GaAs (either n-type or p-type) [32]. When semi-insulating material exists, no current flows and chemical etching proceeds in the usual manner. When conductive material is reached, electric current flows and provides electrons which satisfy the demand of the oxidizing agent for electrons; etching then ceases. Another way of

TABLE 5.4

GaAs Etchants and Etch Rate at Room Temperature Without Stirring [6]

(proportions are equal volumes unless otherwise noted)

Etchant	Etch Rate (μm/min)
$HCl/CH_3COOH/H_2O_2$	2.0
$HCl/CH_3COOH/(1N-K_2Cr_2O_7)$	0.40
$HCl/H_3PO_4/H_2O_2$	0.75
$HCl/H_3PO_4/(1N-K_2Cr_2O_7)$	0.04
HCl/HNO_3 (1:1)	0.50
(1:2)	0.75
(2:1)	0.25
$HCl/HNO_3/H_2O$	0.83
$HCl/HNO_3/H_2O_2$	1.0
$HCl/HNO_3/CH_3COOH$	1.3
HNO_3/H_2O_2	7.0
HNO_3/CH_3COOH	-
$HNO_3/CH_3COOH/H_2O_2$	4.5
HNO_3/H_3PO_4	10.0
$HNO_3/H_3PO_4/H_2O_2$	3.5
$H_2SO_4/H_2O_2/H_2O$	5.0
$H_2SO_4/CH_3COOH/H_2O$	2.5
$H_2SO_4/H_3PO_4/H_2O$	3.0
$H_2SO_4/HCl/(1N-K_2Cr_2O_7)$	0.75
$H_3PO_4/H_2O_2/H_2O$	4.0
$H_3PO_4/CH_3COOH/H_2O_2$	2.0
$H_3PO_4/CH_3OH/H_2O_2$	2.5
$H_3PO_4/C_2H_5OH/H_2O_2$	2.0
$HF/HNO_3/H_2O$	10.0
$HF/HNO_3/H_2O_2$	8.0
$HF/HNO_3/CH_3COOH$	80.0
$HF/HNO_3/H_3PO_4$	50.0
$HF/H_2SO_4/H_2O_2$	21.0
HBr/HNO_3	0.75
$HBr/NHO_3/H_2O$	0.05
$HBr/CH_3COOH/(1N-K_2Cr_2O_7)$	1.5
$HBr/H_3PO_4/(1N-K_2Cr_2O_7)$	1.0
4% Br/CH_3OH	6.0
1% Br/CH_3OH	0.67
(1% Br/CH_3OH)/CH_3COOH	0.20
(1% Br/CH_3OH)/H_3PO_4	0.02
NaOCl	1.8

NaOCl/HCl (5:1)	0.75
(1N-NaOH)/H_2O_2/H_2O (1:1:10)	0.38
(1N-NaOH)/H_2O_2/NH_4OH (5:1:1)	1.1
NH_4OH/H_2O_2/H_2O (1:1:5)	1.8
(1N-KOH)/H_2O_2/H_2O (1:1:10)	0.5
(1N-KOH)/H_2O_2/NH_4OH (5:1:1)	1.4

TABLE 5.5

Simple Etchant Compositions for Various Materials
(Most material from reference [5])

(The following list is not meant to be exhaustive, but to list basic compositions that will etch the material. Commercially available proprietary etchants exist for many of these; they generally work well and are recommended. Etching speed will depend on temperature and the extent of dilution with water.)

METALS:

Aluminum	HCl/H_2O (1:4)
	NaOH solutions
	H_3PO_4/HNO_3/H_2O (74:7:19) (rapid)
Chromium	HCl
	HNO_3
Gold	HCl/HNO_3 (3:1)
	KI/I_2/H_2O (4g:1g:40ml)
Molybdenum	H_3PO_4/HNO_3/H_2O (5:3:2)
	H_2SO_4/HNO_3/H_2O (1:1:1-5)
Nickel	HNO_3/CH_3COOH/H_2SO_4 (5:5:2) plus water
	HNO_3 HCl/H_2O (1:1:3)
Platinum	HNO_3/HCl/H_2O (1:7:8)
Silver	HNO_3/H_2O (5-9: 1-5)
	Fe$(NO_3)_3$/H_2O (11 g: 9 ml)
Tantalum	H_2SO_4/HNO_3/HF (5:2:2)
	HNO_3/HF/H_2O (1-2:1:1-2)
Tin	HNO_3/C_2H_5OH (1:49)
	$FeCl_3$ solutions
Titanium	HF/H_2O (1:9)
	HNO_3/HF/H_2O (2:1:7)
Tungsten	KH_2PO_4/KOH/$K_3Fe(CN)_6$ (34g:13.4g:33g)
	+water to make 1 liter

(Table continued next page)

TABLE 5.5
(continued)

DIELECTRICS:

SiO_2	HF, BHF (buffered HF)
Si_3N_4	HF, BHF, H_3PO_4
$Si_xO_yN_z$	HF, BHF
$Si_xN_yH_z$	HF, BHF, H_3PO_4
TiO_2 (amorphous)	HF, BHF, H_3PO_4, H_2SO_4
Ti_2O_5 (amorphous)	HF, BHF, KOH/H_2O_2/H_2O
Al_2O_3 (amorphous)	HF, BHF, H_3PO_4

describing this is anodic etching (above) supplies holes to enhance or allow etching; cathodic etching supplies electrons to suppress or prevent etching. Indeed, the etch rate of conductive material during cathodic etching is a function of the current density [32]. The procedure works equally well with both p-type and n-type conductive material. Although p-GaAs formally results in a reverse-biased junction, Schottky barriers on p-GaAs are generally leaky and allow sufficient current flow. The technique can be used to remove semi-insulating substrates or material that is rendered insulating by ion implantation. The latter process would be the reverse of the anodic case above, in which the ion-implant damaged layer remained.

5.4 WET ETCHING OF OTHER MATERIALS

Many other substances, mainly metals and dielectrics, may require wet etching during GaAs processing. Table 5.5 lists a number of these and some basic etchants. In many cases, more complex compositions and/or commercial etchants of proprietary composition are available. Usually, the major ingredients of the commercial products are known, but other components may have been added to improve results. These commercial products generally perform well and are worth using.

REFERENCES

[1] A.H.P Skelland, *Diffusional Mass Transfer*. New York: Wiley-Interscience, 1964.

[2] B. Tuck, *Journal of Material Science*, 10, 1975, p. 321.

[3] D.J. Stirland and B.W. Straughan, *Thin Solid Films*, 31, 1976, p. 139.

[4] W. Kern, *RCA Review*, 39, 1978, p. 278.

[5] W. Kern and C.A. Deckert, in *Thin Film Processes*, J.L. Vossen and W. Kern, editors. New York: Academic Press, 1978, p. 401.

[6] S. Adachi and K. Oe, *J. Electrochem. Soc.*, 130, 1983, p. 2427.
[7] D. R. Turner, *J. Electrochem. Soc.*, 107, 1960, p. 810.
[8] F. Kuhn-Kuhnenfeld, *J. Electrochem. Soc.*, 119, 1972, p. 1063.
[9] R.W. Haisty, *J. Electrochem. Soc.*, 108, 1961, p. 790.
[10] D.L. Rode, B. Schwartz, and J.V. Dilorenzo, *Solid-State Electronics*, 17, 1974, p. 1119.
[11] P.D. Greene, *Solid-State Electronics*, 19, 1976, p. 815.
[12] Y. Tarui, Y. Komiya, and Y. Harada, *J. Electrochem. Soc.*, 118, 1971, p. 118.
[13] D.N. MacFadyen, *J. Electrochem. Soc.*, 130, 1983, p. 1934.
[14] S. Iida and K. Ito, *J. Electrochem. Soc.*, 118, 1971, p. 768.
[15] D.W. Shaw, *J. Electrochem. Soc.*, 128, 1981, p. 874.
[16] D.W. Shaw, *J. Crystal Growth*, 47, 1979, p. 509.
[17] H.K. Kuiken and R.P. Tijburg, *J. Electrochem. Soc.*, 130, 1983, p. 1722.
[18] D.W. Shaw, *J. Electrochem. Soc.*, 113, 1966, p. 958.
[19] L.A. Koszi and D.L. Rode, *J. Electrochem. Soc.*, 122, 1975, p. 1676.
[20] S. Adachi and K. Oe, *J. Electrochem. Soc.*, 131, 1984, p. 126.
[21] I. Shiota, K. Moloya, T. Ohmi, N. Miyamoto, and J. Nishizawa, *J. Electrochem. Soc.*, 124, 1977, p. 155.
[22] T. Kobayashi and K. Sugiyama, *Jpn. J. Appl. Phys.*, 12, 1973, p. 619.
[23] M.M. Fakter, D.G. Fiddyment, and M.R. Taylor, *J. Electrochem. Soc.*, 122, 1975, p. 1566.
[24] C.J. Nuese and J.J. Gannon, *J. Electrochem. Soc.*, 117, 1970, p. 1094.
[25] J.J. Gannon and C.J. Nuese, *J. Electrochem. Soc.*, 121, 1974, p. 1215.
[26] J.C. Dyment and G.A. Rozgonyi, *J. Electrochem. Soc.*, 118, 1971, p. 1346.
[27] M. Otsubo, T. Oda, H. Kumabe, and H. Miki, *J. Electrochem., Soc.*, 123, 1976, p. 676.
[28] J.L. Merz and R.A. Logan, *J. Appl. Phys.*, 47, 1976, p. 3503.
[29] C.J. Nuese and J.J. Gannon, *J. Electrochem. Soc.*, 117, 1970, p. 1094.
[30] B. Schwartz, CRC Critical Rev., *Solid State Sci.*, 5, 1975, p. 609.
[31] K.C. Lee, J. Silcox, and C.A. Lee, *Appl. Phys. Lett.*, 43, 1983, p. 488.
[32] P.D. Greene, *Solid State Electronics*, 19, 1976, p. 815.

CHAPTER 6

PHOTOLITHOGRAPHY

6.1 INTRODUCTION

Virtually all patterning techniques used in semiconductor processing employ energy-sensitive chemical substances called resists. These are applied to the slice as thin-film coatings and then selectively exposed to an energy pattern (light, electrons, etc.) that creates exposed areas. The resist film is then subjected to a development process that selectively removes either the exposed or the unexposed resist. The remaining pattern can then be replicated in other materials using techniques such as etching, evaporation, or plating. The exposure may be accomplished using light (photolithography), electron beams (e-beam lithography), x-rays (x-ray lithography), or even ion beams (ion beam lithography). In *positive* resists, the development step removes the exposed resist —*negative* resists are those in which development removes the unexposed resist. Although nonoptical lithography techniques are very important for some uses, photolithography is the dominate technique used in semiconductor processing in general and in GaAs processing in particular; it is the subject of this chapter. E-beam lithography, used to define submicron geometries, will be discussed in the following chapter. X-ray and ion beam lithography will be mentioned briefly in that chapter as well; they are not yet in general use.

Although lithography techniques are well-known to silicon process engineers, GaAs processing has some unique distinctions. The basic silicon process uses resist as an etch mask. Assuming that a metal is being patterned, the process (Figure 6.1 (a)) consists of applying metal on the slice, spinning resist over the metal, exposing and developing the resist to define the pattern, etching away all exposed metal, and then removing the remaining resist. The name "resist" derives from its use to resist the action of the etchant. This procedure leaves the metal on the slice in the desired pattern. This etching process is used in GaAs fabrication for patterning dielectric films, but rarely in patterning metals. There are several reasons for this: many metal etchants will also attack the GaAs substrate; silicon processing often uses aluminum metalization which etches in relatively innocuous etchants; GaAs processes often include metals such as gold that require far stronger etchants; GaAs processing also tends to use composite metalization, such as AuGeNi for ohmics and TiPtAu for gate or overlay metals, which are difficult to etch regardless of the substrate.

Metal patterning in GaAs processing usually uses a procedure called the *liftoff process*, illustrated in Figure 6.1 (b). Resist is applied to the slice, exposed, and developed to define the desired pattern. Metal is next applied to the slice, usually by evaporation. Then a solvent is used which dissolves the resist. This results in "lifting off" the metal which is on top of the resist. Unlike etching processes, the success of the liftoff process is highly sensitive to the edge profile of the patterned resist. A protruding lip (Figure 6.1 (b)) at the top edge of the resist profile is desirable. Most of the lithography techniques used in GaAs processing were developed to provide such an undercut resist edge profile.

Many papers are published describing high resolution photolithography, some using rather exotic procedures. Most of these never mature into full-scale production use. This chapter is confined to the proven, commonly used lithographic techniques and does not attempt to review the numerous alternative approaches that are under investigation. To do so would tend to obscure the major points for readers inexperienced in photolithography as it pertains to GaAs processing.

Section 6.2 reviews the different categories of optical lithography; a separate section is devoted to optical steppers. Resist characteristics are reviewed in section 6.3. Edge profile and multilevel resist techniques are considered in section 6.4.

6.2 TYPES OF PHOTOLITHOGRAPHY AND EQUIPMENT

Photolithography exposes resist in a pattern matching that on a photomask. These masks are usually generated using electron beam lithography (Chapter 7). They consist of sheets of glass or quartz with the desired pattern defined on them in thin-film materials such as chromium, iron oxide, or silicon. The latter two are used to make "see-through"

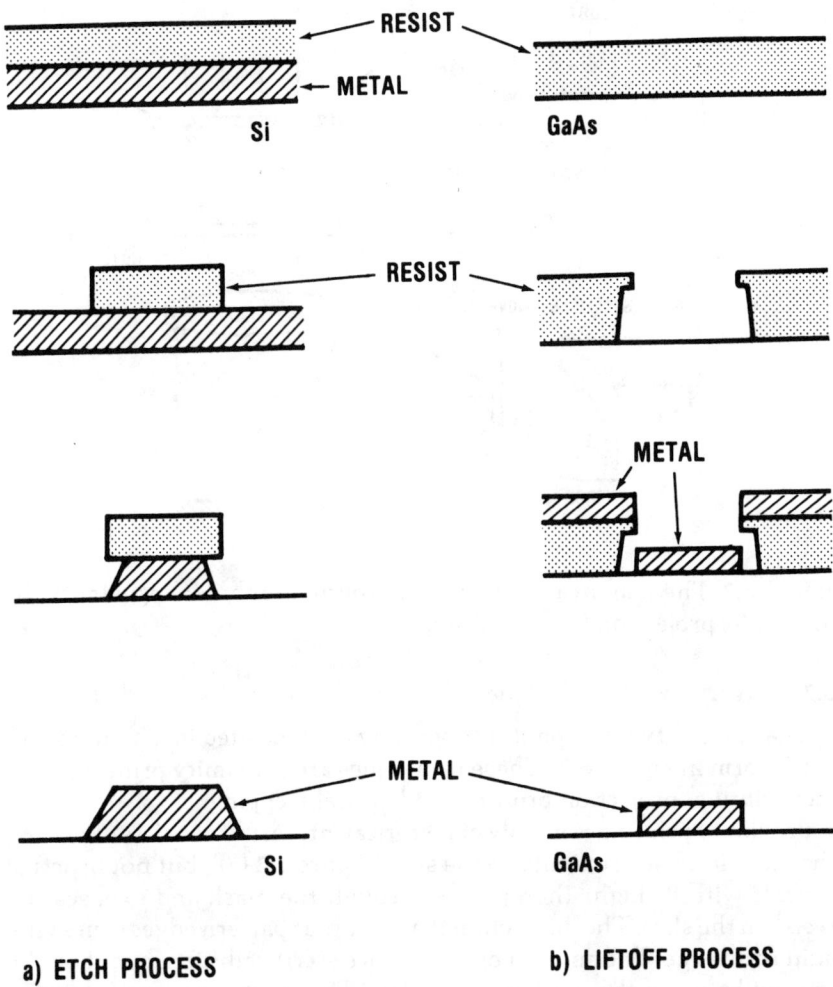

Figure 6.1 The two basic resist processes: (a) the etch masking process typically used in silicon processing (b) the liftoff process typically used in GaAs processing.

masks. Photoresist is sensitiive only to the high frequency end of the optical spectrum and substances such as iron oxide are opaque at these wavelengths and can function as masks. But iron oxide is transparent at longer wavelengths, allowing the operator to "see through" the entire mask when aligning it to the pattern on the slice; then the appropriate wavelength of light is used to make the exposure. Photolithography is sometimes referred to as printing, and photolithographic machines are sometimes called printers or exposure towers.

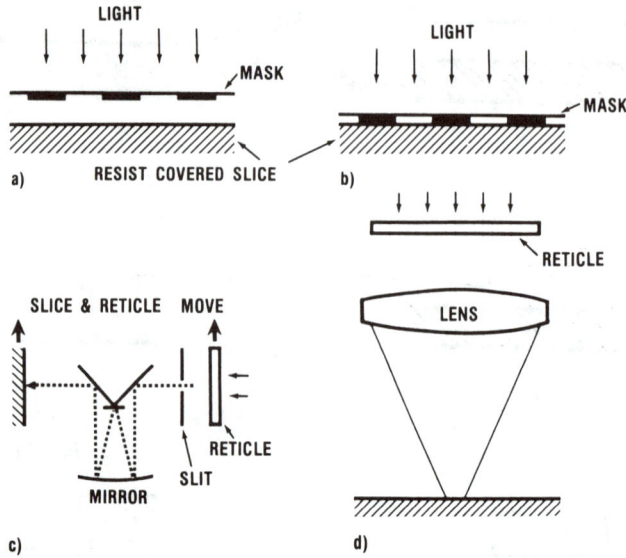

Figure 6.2 The major categories of photolithography: (a) proximity (b) contact (c) projection (d) optical stepping.

6.2.1 Overview of Techniques

The various types of photolithography are indicated in a highly schematic form in Figure 6.2. These techniques are proximity printing, contact printing, projection printing, and optical stepping.

Proximity printing is mainly of historical interest. It consists of placing the mask in close proximity to the slice (Figure 6.2 (a)), but not in actual contact with it. Light then passes through the mask and exposes the resist on the slice. The diffraction that occurs at pattern edges causes the light to diverge. The amount of divergence is critically dependent on the spacing between the mask and the slice. This divergence, and its non-uniformity across a slice, makes this technique incapable of achieving the resolution and reproducibility required in modern photolithography.

Contact printing is a popular approach to GaAs device fabrication; it is illustrated in Figure 6.2 (b). In this technique the mask is placed in proximity to the slice for alignment; then it is vacuum clamped directly against the slice for exposure.

However, several difficulties arise using contact printing. The actual contact between slice and mask tends to damage the mask, both from abrasion during the printing operation and from the frequent cleaning required to remove resist that gets attached to the mask. These defects are then replicated in all subsequent uses. Such problems make it necessary to replace masks after approximately one hundred uses. A second disadvantage of contact printing is "runout". This occurs because the mask and the slice are never perfectly flat. Any curvature results in imperfect registration between the pattern on the mask and the pattern on the slice to which it is aligned. That is, it will be possible to align the mask to the slice accurately in some areas, but not simultaneously in all areas. Considering that some devices may require overlay registration tolerances less than a micron, minimal curvature can cause substantial problems. These difficulties have made contact lithography unsuitable for many of the complex devices made in silicon processing.

But contact lithography is still a mainstay in GaAs processing for several reasons. First, GaAs microwave devices and MMICs generally employ only a few gates or critically active areas (as opposed to silicon digital devices that may have tens of thousands of gates) and can therefore tolerate a higher defect density. Critical geometries, such as FET gates, can be exposed by other means, such as e-beam lithography. These considerations mean that mask defects have minimal consequences. Second, the overlay requirements for analog GaAs microwave devices are usually not severe. The gate pattern is an exception to this, but as noted above, it may be fabricated to a high degree of accuracy using e-beam lithography. Also, GaAs slices, at this writing, are 2 inches or 3 inches and hence, are not as big as silicon slices; so runout is not as serious a problem as with the larger silicon slices.

The resolution achievable with contact photolithography is approximately one micron. Better results have been reported, but these are difficult to achieve on a reliable, reproducible basis over a full slice. Slightly higher resolution can be achieved using *deep ultraviolet* (DUV) light because a shorter wavelength allows greater resolution (section 6.2.2). Contact printing equipment exists that is specifically designed for DUV. Xenon-mercury lamps are suitable light sources and emit in the 220-240 nm range. Glass is highly absorptive at these wavelengths and hence, quartz masks are employed. Usually different types of resists are required. Resists developed for electron beam lithography, such as PMMA (Chapter 7), tend to be sensitive to DUV light and are used for this purpose.

Intimate contact between mask and slice is absolutely essential to obtaining good resolution and pattern definition in contact printing. Optical interference fringes should be visible (if using see-through masks) when the mask is clamped to the slice. If technical reasons do not eliminate contact photolithography for a given application, it is an attractive technique on the basis of its speed and cost. Because the entire slice is exposed at one time, it is fast in contrast to the field-by-field exposure typical of optical steppers or e-beam machines. The cost of contact printers is also much less than the alternative choices. Contact photolithography is very appropriate for fabricating GaAs analog devices and MMICs. GaAs digital circuits require higher-yield techniques.

A third type of photolithography is projection printing as illustrated in Figure 6.2 (c). In this case the mask is completely separated from the slice and optics are used to image the mask pattern onto the slice. Projection printers are also called scanners because the optics image a line (usually curved for technical reasons) and this line is scanned across the mask and slice to form the exposure. The slice must be very flat because depth of focus of such systems is only a few microns at most. Projection printers are losing popularity to optical steppers.

The optical stepper is becoming the dominant machine for high-yield photolithography. This technique is indicated in Figure 6.2 (d). A mask, called a reticle, contains the image of one field. This pattern is imaged onto the slice, then the slice is moved and the exposure repeated. This "step and repeat" procedure continues until the entire slice is exposed. The pattern on the reticle may be the same size as the pattern on the slice, five times as large, or ten times as large. Steppers are very popular for low-defect photolithography, which usually is required for fabrication of digital circuits. Steppers will be considered in more detail in section 6.2.2. Both projection printers and steppers share the advantage that the mask, or reticle, never touches the slice. Correct handling procedures can result in virtually damage-free usage.

All of the above techniques require excellent collimation of the light source, and uniform and constant intensity over the mask area. Collimation is important so that all light impinges on the mask at normal incidence. Some lithographic machines continuously monitor the light output to maintain constant intensity. Nevertheless, it is wise to routinely monitor the light intensity using an independent meter. The sensing head should be filtered to respond only to the wavelength or wavelengths to which the resist is sensitive. In the case of contact printers, the measurement may be made at numerous locations across the slice stage to assure uniform illumination. All lithographic techniques are also very sensitive to vibration. Relative movement of even a fraction of a micron between slice and mask can cause problems. Hence, printers are placed on special foundations and/or tables that isolate them from vibration.

In summary, contact photolithography is popular for analog GaAs devices. Critical geometries may be defined using e-beam lithography. Optical steppers will be required for the growing GaAs digital efforts.

6.2.2 Optical Steppers

Because steppers are complex instruments and will be employed for production of GaAs digital circuits, it is useful to review the considerations involved. The optical imaging can be performed using either reflective optics (mirrors) or refractive optics (lenses); most use lenses. In either case, the required accuracy demands that the optical components be diffraction limited. This means that the surfaces are so perfect that performance is limited by the physics of wave optics (diffraction) rather than by defects in the optics. This requirement generally means that surfaces must be accurate to at least one-quarter wavelength, or about 0.1 micron. Fortunately, optical fabrication techniques, driven by telescopic and space applications, make such perfection achievable, although expensive. By comparison, modern camera lenses are marvels of performance, but (with a few special exceptions) are far from diffraction limited. This is no problem in photography because the lenses are often better than the film which can be the limiting element for resolution in common photography.

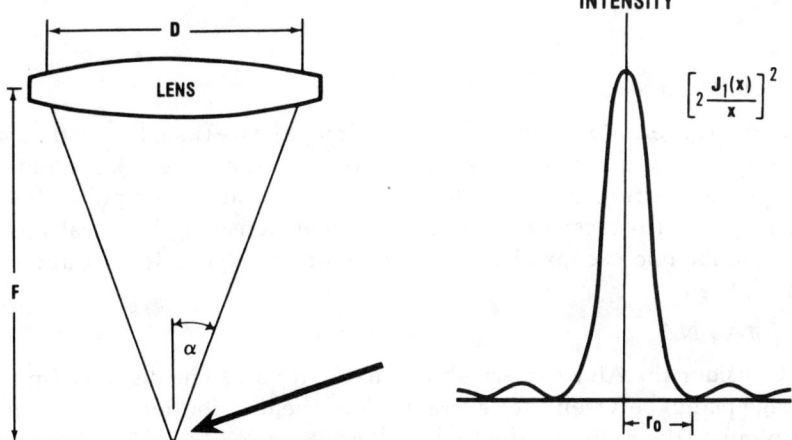

Figure 6.3 Imaging in an optical system showing the Airy pattern that results from imaging a point.

In addition to expensive optics, the accuracy required of steppers requires extremely precise construction, low thermal expansion of critical elements, temperature compensation, vibration isolation, highly accurate stage stepping, and complex alignment and focusing systems. These considerations make steppers expensive. They are ten to twenty times as expensive as the very best contact exposure towers, and in 1984 may cost a significant fraction of a million dollars.

Several relevant aspects of imaging are represented in Figure 6.3. If a point is imaged onto a surface by a perfect optical system, the wave nature of light will result in the image being not a point but a more complex pattern, as illustrated in Figure 6.3. It is called an Airy pattern and the intensity on the image plane is represented by the equation

$$I(x) = K \frac{J_1^2(x)}{x^2}$$

where J_1 is a Bessel function. (A rectangular aperture yields the form $I(x) \propto \mathrm{sinc}^2(x) = \sin^2(x)/(x^2)$)

The central region inside the first null contains 83.8% of the light. (This and other optical information may be found in almost any college optics textbook, such as reference [1] or [2]. The relationship between the aperture, D, and the focal length, F, of an optical system is expressed either by the F-number (F #) or the numerical aperture (NA). These are defined as follows (also see Figure 6.3):

$$F\# = F/D$$

and

$$NA = n \sin(\alpha)$$

where n is the index of refraction in the image space (usually air, then $n = 1.00029 \approx 1$) and α is the half angle of the maximum cone of light coming from the aperture of the system and reaching an image point. F # is usually used to describe telescopes and camera lenses. NA is usually used to describe microscope objectives and optical steppers. If $n = 1$ and α is small, then

$$F\# = \tfrac{1}{2}NA$$

Consider the Airy pattern shown in Figure 6.3. The distance (in the image plane) between the central peak of the distribution and the first minimum (the radius of the first null) is

$$r = \frac{1.22\,\lambda\,F}{D} \approx \frac{1.22\,\lambda}{2\,NA} \tag{6.1}$$

where λ is the wavelength of the light (this result assumes incoherent light — see below for discussion of coherence). This value is commonly taken to be the resolution of the system (it is called the Rayleigh criterion). In practice, slightly higher resolution can be obtained in optical systems with great care, but this definition of resolution is easily specified and is close to what can be achieved in most systems. Often the coefficient 1.22 in equation (6.1) is taken as 1.0 and the resolution, w, of optical steppers is quoted as

$$w = \frac{\lambda}{2\,NA} \tag{6.2}$$

Similar considerations lead to an expression for the depth of field, d, of optical steppers being

$$d = \frac{\lambda}{2\,(NA)^2} \tag{6.3}$$

In practice, changes in focus lead to line-width variations and it is these that set the practical limit on depth of field. In either case, the depth of field is only a few microns at most (usually less) and hence, steppers require very flat slices and/or automatic focusing ability. Such automatic focusing can be achieved using capacitance techniques if the slice is locally flat.

Another important criterion for all optical systems including steppers is the *modulation transfer function* (MTF). It is analogous to the frequency response used to assess audio systems. Audio systems are judged by their ability to reproduce audio frequency, with response falling off at sufficiently high frequencies. In an analogous manner, optical system resolution can be assessed by the ability to image (reproduce) spatial frequencies consisting of a sinusoidal variation between complete transparency and complete darkness. The spatial frequency is expressed as cycles/mm or lines/mm. (In practice, such a sinusoidal variation is difficult to produce and use. Consequently, the technique usually is modified slightly to employ equally-wide lines and spaces.) For large spacings (low spatial frequency) the optical system will accurately reproduce the line-and-space pattern in the image; for sufficiently small spacings the lines and spaces will blur in the image and not be accurately reproduced. The audio analogy is not exact because a perfect audio system would accurately reproduce all frequencies. However, a perfect optical system is still limited by the wave nature of light and will not accurately image all spatial frequencies. The modulation, M, in the image is defined to be

$$M = \frac{I_{max} - I_{min}}{I_{max} + I_{min}} \tag{6.4}$$

where I_{min} and I_{max} are the minimum and maximum intensity in the image. The modulation transfer function is M as a function of spatial frequently (lines/mm). If λ is the wavelength of the (incoherent) illumination, and ν the spatial frequency, a perfect optical system would have an MTF given by [3,4]

$$\text{MTF} = (2/\pi)\,[\arccos(y) - y\,(1 - y^2)] \tag{6.5}$$

where $y = \lambda\, F\#\, \nu$

Equation (6.5) is plotted in Figure 6.4 for two wavelengths and with $F\#$ 3.0 Perfect focus is assumed; deviations from perfect focus will rapidly degrade the MTF. The maximum cutoff frequency (incoherent illumination) is

$$\nu_{max} = \frac{1}{\lambda F \#}$$

This cutoff frequency is for MTF = 0. Most resist exposure requires MTF > 60%.

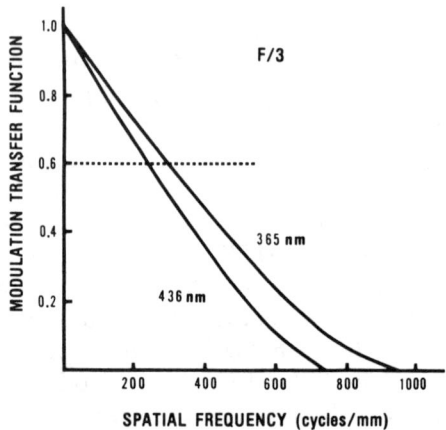

Figure 6.4 Modulation transfer function (MTF) of a diffraction limited F/3 lens at two wavelengths. The 60% MTF line represents the minimum MTF required by most resists for suitable results.

Optical systems have aberrations which limit off-axis performance. Possible aberrations are spherical aberration, coma, astigmatism, field curvature, and distortion [1,2]. The amount of aberration in an imaged point is a function of its distance from the optical axis of the system. Even diffraction-limited systems exhibit degradation of the image as it moves off-axis. Specifically, the modulation transfer function will itself be a function of location in the image plane; it will be best on-axis. This effect limits the useful field over which patterns can be imaged with suitable resolution. The greater the numerical aperture (NA) of the lens, the more rapidly the image degrades off-axis. Hence, improving resolution by increasing NA (equation 6.1) means reducing the useful field size (and reducing the depth of the field — equation 6.3). High NA lenses are also more difficult to fabricate, assuming diffraction-limited performance.

Another issue important to optical steppers is optical coherence. The coherence of the illumination affects MTF and the resulting resist exposure. It should be noted that there are two types of optical coherence: temporal and spatial. Temporal coherence refers to the wavelength content of the light. Broadband sources are incoherent; light that consists of vary narrow range of wavelengths is temporally coherent or monochromatic. Examples of temporally coherent light are laser light and molecular emission lines in low-pressure gaseous discharges. Optical steppers (especially those using lenses rather than mirrors) must use one or at most two distinct wavelengths (the optics are not corrected for multiple wavelengths). Usually the h or g lines of mercury (405 nm and 436 nm) are used (see section 6.3) Hence, temporal coherence is always present. The word coherence used in the context of optical steppers always refers to spatial coherence; it will be used here in that sense in all further discussion. The light reaching any two points in the image may have different relative phases, but if the phases vary with time in an identical fashion, the illumination is spatially coherent. Coherent illumination is obtained when the light appears to arise from a single point (actually, this is a sufficient but not a necessary condition [3]).

A complete discussion of coherence is beyond the scope of this book. A few generalizations follow; more information may be found in reference [5]. Light in steppers is gathered and collimated by a condenser system which has its own numerical aperture, NA_c. The numerical aperture of the objective lens is designated NA_o. The degree of optical coherence may be taken as [6]

$$\sigma = \frac{NA_c}{NA_o}$$

Complete coherence would be represented by $\sigma = 0$. Complete incoherence would be represented by $\sigma = \infty$. Any intermediate condition is partial coherence. The degree of coherence affects several optical parameters including the resolution and the modulation transfer function. All definitions and results given above were for complete incoherence. For example, the resolution as represented by the first null in the Airy pattern (equation (6.1)) becomes [1]

$$r = \frac{1.54 \, \lambda \, F}{D} \quad \text{(coherent light)}$$

for complete coherence. The modulation transfer function is also affected in that any degree of partial coherence decreases ν_{max}. For complete coherence, ν_{max} is reduced by a factor of two:

$$\nu_{max} \text{(coherent)} = \tfrac{1}{2} \, \nu_{max} \text{(incoherent)}$$

It should therefore be clear that results obtained using optical steppers depend on the amount of partial coherence as well as the numerical aperture. For complex reasons, line size control is better with some amount of partial coherence [5]. However, σ values less than 0.7 lead to only marginal improvements while requiring longer exposure times [5].

6.3 RESIST PROPERTIES

As noted in the introduction, photoresists are either positive or negative depending on whether the exposed portion is removed or remains during development. As a general rule, positive resists have more advantageous properties. They offer higher resolution and allow more appropriate edge profiles for liftoff processes. Positive resists are normally the preferred choice for most aspects of GaAs processing. There are many types of resist available, with a variety of characteristics, and manufactured by many different companies. These companies include Shipley, Philip A. Hunt, Eastman Kodak, KTI Chemicals, J.T. Baker, Allied Chemical, Macdermid, Azoplate, Tokyo Ohka, and Micro Image.

6.3.1 Positive Resists

A typical positive resist (such as Shipley's AZ1350J) is composed of three major components: a photoactive compound (inhibitor), a base resin, and a suitable organic solvent system. It will also contain other proprietary ingredients. All or most of the solvent is lost after spin and bake. The base resin alone is moderately soluble in the aqueous alkaline developer, the removal rate being approximately 15 nm/s [7]. When the photoactive compound is present (usually to about 25-30 wt %), the removal rate of unexposed resist is reduced to approximately 0.1 nm/s. This is why the photoactive component is sometimes referred to as an inhibitor. Radiant energy in the wavelength range of 300-450 nm destroys the photoactive component and results in an increased removal rate, up to 100-200 nm/s. A more detailed explanation of each part of the resist process follows below. A good understanding of the principles involved in each aspect makes resist process development merely difficult to accomplish. Without such understanding, it is impossible.

Resist is usually applied to the slice by spinning. That is, an appropriate amount of resist is placed on the slice and the slice is spun at a specified speed (2000 to 8000 rpm) for a specified time (20 to 40 seconds) to provide a uniformly thick coating. The spinning operation also pre-drys the resist so that it stays in place after spinning ceases — the thickness of the resist film is a function of the spin speed. Final thickness is also affected by subsequent heating steps. Resist manufacturers supply curves showing thickness as a function of spin speed. The resist will require a post-spin bake to harden it by removing more of the remaining solvents. Resist films are usually chosen to be between 0.3 μm and 2.5 μm thick depending on the application. The spinning procedure can coat a flat slice to a highly uniform thickness. But substantial local variations in the thickness of the resist film can occur if the slice has topographical features [8], such as etched mesas (for device isolation) or metal patterns. Resist thickness variations translate directly into linewidth variations during exposure. The variations are worse for high-reflectivity substrates. Such variations may or may not be a problem for a given application. Multilevel resist techniques (section 6.4) can be used to alleviate the nonuniformity of topographical features.

Both the resist and the slice must be exceptionally clean. Any particle contamination will result in poor imaging or defects. It is common to filter resist to remove particles as small as 0.2 μm. *Relative humidity* (RH) must be also controlled, generally below 50% RH. The slice has to be completely dry before resist is applied. A bake step or hot plate step should be included to assure complete dryness. If resist is being applied over an oxide, or any dielectric, adhesion can be improved by using an adhesion promoter prior to resist application — resist manufacturers supply commercial compositions. Also, the chemical HMDS (hexamethyldisilazane) is often used for this purpose. Such adhesion promoters have limited use when the resist is applied over metal or GaAs material.

Most positive resists are sensitive to light between 300-450 nm. Absorption is low below 200 nm or above 500 nm. Mercury lamps are a common light source — they have copious output in the 200-600 nm range. The major lines in the Hg spectrum are designated as follows:

e line   546 nm
g line   436 nm
h line   405 nm
i line   365 nm

The g and/or h lines are most commonly employed in optical steppers. Reproducible processes require that the intensity of the light and its uniformity be closely controlled. This applies to all types of lithographic equipment.

Light is absorbed as it travels through the resist. To take a specific example, the index of refraction of Shipleys AZ1350J resist is approximately [5]

$$n = 1.65 - j0.02$$

which gives an attenuation of about 0.6/μm before exposure. The attenuation changes during exposure and can decrease to 0.1/μm [5]. The light traverses through the resist film and is reflected at the bottom interface of the resist film, and again at the top interface. Depending on the substrate material (GaAs, oxide, metal), there may be a phase change of the light upon reflection. These multiple reflections and attenuations create interesting problems. The reflections generate a standing wave pattern that causes differential exposure of the resist in the vertical direction. These effects may be seen in the edge profile of the resist pattern shown in Figure 6.5. The exact standing wave pattern that is obtained is a function of the resist thickness and the nature and thickness of the underlying substrate. Dielectrics will give little reflection. Metals will give strong reflections and phase changes. The variegated exposure can be alleviated by the post-exposure bake [9,10] which tends to redistribute the photoactive component. The function of heat treatments will be discussed below.

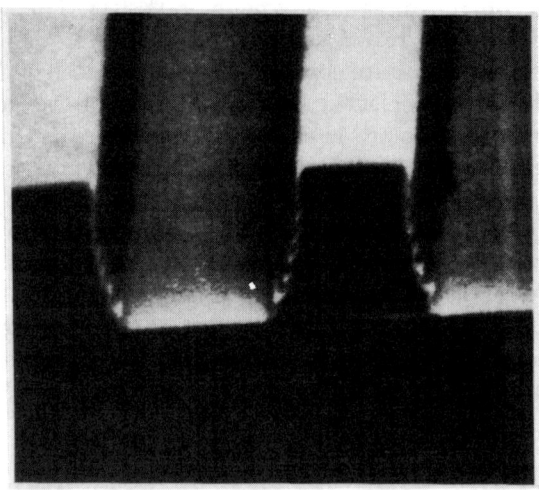

Figure 6.5 Photoresist profile of 1.25 μm lines and spaces without a post-exposure bake. The scalloped edges represent standing wave effects (reference [9], reprinted with permission from *Solid State Technology*, Technical Publishing, a Company of Dun & Bradstreet).

A certain amount of contrast must be present to expose one area of resist while leaving another unexposed. The ideal case would be 100% contrast: full intensity of light in the area to be exposed; no light in other areas. However, diffraction effects make this goal impossible. Diffraction occurs at pattern edges in contact lithography. Diffraction effects lead to blurring in optical steppers as explained in the discussion of the modulation transfer function. This means that some light will be diverted from areas intended to be exposed. Nevertheless, the pattern can be successfully exposed in the resist if there is sufficient contrast between the intensity of light in the two areas (desired patterned and desired unpatterned). A certain minimum MTF, approximately 60% for positive resists, must exist to yield suitable patterning.

The pattern that is replicated in the resist will not necessarily be the exact size of the pattern on the mask. This topic is usually addressed by assuming the pattern on the mask is a line of a given width. The corresponding linewidth produced in the resist is a complicated function of many parameters. For contact lithography, the exposed line can be smaller than, equal to, or larger than the linewidth on the mask. For optical steppers, linewidth will depend on the MTF (which includes focus effects) and the degree of optical coherence. In both cases, variations in resist thickness and/or substrate reflectivity will affect the linewidth. A number of these issues are discussed in articles in the July, 1975 issue of *IEEE Transactions on Electron Devices*, volume ED-22; the entire issue is devoted to pattern generation and lithography. In addition, information on linewidths can be found in references [5,7,11,12, and 13].

The optical exposure process results in chemical changes to the photoactive component of the resist. The resulting chemical reactions depend critically on the water content of the resist film. The water content is usually less than 0.5%, but its presence is essential. When exposed to light, the photoactive component degrades to form a ketene. In the presence of water, this ketene forms an indene carboxylic acid which is highly soluble in the basic developer. But in the absence of water, the degraded photoactive component forms ester linkages with other molecules (either of the photoactive component or of the resin) [9]. This crosslinking causes a decrease in solubility. A relatively small amount of crosslinking can significantly decrease solubility [9]. The critical role played by water makes the resist process very sensitive to humidity control in the room (Chapter 5) and to the baking steps that are used. The photoactive component will also degrade thermally, as well as by exposure to light, and result in crosslinking.

Baking in resist operations serves a number of purposes. Although the term *bake* is used, the actual heating process can be performed in an oven or on a hot plate. Hot plate bakes are generally between 60° and 100° C for periods of 20 to 60 seconds. Oven bakes are generally between 60° and 140° C for periods of 5 to 60 minutes. Two different types of bakes are used. One immediately follows resist application and is sometimes referred to as a *softbake*. The other is a bake following exposure, but before developing — it is called a *postbake*. The softbake serves to remove more of the water and solvents that remained in the film after spin-on. This hardens the resist film and improves adhesion. The removal of solvents also decreases the film's thickness. Figure 6.6 presents typical data showing the decrease in resist thickness as a function of hot plate temperature for a 45-second softbake [9]. Note that resist thickness decreases rapidly over the lower temperature range, making the final thickness especially sensitive to temperature variations. Greater process latitude is obtained with higher softbake temperatures. As indicated above, the amount of water left in the film affects exposure and development, so these properties are directly affected by the softbake. Typical data showing the dissolution rate of unexposed resist as a function of bake temperature is shown in Figure 6.7. Some amount of softbake (even a "room temperature" bake) is necessary to remove enough solvents to harden the resist. Excess softbake impedes easy exposure and development by promoting crosslinking. Subsequent removal of the resist is also affected by baking. The higher the bake temperature, the more difficult it will be to strip the resist from the slice.

The second possible bake is after exposure but before development. This postbake can result in several advantages [9,10]. These include improved linewidth control, elimination of standing wave effects, increased contrast, increased edgewall angle, reduction of scumming (see below), and improved adhesion. Postbakes can also form a crust on the

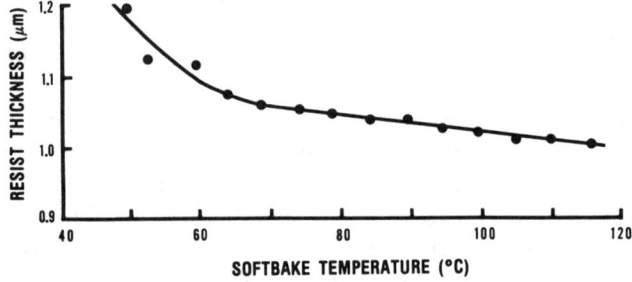

Figure 6.6 Resist thickness after a 45 second hotplate softbake as a function of bake temperature (after reference [9]).

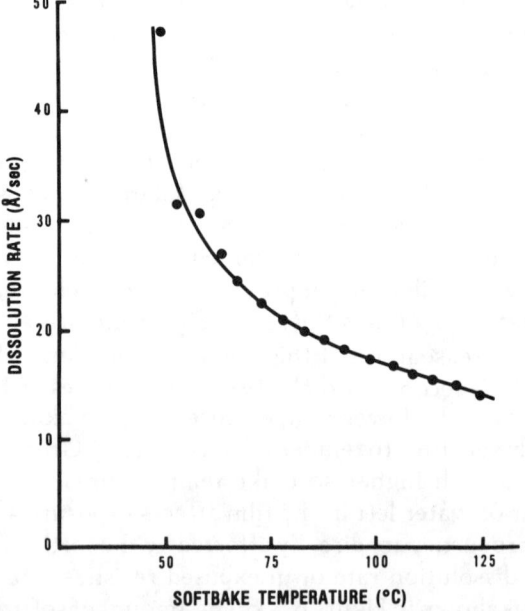

Figure 6.7 Average dissolution rate of unexposed photoresist as a function of softbake temperature (after reference [9]).

upper surface of the resist giving it a very slight overhang after development — this is desirable for liftoff. A postbaked edge profile, shown in Figure 6.8, illustrates several of these features, especially the absence of standing wave effects. The postbake drives out most of the remaining solvents intentionally left after softbake to aid exposure and development. The complete removal of all remaining solvents results in more uniform development rates as a function of resist thickness (the softbake can leave a nonuniform distribution of solvents in the film in the vertical direction). The complete absence of water, combined with the thermal degradation of the photoactive compound, results in more crosslinking. This crosslinking is what makes the hardbake effective; it also makes the resist more difficult to strip. The presence of the crust results in lower solubility rates in the upper part of the film. This effect, combined with the uniform solubility of the lower parts of the film, results in the steep slope and undercut profile evident in Figure 6.8.

The postbake can also reduce the phenomenon known as scumming. Even unexposed resist will slowly dissolve in the developer. This unexposed resist can be redeposited in developed-out spaces. Such material must then be removed by a "descumming" operation such as etching in oxygen plasmas (*ashering*). The hardening of the resist surface that occurs during postbake results in decreased solubility and hence, decreased scumming. Baking operations also make the resist more difficult to remove or lift off, and extremes should be avoided. Higher bake temperatures can result in resist flow.

Figure 6.8 Photoresist profile of 1.25 μm lines and spaces with a post exposure hotplate bake. Note the surface crust and the absence of standing wave effects (after reference [9], reprinted with permission from *Solid State Technology*, Technical Publishing, a Company of Dun & Bradstreet).

Resist patterns are developed either by immersing the slice in the developer or by spraying the developer onto the slice. Spray developing may be performed in a highly reproducible manner using commercial machines which spin the slice while spraying various chemicals (developer, water, etc.) on it under microprocessor control. Some form of agitation is usually used in immersion developing. The slice must be rinsed to remove the developer, usually using DI water, and then dried. Spin drying is effective in commercial machines.

After development, it may be useful to subject the entire slice to intense light (*blanket exposure*). This will expose the remaining resist, making subsequent liftoff or stripping operations easier. It can also remove nitrogen from the film which can cause blistering of sputtered or evaporated metals. Such blanket exposures are usually several times longer or more intense than the normal, patterning exposure. Note that some water must be present in the film or the exposure will result in increased crosslinking and make resist removal more difficult rather than easier. A blanket exposure before an extremely high temperature bake (greater than 140°C) will allow the bake-hardened resist to be removed by subsequent chemical means. Otherwise, such bakes would result in resist films that are almost impossible to remove.

Acetone is the major solvent used to remove positive resist either in liftoff or stripping operations. It can be sprayed on the slice, or the slice can be immersed in hot acetone. Alcohol and water are used to remove the acetone. Sometimes, stubborn portions of the resist may remain. Stronger solvents, such as J-100 (Chapter 5) may be used to remove such material, or it may be removed by plasma etching (ashering — see Chapter 9).

It should be clear that results obtained in resist processes depend on many variables, including resist type, spin speed and time, baking method, bake time, and bake temperature. Many of these variables will also affect the final linewidth obtained. For example, plasma descumming will remove some resist from all surfaces and increase linewidth. Plasma etching can also change the resist profile by etching protruding lips faster than other features.

No specific, detailed recommendations were given above for a complete resist process. This is partly because these processes tend to be proprietary. But more importantly, the many types of resist available and the many ways in which they are used make general recommendations useless. For a specific resist, the manufacturer's data sheet can serve as an initital guide. Then resist processes must be developed by experiments geared toward the specific application being considered. Good results in resist operations depend on careful experimental development, scrupulous attention to detail, and stringent process controls.

6.3.2 Negative Resists

The above discussion concentrated on positive resists. Negative resists may also be used in some applications in GaAs processing. They generally give worse resolution, worse edge profile, and worse linewidth control than positive resists. These undersirable features arise partly from a tendency of negative resist to swell in the presence of solvents. The poor edge profiles make negative resists unsuitable for liftoff operations. Negative resists are also generally more difficult to remove from the slice. J-100 (Chapter 5) is commonly used to remove negative resists. Negative resists do exhibit good resistance to liquid etchants and plating baths, but many modern positive resists also exhibit this good resistance.

One negative resist worthy of special mention is Riston (a product of Dupont). This is a thick (18-50 μm), sturdy resist that comes in sheet form, protected between cellophane-like sheets on both sides. One protective sheet is removed, allowing the material to be placed on a slice. Heating to about 60°C adheres the Riston to the slice. It is then exposed, the top protective sheet removed, and developed. The presence of the top protective sheet during exposure limits the resolution to approximately 20 μm. Riston is very tough, and will stand up to conditions that rapidly attack most other resists. These include plating operations and even sandblasting. Riston is soluble in methylene chloride.

Finally, it should be emphasized that there are many types of both positive and negative resists, each with distinguishing characteristics including resolution, exposure speed, resistance to flow under temperature, and sensitivity to various treatments, such as plating. It is generally desirable to make use of this diversity and employ several types of resist in a complete process flow.

6.4 EDGE PROFILE AND MULTILEVEL TECHNIQUES

Liftoff techniques, commonly used in GaAs processing, require proper resist edge profiles to yield reliable results. This requirement, in combination with others discussed below, has driven the development of special techniques. These techniques fall into two categories: edge profile modifications (single coating of resist), and multilevel resist procedures (two or more resist layers). Each is discussed below.

The liftoff process works best when the resist edge profile is undercut and the applied metal thickness is less than the resist thickness, as indicated in Figure 6.9(a). However, successful liftoff can be accomplished even if the resist thickness is less than the metal, as indicated in Figure 6.9(b). This is possible because metal grain growth occurs as successive layers of evaporated metal reach the slice. This results in film growth in the lateral direction in addition to the usual vertical accumulation. As indicated in the figure, the metal on the slice tends never to

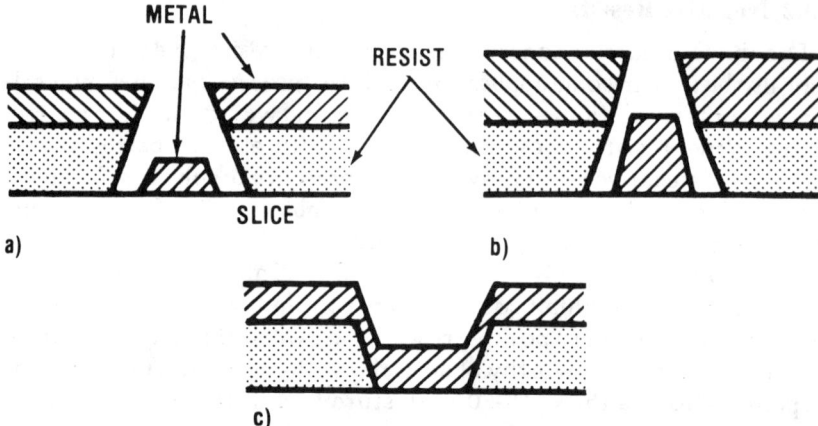

Figure 6.9 Liftoff technique with: (a) resist thicker than metal, undercut resist profile (b) resist thinner than metal, undercut resist profile (c) resist thicker than metal, non-undercut resist profile.

connect with the metal on the resist. The amount of lateral spread depends on the metal and the deposition parameters. Under proper conditions, this effect can yield successful liftoff for metal films approaching twice the resist thickness. Although this phenomenon is beneficial to liftoff, it makes it difficult to achieve step coverage of a permanent, vertical feature on the slice if the metal arrives perpendicular to the slice (as is characteristic in evaporation). Sputtering (Chapter 13) results in metal impinging on the slice from many directions, not just the vertical. That is why metal application by sputtering is very good for step coverage, but very bad for liftoff techniques.

Some success can occur even if the resist edge profile is not undercut as shown in Figure 6.9(c). Too much of this effect will yield a strong, continuous film over the resist edge and result in failure to achieve liftoff. However, if the connecting film is very thin, the liftoff procedure can tear the connection and leave the desired metal pattern on the slice. To achieve this the metal in contact with the slice must have sufficient area to allow its adhesion to resist the tearing action. Metal etchants can be used to reduce or eliminate the connecting film before liftoff. The exact conditions for success are complex and not easily quantified. As with most aspects of lithography, experiments are needed to determine the success of any given process. In this regard, it is exceedingly difficult (if not impossible) to engage in lithographic process development without the use of a *scanning electron microscope* (SEM). This instrument is essential for inspection of modern GaAs devices and process features.

Optical microscopes are simply not sufficient. Fortunately, the cheaper SEMs are not much more expensive than top-of-the-line optical microscopes.

6.4.1 Edge Profile Modification

It was noted in section 6.3 that a post-exposure bake tends to form a crust on the resist and provide a lip suitable for liftoff (Figure 6.8). All such processes work by retarding the action of the developer on the upper layer of the resist. Lower layers are developed more vigorously and result in an undercut profile. Exposure to a plasma, or even ion implantation, can harden the upper surface by promoting crosslinking and can achieve the same effect.

A very popular process to achieve the undercut profile is the *chlorobenzene method* [14, 15]. Soaking the resist in the organic solvent chlorobenzene removes residual solvents and low molecular weight resin from the upper layers of the resist film. This retards developing action in these layers and therefore yields an undercut profile. Other solvents, such as benzene or trichloroethylene, do not yield the same effect [15]. The slice can be soaked in the solvent before or after exposure, but pre-exposure seems to work more rapidly. The greater the soak time, the greater the thickness of the overhanging layer. Typical soak times are five to fifteen minutes. The size of the overhanging layer is also a function of the softbake; higher temperatures yield smaller overhanging layer thickness for constant soak times. Clearly, the chlorobenzene process will affect developer action and linewidth. And so, as always, some experimentation and process development is necessary to arrive at a suitable process for a specific application.

A good, reliable process will not just happen by itself. The author recalls an early attempt at the chlorobenzene process in which SEM inspection revealed beautiful, well-defined overhangs in some areas, but the complete absence of overhangs in an area less than 1 mm away — this was ascribed to mask contact problems. Also, different applications (required resist thickness, linewidth, etc.) may require different parameters in the chlorobenzene process. A colleague once told the author that his company used three different chlorobenzene processes, each optimized for different applications. He also stated that their required process experimentation and development took about six months. This is not unreasonable.

6.4.2 Multilevel Resist Techniques

Multilevel resist techniques use two or more layers of resist. These are relatively complex resist processes requiring several spin-on steps, bake steps, or other processing. This complexity would tend to make such procedures unattractive were it not for their significant benefits. These include planarizing the surface, easy production of an overhang resist profile, exposure in thin resist, elimination of standing wave effects, and better linewidth control (especially over topography). Only the major two categories of multi-resist techniques will be considered here. These are bi-level and tri-level techniques and each is illustrated in Figure 6.10. A general overview of these techniques may be found in reference [16], with references therein giving details for specific implementations.

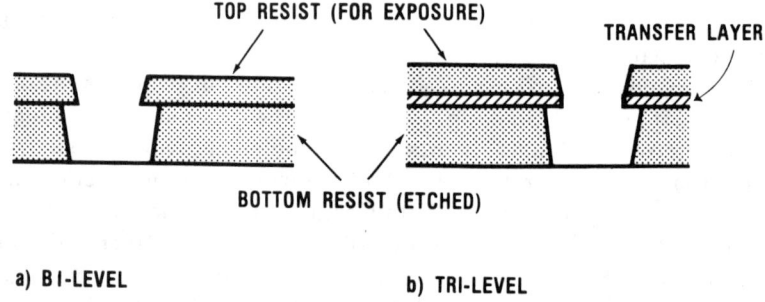

Figure 6.10 Multilevel resist techniques: (a) bi-level (b) tri-level.

The lowermost resist level in multilevel processes is relatively thick and serves to planarize the surface topography. Of course, other materials such as polyimides may be used for the first level. The topmost layer is very thin and is used for the optical exposure. The exposed pattern in the top resist layer is then replicated in the lower level(s) by various means. The ability to expose the pattern in a very thin, highly planar resist layer creates optimum conditions for photolithography, eliminating problems arising from topography or standing wave effects. The ability to separately etch the lower layer allows the undercut edge profiles needed for liftoff techniques. Because the top level of resist acts as a mask for the lower level(s), the term *portable conformable mask*, or PCM has been applied to the multilevel process. The term MLR for *multilevel resist* is also commonly used in the literature.

Only two resist layers are used in bi-level methods. The major problem in the bi-level technique is to prevent the two resists from intermixing. The solvent in the top resist tends to dissolve the lower resist. If the lower resist is baked an amount sufficient to prevent intermixing, it will be virtually impossible to remove. An exception to this is the use of an electron-beam resist such as PMMA (see Chapter 7) as the lower resist. These resists can be baked at a high temperature, around 180°C, and can still be chemically removed. They are not sensitive to ordinary light, but usually are sensitive to deep ultraviolet (DUV) light. Therefore, one bi-level process consists of spinning PMMA onto the slice, baking it, spinning a positive resist over it, exposing and developing the upper resist in the usual manner, exposing the entire slice to DUV (the upper resist acts as a mask for the PMMA), and then using a developer which disolves the exposed PMMA but leaves the upper resist. Alternatively, the PMMA could be blanket exposed before applying the upper resist. The details depend on the choice of developers.

When using ordinary (not e-beam) positive resists, the surface of the first layer may be hardened by exposure to a plasma. Then the second layer may be applied. After the top layer is exposed and developed by ordinary methods, dry etching techniques (Chapter 9) can be used to open up the bottom layer. Alternatively, the bottom layer could have been blanket exposed before the top level was applied. Then developing action, after opening the exposed pattern in the top layer, can continue to dissove material in the lower layer, yielding an undercut profile. Obviously, the amount of overhang obtained in all these cases is very sensitive to the exact process chosen. Although bi-level processes are used, they are rather limited by the conditions that must be applied to prevent layer intermixing. And the practical application of these techniques often proves more difficult than the above descriptions imply. Production-worthy uniformity and reproduciblity may be difficult to achieve. These difficulties are eliminated by the tri-level procedure.

Tri-level techniques offer the greatest advantages. These use the same two layers as the bi-level process, but include a third, intermediate layer between the lower and upper resists (Figure 6.10(b)). This transfer layer is very thin (400 to 2000Å) and is usually a metal or dielectric film. The transfer layer completely separates the two resist layers so that virtually any pair of resists can be employed without intermixing. Hence, the characteristics of each resist can be chosen without any restriction imposed by the properties of the other. After exposing and developing the top layer, the pattern can be replicated in the two lower levels by any of several techniques including wet etching, plasma etching, or reactive ion etching. As is always the case, the exact details are best worked out by each user for his particular equipment and application.

REFERENCES

[1] M. Born & E. Wolf, *Principles of Optics*, 4th ed. London: Pergamon, 1970.

[2] F.A. Jenkins and H.E. White, *Fundamentals of Optics*, 3rd ed. New York: McGraw-Hill, 1957.

[3] J.W. Goodman, *Introduction to Fourier Optics*. New York: McGraw-Hill, 1968.

[4] W.T. Welford, *Aberrations of the Symmetrical Optical System*. New York: Academic Press, 1974.

[5] J.D. Cuthbert, *Solid State Technology*, 20(8), 1977, p. 59.

[6] R.E. Swing and J.R. Clay, *J. Opt. Soc. Am.*, 57, 1967, p. 1180.

[7] F.H. Dill, W.P. Hornberger, P.S. Hauge, and J.M. Shaw, *IEEE Trans. Electron Devices*, 22, 1975, p. 445.

[8] D.W. Widmann and H. Binder, *IEEE Trans. Electron Devices*, 22, 1975, p. 467.

[9] T. Batchelder and J. Piatt, *Solid State Technology*, 26(8), 1983, p. 211.

[10] E.J. Walker, *IEEE Trans. Electron Devices*, 22, 1975, p. 464.

[11] D.A. McGillis and D.L. Fehrs, *IEEE Trans. Electron Devices*, 22, 1975, p. 471.

[12] T.S. Chang, D.F. Kyser, and C.H. Ting, *Solid State Technology*, 25(5), 1982, p. 60.

[13] W. Arden, H. Keller, and L. Mader, *Solid State Technology*, 26(7), 1983, p. 143.

[14] B.J. Canavello, M. Hatzakis, and J.M. Shaw, *IBM Tech. Disclosure Bull.*, 19, 1977, p. 4048.

[15] M. Hatzakis, B.J. Canavello, and J.M. Shaw, *IBM J. Res. Develop.*, 24, 1980, p. 452.

[16] B.J. Lin, *Solid State Technology*, 26(5), 1983, p. 105.

CHAPTER 7

NON-OPTICAL LITHOGRAPHY

7.1 INTRODUCTION

Although optical lithography (Chapter 6) remains a mainstay of semiconductor processing, it has one unsurpassable limitation: the wavelength of light. This limits the resolution of photolithography to approximately one micron. The exact limit is arguable. Advanced techniques using multilevel resist and/or procedures such as angle evaporation can achieve submicron resolution over a small area. In fact, one-half micron FET gates have been fabricated using contact photolithography in conjunction with deep UV exposure. But processes which press the very limits of optical lithography tend to be difficult to use and have a low yield. Reliable, reproducible, high yield photolithographic procedures suitable for production tend to be limited to resolution over one micron. Regardless of the exact value of this limitation, optical lithography is not sufficient to pattern the smallest geometries used in many modern GaAs devices. Features such as the gates of GaAs field effect transistors may require lithographic resolution of 0.5 μm or less. Not only must such small geometries be defined, but often they must also be placed to high accuracy. Registration requirements (the ability to define the pattern at a precise location) may be as stringent as 0.1 micron.

Such severe requirements demand other forms of lithography be used. These are electron beam (e-beam) lithography, x-ray lithography, or ion beam lithography. E-beam lithography is by far the most popular and is the main topic of this chapter. The other two techniques are not in general use. Following the intent of this book to restrict itself to discussing only proven procedures, they would not be included in the discussion. However, their potential is enormous and both (especially x-ray lithography) could experience explosive growth in the near future. This fact, combined with the growing need for smaller and smaller features on GaAs devices, makes it appropriate to include a description of these techniques. In fact, x-ray lithography has already been used in limited production of devices by Bell Laboratories.

7.2 ELECTRON BEAM LITHOGRAPHY

Electron beam (e-beam) lithography is used to form patterns on photomasks used in optical lithography, and for exposing patterns directly on slices. The latter is known as *direct slice writing* (DSW) and is often used in GaAs device processing to define small features such as gate patterns on FETs. The patterns required on GaAs devices are often in the micron or submicron range. It should be emphasized that this application of e-beam technology is more demanding than that historically required in silicon-related processing to generate photomasks using resolutions above one micron and often above two microns. Some of these machines may not be capable of the sub-micron requirements of direct slice writing as used in GaAs processing.

The basic e-beam lithographic process is similar to optical lithography except that energetic electrons instead of light are used to expose the resist. Special e-beam resists which are usually not sensitive to optical light (although they may be sensitive to deep ultra violet light) are needed. They can be handled under normal lighting conditions. E-beam resists exist in both positive and negative formats (Chapter 6). As was the case with photolithography, the positive resists have higher resolution and more desirable characteristics. Unlike photolithography, no mask is used in e-beam lithography. The electron beam is scanned over the slice under computer control to draw the desired pattern.

7.2.1 Advantages and Disadvantages of E-beam Lithography

The advantageous properties of e-beam lithography include improved resolution, great depth of field, excellent level-to-level overlay tolerance (registration), easy pattern modification, and absence of mask defects (no masks are used). Improved resolution results from the ability to focus electron beams to spot sizes much smaller than optical resolutions, even to diameters below 100 Å. Spot size, however, is not the resolution determining factor. As will be discussed below, electron scattering in the resist layer and backscattering from the substrate are the limiting factors. Nevertheless, the resulting resolution limitations are well below those of optical lithography. Because of the nature of electron beam physics, the practical depth of field is enormous (by optical standards), being approximately ± 25 microns [1]. This means that slice flatness is not a limiting factor.

Sub-micron patterns must often be placed with sub-micron accuracy. Even though optical techniques can be pushed to sub-micron resolution (albeit with difficulty and low yield), these techniques cannot place the pattern to comparable accuracy. But accurate placement can be important in GaAs processing. The gate pattern of FETs is an excellent example. Variations in source-to-gate and gate-to-drain spacing result in variations in FET parameters (such as source resistance) which effect performance. Reproducible FETs require reproducible gate placement. Low noise FETs may have source-drain spacings of 3 μm or less and have the gate offset toward the source (Chapter 3) for minimal source resistance. In some cases, desired registration may be as stringent as 0.1 μm. With appropriate alignment markers on the slice (see below) such accuracy can be obtained.

Because the e-beam process does not use a mask, there can be no mask defects to degrade the image. The pattern is stored in a computer — it can be changed or modified rapidly and with little expense. These latter advantages mean that e-beam lithography can be attractive even for defining geometries within the capability of photolithography.

The major disadvantages of e-beam lithography are complexity, expense, and throughput (writing speed); although writing speed may not be a disadvantage for some patterns (see below). A modern e-beam machine is very complex. Although some machines have been created by modifying scanning electron microscopes, such machines are mainly only suitable for research. Several commercial companies now manufacture e-beam machines and at least one major semiconductor company manufactures its own machine for internal use. The requirements of small spot size and accurate placement make the e-beam column very complex. A high vacuum must be maintained in the column. The stage which holds the slice or mask blank must move rapidly and with high accuracy; it may be controlled using a laser interferometer. The related electronics may include features such as automatic, dynamic correction of the electron beam shape as a function of its position within the field. And of course, there is the computer that controls everything. These complexities translate into expense. In 1984, commercial machines can cost over one million dollars. In addition, skilled personnel must be available for repair and maintenance of these machines.

Throughput can be a disadvantage. This is a function of several parameters including stage stepping time, field alignment time, and writing time. Writing time is usually the limitation. Unlike a masking operation, each pattern must be individually drawn by the e-beam. Patterns using large area require a long writing time. Digital logic circuits will encounter this disadvantage. In this case, the writing time may be significantly longer than that of an optical stepper, perhaps prohibitively longer. But many analog GaAs devices and MMICs have only a few, low-area patterns that require e-beam exposure. For gate patterns, the dominant case, the total exposure time is similar to that of an optical stepper.

Charging is a complication that can arise when exposing patterns on GaAs slices having semi-insulating substrates. If electrons from the electron beam are not drained away, enough charge will build up to deflect the beam. This charging can cause problems both in acquiring alignment marks to high accuracy, and in exposing patterns. Either can result in misalignments. Curved or misplaced patterns can be evidence of severe charging. Because the beam current is usually 50 nA or less, only minimal conductivity is required. Under some conditions a buffer layer on the slice may be sufficiently conductive to drain away the charge. High resistivity semi-insulating substrates are usually not sufficiently conductive. A useful approach is to use the conductive GaAs material to drain the charge away. Such material forms part of the active device. It is often possible to define the isolation level (Chapter 10) so that a grid of the conductive material extends across the slice, interconnecting all devices and alignment markers with the metal slice holder at the slice edge. This approach is especially suitable for discrete GaAs devices, but will not be possible in all applications. GaAs MMICs, for example, may have complex inductor patterns that must rest on semi-insulating material; these prevent the presence of such a continuous grid. In these cases, thin coating of a metal such as Al or Au have been suggested in the literature [2] to drain the charge away. Obviously, such metals must be removed at an appropriate point in the process. If multi-level resist processes are used, the middle layer may be chosen to be a conductive material [2].

Because e-beam machines are usually employed to define micron or sub-micron patterns, the beam spot size must be small. This means that larger patterns must be exposed by rastering the beam back and forth across the desired area. This results in slow writing times when exposing large geometries. One approach to this problem is to use e-beam machines which use beam shaping optics to change the size and shape of the spot under program control [3,4]. The variable shaped beam can be enlarged into a square, for example, to expose large pad areas. A second approach to the problem is to restrict e-beam exposure to only the small, critical patterns (such as FET gates) and expose other parts of the pattern in a subsequent step, perhaps as part of a lithography step that is required for another purpose anyway. The latter approach may seem the least desirable, but can be the most practical. Definition of the smallest geometries requires critical adjustment of spot size, focus, and astigmatism; this is more difficult in machines which must accommodate dynamic adjustments of beam size and shape.

Good control of spot size (focus, astigmatism) of the electron beam can be maintained only for small deflections of the beam from the vertical, even with dynamic correction of astigmatism. Hence, only a small area of the slice, a *field*, can be exposed at a time. The exact size of this field is dependent on the linewidth and placement accuracy required and, of course, on the particular machine. For accurately placed sub-micron patterns, the field size is typically between 1.5 and 3 mm. After one field is exposed, the stage holding the slice is moved to place another field under the e-beam column. Stage movement may be laser controlled and very accurate, but it is not sufficiently accurate for stringent registration requirements approaching 0.1 μm. Hence final alignment in each field is accomplished by using the e-beam to scan alignment markers on the slice. A detector monitors either backscattered or secondary electrons. This technique allows accurate alignment within that field. Therefore, e-beam exposure consists of a series of stage movements, alignment mark acquisitions, and exposures. The ability to align to each field means that e-beam patterns can be placed accurately in spite of minor variations that always exist in the location of patterns already on the slice. Such variations can be caused by slice curvature or the unavoidable small errors that occur in generating photomasks. Sophisticated e-beam machines have the capability of exposing different patterns at different locations on the slice. This allows the use of *plug bars* — special fields which contain diagnostic patterns and which are placed in a few locations across the slice (Chapter 17).

The exact configuration of the alignment markers (shape and size) will depend on the requirements of the specific e-beam machine being used. All methods will use some form of edge detection. The resulting placement accuracy can be no greater than the tolerance to which the markers are detected. Therefore well-defined, smooth edges are essential. For highest accuracy, the alignment markers must be part of the pattern to which the e-beam pattern is being aligned. For example, e-beam gate patterns of FETs should be aligned using markers which are part of the source-drain pattern. The corners of each e-beam field are convenient places to locate alignment markers. They are especially appropriate for fields of discrete devices. However, MMICs may be large enough that corner placement of the markers would result in too great a space between them. If this is the case, the markers must be located within the MMIC and therefore must be considered in the initial layout of the *integrated circuit* (IC). Larger MMICs may require more than one set of markers within the IC.

7.2.2 E-Beam Resist and Exposure Characteristics

A focused beam of high energy electrons is used to expose e-beam resists. The electron energy is generally between 10 to 25 keV, the beam current less than 50 nA, and the spot size can be well under one micron. Note that although the beam current is small, the current density can be appreciable, on the order of 1 to 10 A/cm^2. The exposure is controlled by adjusting the spot size, the beam current, and the speed that the beam is moved across the slice. When a high energy electron enters the resist, it undergoes electromagnetic interactions with the electrons of the resist atoms and rapidly loses energy. This energy is transferred to molecules in the resist, resulting in either bond breaking or crosslinking. In positive resists, bond breaking occurs and results in greater solubility of the resist polymer by the developer. In negative resists, crosslinking occurs and results in decreased solubility. The resist will become exposed over an area wider than that of the incoming electron beam. This is caused by three effects: electron scattering within the resist, secondary electrons generated in the resist, and backscattered (or secondary) electrons from the substrate. These effects are shown in Figure 7.1. This lateral exposure gives rise to what is called the proximity effect; if two lines or patterns are exposed very close to each other, there will be some exposure of the resist between them, as illustrated in Figure 7.2. This undesired, lateral exposure can extend as far as 50 μm from the pattern edge, depending on resist thickness and substrate material.

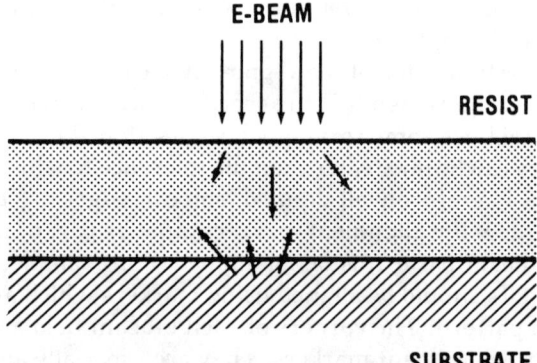

Figure 7.1 Lateral exposure of e-beam resist resulting from forward scattering in the resist, secondary electron emission, and backscattering from the substrate.

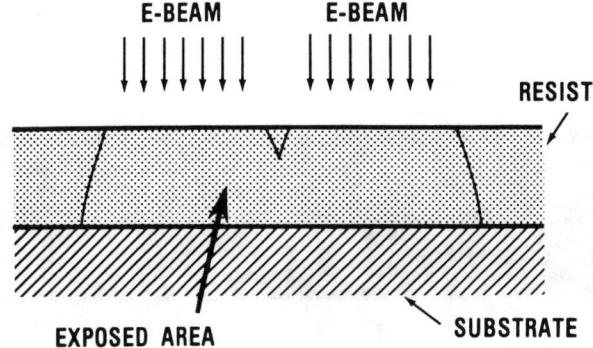

Figure 7.2 Proximity effect resulting from electron scattering.

The backscattering of electrons from the substrate dominates over forward scattering in the resist. Assuming sufficiently intense exposure and reasonably thick resist, this effect will result in greater exposure at the bottom of the resist film than at the top. Figure 7.3 contains a series of SEM photographs showing patterns resulting from increasing exposure. Low exposure (electron dose) results in patterns that do not reach the substrate; high doses exhibit the undercut profile caused by substrate backscattering. The intermediate cases are also shown. Such an exposure series can be used during process development to determine correct exposure (of course, it is not necessary to examine cleaved cross-sections for this purpose). The undercut profile is perfect for lift off purposes. But as can be seen from the figure, the smallest lines have vertical or non-undercut profiles (all other parameters being held constant). These differences in exposure between the top and bottom layer of a resist are a distinguishing characteristic of e-beam lithography [5]. As evidenced in the figure, variations in electron dose have less effect on the line size at the top of the resist. The variations have greater effect at the bottom of the resist. This feature is beneficial for lift off, in which minor exposure variations yield nearly the same resulting linewidth. However, the size of the pattern at the bottom of the resist level can be important. For example, in the fabrication of recessed gates on FETs, the opening at the top of the resist will determine the gate length but the opening at the bottom of the resist will determine the size of the recessed slot produced by wet etching (Figure 7.4). The size of the recess in relation to the size of the gate strip can be important. Therefore, the amount of reverse slop can be important in some applications.

156   GaAs Processing Techniques

Figure 7.3 SEM photographs showing exposed line profiles as a function of electron dosage (exposure).

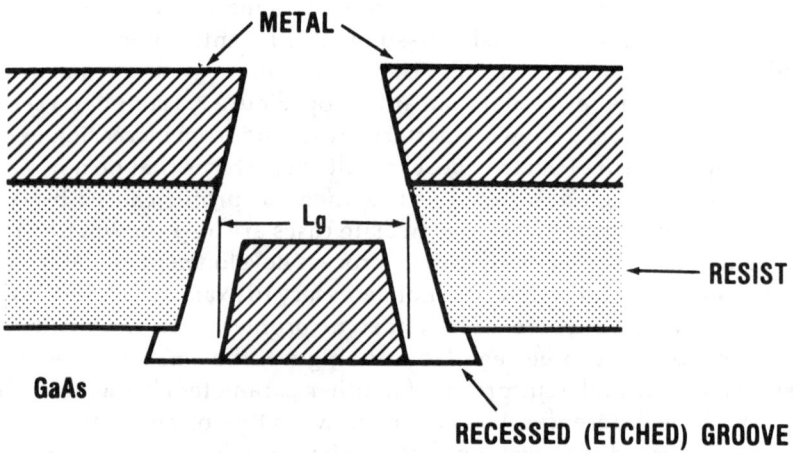

Figure 7.4 Consequences of exposure disparity at top and bottom of e-beam resist, as exemplified by formation of a recessed gate on a field effect transistor. The opening at the top determines the metal line size (using lift off processes); the opening at the bottom of the resist determines the size of the etched recess slot.

The amount of backscattering from the substrate is approximately proportional to the atomic number of the scattering material [2]. There is less scattering in resists than in semiconductor substrates or dielectric films. Differences in substrate material therefore cause apparent differences in exposure. An e-beam line that extends across a GaAs slice and onto a metal film will show this effect clearly, appearing wider (or more open) in one area (usually over the metal) than the other. In summary, linesize will depend on e-beam spot size and lateral exposure from electron scattering. This makes linesize dependent on electron energy, electron current, resist thickness, and substrate type. Half-micron lines can be exposed in resist sufficiently thick (~ 0.8 μm) for many lift off requirements without inordinate difficulty. Significantly smaller lines can be defined using combinations of low beam currents, low energy, thin resists, and/or low-scattering substrates (using multilevel resist techniques). Such techniques have been used to expose lines as small as 10 nm in thin resist [6,7]. The minimum spot size is limited by electron-electron repulsions in the beam — so it is a function of beam current. Low beam currents allow smaller spot size. There is generally a Gaussian profile to the beam intensity when it is focused near its minimum diameter.

The most common e-beam resist is the material PMMA, (polymethylmethacrylate) although many other e-beam resists exist. Properties of some modern e-beam resists are compared in reference [8]. Required electron dosage for exposure is generally between 2×10^{-7} and 8×10^{-7} C/cm^{-2} [8]. PMMA is a positive resist having good resolution. It is interesting to note that PMMA is essentially plexiglass. It can be developed using MIBK (methylisobutylketone). Development time will also affect the resulting profile. Given a constant exposure, longer development times will tend to produce the undercut profile; shorter development will not have time to dissolve all the lower material.

Multilevel resist techniques, described in Chapter 6 for optical lithography, are also used in e-beam lithography. These techniques are most relevant for very small linewidths (sub-halfmicron) or linewidth control over topography. Because the lower resist level acts as the substrate, backscatter and proximity effect is greatly reduced. As noted above, the e-beam process naturally results in an undercut profile, and multilevel techniques are not required to obtain these. Several bi-level processes have been reported. One uses conventional PMMA as the lower level resist. After baking, a copolymer (methylmethacrylate and methacrylic acid) is applied as the upper resist layer [9,10]. After e-beam exposure, the two layers are developed using different developers, such as ethoxyethanol for the top layer and chlorobenzene for the bottom layer [9]. A second approach uses different compositions of PMMA [11]. Low molecular weight PMMA is applied as the first level and baked. High molecular weight PMMA is then used for the top level. After exposure, the composite resist layer is developed using MIBK. As with optical lithography (Chapter 6), tri-level techniques are more flexible and reliable than bi-level techniques and have been utilized in e-beam lithography [12]. In this case, the top resist can be very thin and consequently it is easier to define sub-micron or sub-halfmicron lines. Use of thin resist also eases inspection. Inspection of near-halfmicron lines, to assure they are open, clearly pushes the limits of optical microscopy. Patterns in thin resist are easier to assess.

7.3 X-RAY LITHOGRAPHY

X-ray lithography has been a theoretically attractive technique for years. A number of practical problems (discussed below) have delayed progress, but x-ray lithography has been used for limited production. Institutions that either offer prototype commercial systems, or are actively engaged in internal development of x-ray lithography, include Bell Labs, Perkin-Elmer, Karl Suss, Nippon Kogaku (Nikon), Bakish Materials, Eaton SEO, Hewlett-Packard Labs, IAM, Micronix, Musashino ECL (NTT), and Westinghouse. This technology is discussed in many places [13-19].

The basic process is in theory, very attractive. Soft x-ray (2-10 Å) are used to expose appropriate resists. The low wavelength (a few Angstroms) makes diffraction effects essentially nonexistent for lithography purposes; masks can be placed in proximity to the slice rather than in contact with it. Backscattering or reflection from the substrate is also almost nonexistent. These features mean that x-rays can expose very small lines in very thick resist and produce patterns with vertical walls. Topographical variations on the slice, or resist thickness variations would be irrelevant. In addition, most dust is transparent to x-rays and hence dust on the mask (or possibly even on the slice) will not effect exposure. Although this technique may sound exotic, in principle it should be much simpler (and therefore less expensive) than either optical or e-beam lithography. There are no complex, highly accurate lenses, mirrors, or electron beam lens required; just an x-ray source and mask. In principle, the system could expose an entire slice at one time or, for more accurate alignment, proceed in a field-by-field exposure mode. However, these fields might be significantly larger than those required in e-beam lithography or optical stepping, up to even a significant fraction of an inch.

The virtual absence of diffraction limitations would seem to make x-ray lithography capable of exceedingly small resolution and almost limitless depth of field. But a practical limitation is illustrated in Figure 7.5. With one exotic exception (see below), the x-rays are not collimated and do not originate from a point source. They issue from an extended, finite source. This gives rise to the penumbra effect as illustrated in the figure. The penumbra effect limits resolution. The extent of this effect clearly depends on the size of the source and the spacing between the mask and the slice. Nevertheless, resolution can be significantly greater than what is obtained by optical methods; although it may not yield better results than using e-beam techniques. But the attractiveness of x-ray lithography is not necessarily miniscule resolution, but excellent resolution combined with other technical advantages such as depth of field, vertical-walled patterns, etc., as well as system simplicity.

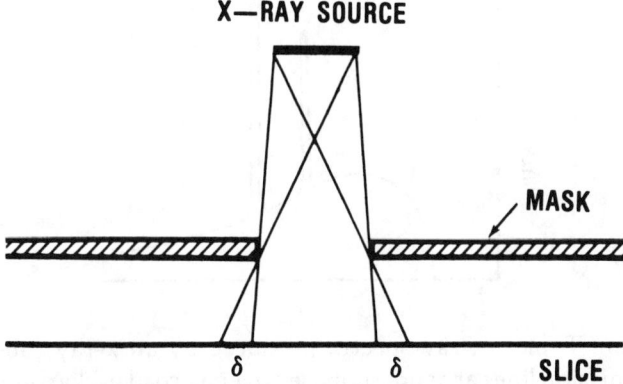

Figure 7.5 The penumbra effect in x-ray lithography arising from non-collimated x-rays issuing from an extended source. This effect limits resolution.

There are, however, aspects of x-ray lithography that have impeded widespread use of this technology. Difficulties have occurred in developing x-ray sources, masks, and suitable alignment procedures. None of these are presently fatal. In fact, as noted above, x-ray lithography is in actual use. But these subjects do represent difficulties which are discussed below.

X-ray sources fall into three general categories: x-ray tubes, plasma sources, and synchrotron radiation [19]. X-ray tubes are the method used to generate x-rays for medical purposes. High energy electrons bombard metal targets and excite electrons in the atoms of the target which decay by emission of an x-ray. The x-ray spectra of such a source is characterized by a low intensity, broad spectrum with major emission lines (Figure 7.6). The broad spectrum is called bremsstrahlung and is caused by the acceleration and deceleration of electrons. The wavelength of the emission lines is characteristic of the target material. The major problem with these types of sources for lithographic purposes is the lack of sufficient intensity. X-ray resists are not very sensitive and need a large flux. The required x-ray dose is 5 mJ/cm^2 or greater [2]. Anodes that are cooled with high pressure water, and/or rotate rapidly, are used to dissipate heat and allow greater electron bombardment and hence greater x-ray intensity. Plasma sources operate by magnetically compressing the plasma to high density with consequent emission of x-rays. These sources work in a pulsed mode. In principle, they are capable of great intensity. But these sources are still largely in the research stage.

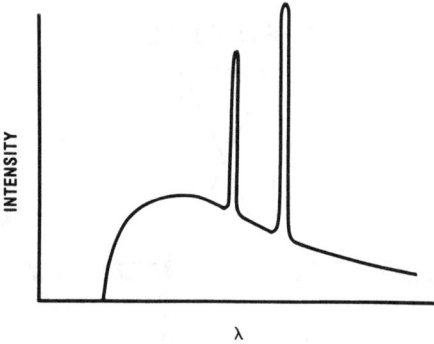

Figure 7.6 Typical x-ray spectra produced by an x-ray tube source. Strong emission lines are superimposed over a broad background spectra called bremsstrahlung.

The optimum x-ray source is a synchrotron [15]. When charged particles, such as electrons, are forced to move in a circular orbit, they emit radiation. This is a consequence of the fundamental laws of physics (Maxwell's laws of electrodynamics). At relativistic speeds this energy is in the x-ray range. More importantly, relativistic effects make the radiation highly collimated, eliminating the penumbra effect. The radiation is also very intense. The difficulty of using synchrotron radiation is obvious; there are few of them around and they cost tens of millions of dollars. Nevertheless, time on several synchrotrons, including the installation at Brookhaven, is being dedicated to x-ray lithography. Several institutions have established user privileges at such machines solely for lithography. There have also been proposals for table-top synchrotrons that use superconducting magnets and which would be reasonably inexpensive (under half of a million dollars).

X-ray masks are another difficulty. The mask substrate must offer low density to x-rays; the masking material must be opaque to x-rays. Because of its high atomic number, gold is common material for x-ray masking. The gold pattern is 0.5 − 1.0 μm thick. Vertical edges, which are not easy to produce, are needed to realize the full resolution offered by x-rays. Other problems arise from the choice of a suitable substrate. These are thin membranes made of low atomic number materials and they must be dimensionally stable. Materials such as boron nitride, silicon carbide, and polyimide have been used. The supporting membrane is usually only microns thick (1-15 μm). The basic mask fabrication process typically proceeds by fabricating the membrane and the gold patterns on a silicon slice and then completely etching the silicon away (any stresses in the membrane can result in pattern distortions). The resulting mask is very fragile and must be handled with care.

The final major difficulty is alignment technique. This is not difficult if overlay tolerance typical of optical techniques are acceptable. But greater registration tolerances, to match or exceed those of e-beam lithography, are difficult to achieve. The fragile mask must be placed in close proximity to the slice and then aligned to markers on the slice. Techniques using Frenel patterns have been investigated, but there is not yet a definitive answer to the alignment problem.

X-ray resists are yet another problem. Optical resists can be relatively insensitive to light if they have other beneficial properties; this is because light can be generated with great intensity and collimated or focused to compensate for low resist sensitivity. With the exception of synchrotron radiation, x-rays cannot be produced at high intensity and collimation. Further, the absorption of a given resist should be made to match the emission line of the x-ray source. Negative resists exist with sensitivities on the order of 10 to 25 mJ/cm^2, but these exhibit the undesirable properties common to all negative resists — swelling and poor resolution. Positive resists are not as sensitive. Potential x-ray resists of both positive and negative format have been reviewed [18] and none are truly satisfactory. Tri-level resist techniques are attractive because the pattern can be exposed in very thin resist, which minimizes the required dose. Such an approach also ameliorates a difficulty illustrated in Figure 7.7. An uncollimated x-ray source will expose lines at an angle corresponding to the distance from the center axis. Tri-level techniques can correct this difficulty by using the thin resist as a pattern, as illustrated in the figure, and then using dry etching techniques to produce a vertical pattern.

The difficulties described above may appear discouraging, but active research is being conducted on all these topics and x-ray systems are presently producing devices. The basic simplicity of the system means that it may be possible to build x-ray printers at a fraction of the cost of optical steppers [18], but with superior performance. This technology could very well replace every other lithographic technique within the next ten years.

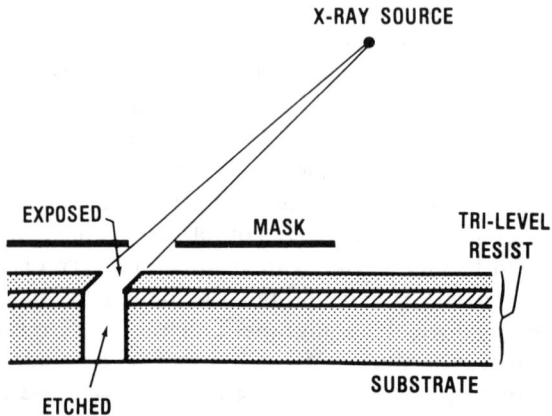

Figure 7.7 Tri-level resist techniques used in x-ray lithography reduce off-axis angle effects caused by a non-collimated beam.

7.4 ION BEAM LITHOGRAPHY

Ion beam lithography is still in the experimental stage. It is similar to e-beam lithography except that ions are used instead of electrons. The heavier ions suffer little scatter (compared to electrons) within the resist or from the substrate. They produce very few, low energy secondary electrons. Therefore resolution should be limited by the beam diameter rather than by the scattering effects characteristic of e-beam lithography. Exposure proceeds by the same mechanism as e-beam lithography; the energetic ions either break bonds (positive resists) or cause crosslinking (negative resists). But an ion can supply much greater energy to the resist than an electron. This results in significantly greater resist sensitivity and should greatly decrease writing time. This basic technology has been reviewed in several places [20-22].

Ion beam lithography is made possible by the existence of suitable ion sources. These are liquid metal sources that typically use gallium, indium, or gold. They can provide ion fluxes of 0.5 A/cm^2; the ion beam can be focused to spot sizes between 0.1 and 3.0 μm. Accelerating voltage is typically 20 kV. The ion sources use a sharpened tungsten needle that is wetted by a thin film of molten metal. A positive potential of approximately 3-6 kV is applied and causes intense ionization. Electrostatic forces cause formation of a liquid cone perched on the tip of the tungsten needle. The radius of curvature of the tip of the molten cone is less than 300 Å and results in large electric fields at that tip, which cause field evaporation. The significant characteristic of such an ion source is that it is essentially a point source of ions (source < 300 Å) [20]. Because of the small source size, the focused spot diameter is limited only by lens aberrations.

Ion beam lithography is the least mature of the lithography techniques discussed in this chapter. But it promises better resolution (no proximity effect) and significantly greater writing speeds than e-beam lithography. A potential difficulty is the effect of ions on semiconductor material if such material is immediately beneath the resist. In such cases, multilevel resist processes would be desirable so that ions come to rest in the lower resist rather than in the semiconductor.

REFERENCES

[1] T.S. Chang, D.F. Kyser, and C.H. Ting, *Solid State Technology*, 25 (5), 1982. p. 60.

[2] B.J. Lin, *Solid State Technology*, 26 (5), 1983, p. 105.

[3] R.D. Moore, *Solid State Technology*, 26 (9), 1983, p. 127.

[4] B.P. Piwczyk and A.E. Williams, *Solid State Technology*, 26 (9), 1983, p. 145.

[5] S.J. Gillespie, *Solid State Technology*, 26 (9), 1983, p. 74.

[6] H.G Craighead, R.E. Howard, L.D. Jackel, and P.M. Mankiewich, *Appl. Phys. Lett.*, 42, 1983, p. 38.

[7] J.L. Jackel, R.E. Howard, E.L. Hu, P.M. Tennant, and P. Grabbe, *Appl. Phys. Lett.*, 39, 1981, p. 268.

[8] G.W. Martel and W.B. Thompson, *Semiconductor International*, Jan.–Feb., 1979, p. 69.

[9] M. Hatzakis, *Solid State Technology*, August, 1981.

[10] M. Hatzakis, *J. Vac. Sci. Technol.*, 16, 1979, p. 1984.

[11] D.J. Elliott, *Solid State Technology*, 25 (12), 1982, p. 91.

[12] J.A. Oro and J.C. Wolfe, *J. Appl. Phys.*, 53, 1982, p. 7379.

[13] E. Spiller and R. Feder, "X-Ray Lithography," in *X-Ray Optics*. H.J. Queisser (ed.), Springer-Berlin, 1977, p. 35.

[14] W.D. Grobman, "Status of X-Ray Lithography," *Proc. of the Int'l Electron. Dev. Conf.*, Washington, D.C., Dec., 1980.

[15] E. Spiller, D.E. Eastman, R. Feder, W.D. Grobman, W. Gudat, and J. Topalian, *J. Appl. Phys.*, 47, 1976, p. 5450.

[16] M. Lepsetter, *IEEE Spectrum*, 18 (5), May, 1981, p. 26.
[17] J.K. Hassan and H.G. Sarkary, *Solid State Technology*, 25 (5), 1982, p. 49.
[18] P.S. Burggraaf, *Semiconductor International*, September, 1983, p. 60.
[19] S. Harrell, *Semiconductor International*, September, 1983, p. 74.
[20] A. Wagner, *Solid State Technology*, 26 (5), 1983, p. 97.
[21] W.L. Brown, T. Venkatesan, and A. Wagner, *Solid State Technology* 24, 1981, p. 60.
[22] R.L. Seliger, and P.A. Sullivan, *Electronics*, March 1980, p. 142.

CHAPTER 8

PLASMA ASSISTED DEPOSITION

Plasma processing is used extensively in the semiconductor industry for growth of thin-film materials and etching. These plasma assisted procedures are sometimes referred to as dry processing to distinguish them from liquid-based processing known as wet processing. Growth or deposition is described in this chapter; etching techniques are described in Chapter 9. Thin-film materials are used in semiconductor processing for encapsulation, scratch and particle protection, environmental protection, interlevel dielectrics, capacitor dielectrics, transfer layers in multilevel resist procedures, and as etch or ion implant masks.

8.1 INTRODUCTION

Plasma assisted techniques are especially important in GaAs processing because they can be accomplished at relatively low temperatures. Deposition of thin-film materials is an excellent example of the low temperature requirement. Such materials can be deposited on slices using *chemical vapor deposition* (CVD) in which the reactant species are introduced into the vicinity of a hot slice by gaseous flow and appropriate reactions take place to grow the material on the slice. The temperatures required to drive the chemical reactions are often 700° to 1000°C. GaAs simply cannot stand up to such temperatures. Not only does arsenic evolve in this temperature range, but metals commonly present on the slice cannot be exposed to these extremes. Alloyed AuGeNi ohmic contacts (Chapter 11) are formed at approximately 450°C. Temperatures close to that, even 300°C, can alter ohmic contact properties. Temperatures above 500°C will rapidly destroy the ohmic contact. Therefore CVD is generally not a useful technique in GaAs processing; low temperature, plasma driven reactions are used instead. The plasma assisted deposition process is referred to as *plasma enhanced chemical vapor deposition* (PECVD).

From the physicist's perspective, the plasma state encompasses a wide range of electron energies and densities and includes such phenomena as flames, low-pressure arcs, solar coronas, and thermonuclear reactions. The area of interest to semiconductor processing is the low-pressure plasma or glow discharge. These are characterized by gas pressures on the order of 0.1 to several Torr, free electron densities of 10^9 to 10^{12} cm^{-3}, and electron energies of 1 to 10 eV.

The plasma is able to generate chemically reactive species at low temperatures because of the nonequilibrium nature of the plasma state. The temperature of the chemical species (atoms, molecules, or radicals) as represented by their translational and rotational energy is generally near ambient. The electrons, however, can exhibit temperatures of tens of thousands of degrees Kelvin. The electron energy is sufficient to break molecular bonds and create chemically active species in the plasma. Any of these species can be excited to higher electronic energy states by further interactions with the electrons. Hence, chemical reactions that usually occur only at high temperatures can be made to occur at low or even ambient temperature in the presence of an activating plasma field. Most of the species remain neutral in glow discharges (the degree of gas ionization is less than 10^{-5}). It is this feature that allows most of the plasma to remain at near ambient temperatures. Although the ionization rate is small, it is enough to provide sufficient numbers of reactive species. The light glow emitted from the plasma is characteristic of the electronic transitions taking place. The light can be used for endpoint detection in etch processes. A concise description of the generation of low temperature plasma species may be found in reference [1].

The plasma conditions result in creation of virtually all possible radicals that can be created from the incoming gases. These may be used to deposit thin film materials such as silicon nitride or silicon dioxide. The slice will generally be heated to aid the deposition process, but only to less than a few hundred degrees Centigrade.

The plethora of chemical species in the plasma, and the nature of plasma state, make processing results extremely sensitive to virtually all possible parameters: gas type, gas flow rate, gas delivery position, pressure, electrode geometry, power, power density, rf frequency, slice temperature, and slice type. The possible chemical reactions can be highly complex because of the presence of so many radicals. These considerations mean that plasma processes are developed more by empirical means than by theoretical analysis. Establishing reliable, reproducible plasma processes is far from straightforward. This problem applies as much to commercially available machines as to "home built" reactors. Substantial effort may be required to determine the proper operating conditions necessary to achieve desired results. The breadth and complexity of reactions among all species defies a comprehensive discussion. The fundamentals of plasma chemistry are treated in several books [1,2,3]. A concise summary of the inelastic collisions and resulting products that occur in flow discharges may be found in reference [4]. Plasma enhanced chemical vapor deposition has been the subject of several reviews [4,5].

The major advantage of PECVD for GaAs processing is the ability to grow films at relatively low substrate temperatures (compared to CVD), usually well under 300°C. This advantage is obtained at a price; compositional control of the the thin-film material is difficult. CVD growth generally results in near stoichiometric composition of the deposited material. PECVD yields films that are amorphous in nature with very little short range structural ordering. Chemical bonding within the film may vary. The plasma assisted deposition process sometimes has been called plasma polymerization to emphasize that the film may be randomly bonded, highly crosslinked, and of variable composition. Species reach the surface in a haphazard manner and may be quickly covered and incorporated into the film. Therefore chemical species other than the desired ones are often included in the film. In this sense, plasma assisted deposition is more complex than plasma assisted etching, in which it doesn't matter what the final products are because they disappear into the pumps. In PECVD films a range of stoichiometry is possible depending on the plasma and operating parameters. This variation in stoichiometry generally results in variations in electrical, mechanical, and chemical properties of the deposited film. The etch rate will be a function of composition. In theory, the ability to adjust physical and electrical properties over wide range could be considered an advantage. In practice, it is not; in almost every situation in GaAs processing, the process engineer would prefer to have the stoichiometric material.

8.2 EQUIPMENT CONSIDERATIONS

Plasma processing occurs in one of two general types of plasma reactors: a barrel (or tube) reactor, or a planar reactor. These are illustrated in Figure 8.1. In barrel systems the plasma is excited using inductive coils or capacitive electrodes outside of the quartz or glass tube. The substrates are generally held in the vertical position by a slice holder and are immersed in the plasma with no electrical bias applied. The slice surfaces are at a floating potential near that of the glow discharge and are subjected to only low energy ion bombardment at probably less than about 30 eV [6]. Uniformity of growth, or etch rate across a slice, is almost impossible to obtain using tube reactors. This follows from non-uniformity in the plasma, the gas flow pattern, and the slice temperature. Barrel reactors generally are used for etching operations (Chapter 9) such as removing resist.

Good uniformity can be achieved using the radial flow planar reactor shown in Figure 8.1(b). The reactant gases are introduced either at the outer radius or on axis, and flow radially between the electrodes. Substrates are placed flat on the lower electrode which is used to heat them. The basic radial flow planar reactor was announced in 1973 [7] and has become enormously popular. Details of industrial implementations of this design tend to be proprietary, but descriptions of such reactors do exist [8,9] and commercial machines are available. As indicated in Figure 8.2, the plasma occupies the region between the two electrodes. It can be excited by any rf frequency, but 13.56 MHz is usually chosen. This frequency has been allocated by international authorities as one which will not interfere with other allocated communication bands. Unfortunately, the highly non-linear plasma effects tend to generate many harmonics of the fundamental signal, so rf shielding is highly desirable to prevent external interference.

The plasma is excluded from the immediate vicinity of the electrode surface by electromagnetic effects. This region is called the plasma sheath, or the dark region, because it does not glow. It is on the order of 0.1 to 10 mm thick. The plasma is generally neutral because the positive species balance the negative species. But the plasma sheath is a region of positive space charge and the electrode surfaces are negative with respect to the plasma. This is due to the higher mobility of electrons which move rapidly to the surface of electrodes. The rf voltage is applied through a large blocking capacitor so that no dc bias is intentionally applied. Most of the voltage between the two electrodes is dropped across the two plasma sheaths. If A_1 and V_1 are the area of the first electrode and the voltage dropped across its plasma sheath, and A_2 and V_2 are the area of the second electrode and the voltage dropped across its sheath, then at low pressure,

$$\frac{V_1}{V_2} = \left(\frac{A_2}{A_1}\right)^4$$

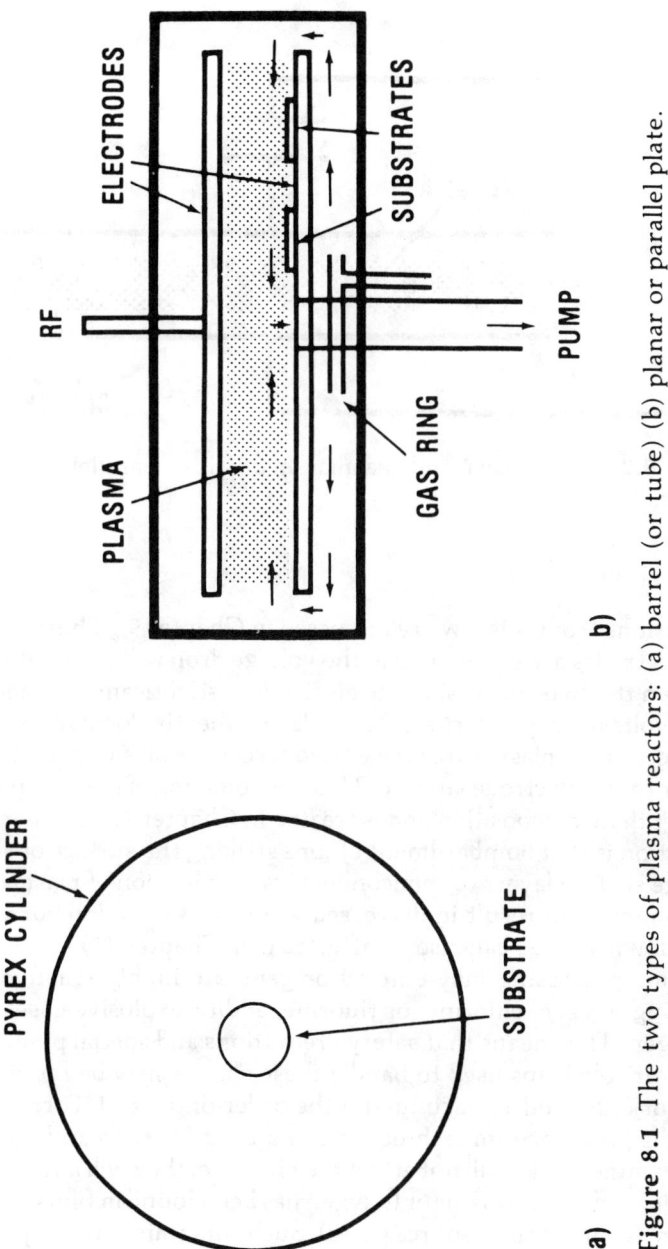

Figure 8.1 The two types of plasma reactors: (a) barrel (or tube) (b) planar or parallel plate.

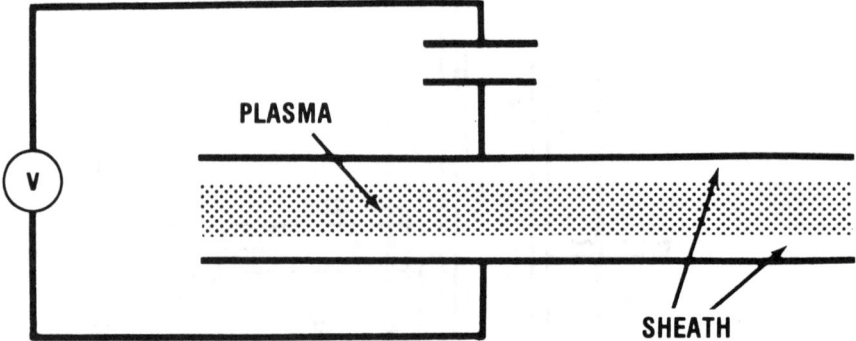

Figure 8.2 The plasma and plasma sheath in a parallel plate plasma reactor.

(Deviations from this law are discussed in Chapter 9). Therefore, if the two electrodes are equal in area, the voltage drop will be equally divided between the two sheaths. If one electrode is significantly smaller, most of the voltage drop occurs across its plasma sheath. Positive species near the edges of the plasma are accelerated across the plasma sheath perpendicular to the electrode surface. This phenomenon is more important in etching than in deposition and is treated in Chapter 9. Its importance to deposition is that bombardment of ions striking the surface of slices can damage surface layers of semiconductors. Application of plasma nitride, for example, can result in decreased values of saturation current [10]. Breakdown voltage can also be affected (**see Chapter 13**).

Plasma processing may employ or generate highly reactive species including oxygen, chlorine, or fluorine, and/or explosive gases such as hydrogen. This means that safety precautions and special pumps, pump fluids, or cold traps used to handle these species may be required. The need for only modest vacuum (on the order of 0.1 to 1 Torr) must not preclude good vacuum techniques being used for seals and pumps. Although minor leaks will not affect the pressure, they will allow contaminants into the plasma reactor (oxygen has been found in films not having oxygen as part of their source gases). Such contaminants can play havoc with processes.

It is necessary to clean reactors regularly. Thin-film materials will tend to deposit on the electrodes and on other parts of the chamber. These materials are most easily removed by operating the reactor in an etching mode and using appropriate gasses (Chapter 9). Higher power is usually used for etching than for deposition. Care should be taken that the gasses used for etching are completely excluded when the reactor is being used for deposition.

8.3 THIN FILM DEPOSITION

A number of materials have been grown using plasma enhanced chemical vapor deposition. The major materials of interest to GaAs processing are silicon nitride and silicon dioxide. These are used particularly for interlevel dielectrics, capacitor dielectrics, and scratch protection. Their use to protect the slice from chemical and mechanical harm has often been referred to as *passivation*. This is unfortunate because the term passivation as used in silicon processing means a treatment that greatly reduces surface states (Chapter 2). SiO_2 performs this function on silicon devices and is crucial to silicon MOS structures. No equivalent material has been found that will accomplish the same thing on GaAs. Hence, many persons, including this author, consider it incorrect to use the term passivation to denote simple environmental sealing. Sometimes the term *glassivation* has been used instead. This seems more appropriate.

Thin-film materials are usually grown to thicknesses of 0.1 to 1.0 micron. Growth rate is on the order of 100 to 500 Å/min. These thin films will usually have pinholes in them. These are small openings, often less than one micron, that expose the underlying material. They can be caused by small particles present on the slice. Such particles can also be formed during deposition if contaminant gasses are present. Pinhole density will also depend on the general operating parameters. Pinhole density can be determined by etching the substrate and film in an etchant that does not attack the film, but etches the underlying material. The number of etched spots can then be counted under an optical microscope.

Much of historical PECVD work and most references are in the context of silicon processing. But virtually all the information applies equally to other substrates such as GaAs. Silicon nitride and silicon dioxide are treated in the two following subsections. The third subsection considers other materials. Such materials may be used as transfer layers in multilevel resist procedures (Chapter 6) or for other purposes.

8.3.1 Silicon Nitride

Silicon nitride (Si_3N_4) is probably the dielectric used most in GaAs processing. It is a better diffusion barrier than silicon dioxide [3] so it is superior for encapsulation or glassivation. Because it also has a higher dielectric constant than silicon dioxide it is preferred for capacitor dielectrics as well (the lower dielectric constant of SiO_2 is best for general interlevel dielectrics where crossover capacitance needs to be minimized). General information about PECVD Si_3N_4 may be found in several references [5,8,11].

Silane (SiH_4) is usually used for the silicon source and nitrogen or ammonia for the nitrogen source. Ammonia is generally preferred because it has a lower ionization potential than N_2. If nitrogen is used, it is difficult to control the refractive index to be in the desired range (~ 2.0) and still flow enough silane to get practical deposition rates [8] (see below). Silane will generally be mixed with an inert carrier gas such as argon. Other reactant gasses may also be mixed with inert gases. Such gas combinations are available commercially at high purity for semiconductor processing. The desired, overall reaction would be

$$3SiH_4 + 4NH_3 \rightarrow Si_3N_4 + 12H_2$$

or

$$3SiH_4 + 2N_2 \rightarrow Si_3N_4 + 6H_2$$

But there are an enormous number of possible intermediate reactions. The plasma conditions make it likely that SiH_4, SiH_3, and SiH_2 are present as well as (for ammonia) NH_3, NH_2, and ionized hydrogen. Oxygen may be present from background gasses or water. Carbon may be present from background hydrocarbons, such as pump oil. The resulting "silicon nitride" may include Si, N, O, H, and C. Even in the complete absence of background components or contamination, hydrogen will be included in the film if ammonia is the source. In fact, plasma Si_3N_4 can contain as much as 20-25% hydrogen [12]. Greater amounts of hydrogen are incorporated at lower growth temperatures. Hydrogen concentration has been correlated with a decrease in the refractive index and an increase in etch rate [12,13]. Some of the hydrogen can be removed by annealing [5]. Although the presence of bonded hydrogen can be determined from the IR spectra, a quantitative determination is more difficult and requires use of such analytical techniques as secondary ion mass spectrometry (SIMS — see Chapter 19) [4]. The Si/N ratio will be a function of the operating parameters and will not necessarily be ¾. The departure of PECVD silicon nitride from true, stoichiometric silicon nitride (grown by high temperature CVD) is indicated by the etch rate. PECVD silicon nitride will etch five to ten times faster in common acid etches than will the CVD nitride [5]. Table 8.1 lists the range of physical and chemical parameters that typically may be obtained in PECVD silicon nitride compared to high temperature CVD nitride [14].

TABLE 8.1
Representative Properties of CVD Silicon Nitride and PECVD Silicon Nitride
(From reference [7], with modifications by this author.)

Property	CVD Si_3N_4	PECVD $Si_xN_yH_z$
Density (g/cm^3)	2.8-3.1	2.5-2.8
Refractive index	2.0-2.1	1-9-2.2
Dielectric constant	6-7	6-9
Dielectric strength (V/cm)	10^7	6×10^6
Bulk resistivity (ohm-cm)	$> 10^{15}$	$10^6 - 10^{16}$
Thermal expansion (/°C)	4×10^{-6}	$4-7 \times 10^{-6}$
H_2O permeability	None	Low-none
Thermal stability	Excellent	Variable $>$ 400°C
Si/N ratio	0.75	0.8-1.0
Buffered HF etch rate room temp (Å/min)	10-15	200-300
Plasma etch rate CF_4/O_2 (Å/min)	200	500

Note: these values are intended to illustrate the difference in the two types of silicon nitride and to list general ranges that are obtained. Specific films can deviate from these.

PECVD silicon nitride may be applied in either a compressive or tensil state. The type can be determined by growing the Si_3N_4 film on very thin substrate material (designed for this purpose) and noting the resulting curvature. These and other physical parameters may be of interest during process development. Detailed analytic procedures such as electron spectroscopy for chemical analysis (ESCA) or Auger characterization (Chapter 19) may be employed to characterize the film, especially as a function of depth. Because of inhomogeneities during growth initiation, the film may not be completely homogeneous as a function of depth. This may be reflected in differing (average) characteristics of very thin films as compared to thicker films. But in routine monitoring of in-process parameters, usually only the film thickness and optical index of refraction are measured. These may be determined simultaneously using an ellipsometer. The optical index of refraction is a general indication of physical characteristics of the film and can correlate with electrical properties such as dielectric breakdown [5], etch rate [8], etc. Usually, an index of refraction near 2.0 is desired. These measurements can be taken from pilot GaAs substrate slices included in the run with production slices.

TABLE 8.2
General Trends in PECVD Silicon Nitride Parameters as a Function of Operating Parameters
(After [4] with additions from [11, 15])

Effect	Freq.	Power*	Press.*	Substrate temperature	Total flow rate	Gas flow pattern	SiH$_4$ conc.	$\frac{NH_3}{SiH_4}$	Addition of N$_2$
Glow discharge stability	+[a]		+[a]						
Disposition rate		±	+[bd]	±	+[bd]		+[d]	−	
Thickness uniformity		−√	+[ac]		±	±[c]			
Si/N atomic ratio		√	±[c]		+	±[c]			
Density		+	±	+			±	±	
Ave. tensile stress	+	−	+		±		±	+	+
Cracking resistance			−	+			±	−	−
BHF etch rate		−	+	−			±		+
Dielectric constant							+		
Resistivity		+[e]					−[e]		

*Interact throughout
(Matching superscripts indicate known interactions)

There are some consistent trends in the characteristics of PECVD silicon nitride as a function of the operating parameters, especially for radial flow planar reactors. The following data will indicate some of these. But it cannot be overemphasized that this data is presented only to illustrate trends. Exact results will depend on the specific machine and the many other operating parameters. It is even possible that some operating conditions will show opposite results. Some of these trends are indicated in Table 8.2, most of which is taken from reference [4]. Data from references [11] and [15] have also been included. A positive sign denotes that the magnitude of the effect increases with the parameter; a negative sign denotes the opposite. A "+/−" sign can denote a minor effect, or one which passes through a maximum or minimum, or conflicting evidence. A check represents a complex effect (usually sensitive to reactor geometry). A blank means no information available, not the absence of an effect. Known interactions are indicated by matching letter superscripts.

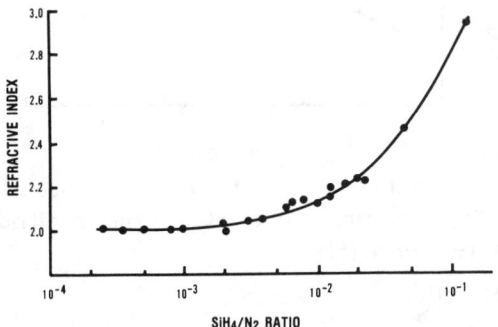

Figure 8.3 Typical data illustrating the dependence of the refractive index of a PECVD silicon nitride film on the SiH_4/N_2 ratio (from reference [16]).

Figure 8.3 illustrates the increase in the index of refraction as the SiH_4/N_2 ratio increases (forming silicon-rich films) [16]. It is clear that this ratio must be kept rather low to obtain the desired (~2.0) index of refraction. The dependence of the index of refraction and the growth rate on pressure and temperature is shown in Figures 8.4 and 8.5 [8]. Growth rate decreases with increasing pressure or temperature. Refractive index increases with increasing pressure or temperature, although the pressure dependence is minimal. The relationship among density, composition (Si/N ratio), and refractive index of plasma nitride films is indicated by the Lorentz-Lorenz relationship shown in Figure 8.6 [17]. Deposition rate increases with silane concentration [8,11]. Growth rate as a function of power is more complex — it increases with power at low power levels. At high power levels the effect is more complicated and depends on gas flow and composition [8]. Nonuniformity of film thickness can be caused by differential depletion of the reactant gasses as they flow across the slices. This can be improved by higher flow rates. For a given temperature, the etch rate of the silicon nitride film in HF is a function of the refractive index as shown in Figures 8.7 and 8.8 [8].

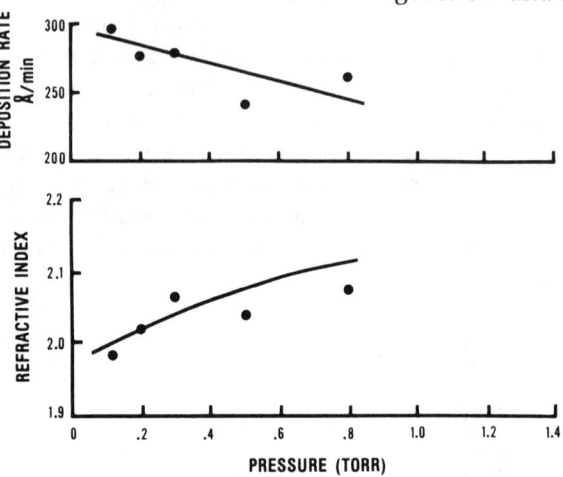

Figure 8.4 Typical data illustrating the dependence of the refractive index and the deposition rate of a PECVD silicon nitride film on total pressure (from reference [8]).

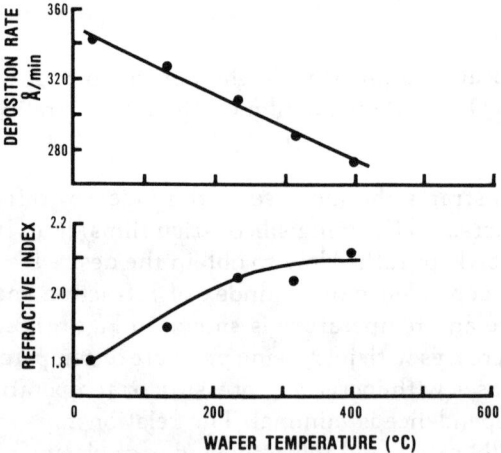

Figure 8.5 Typical data illustrating the dependence of the refractive index and the deposition rate of a PECVD silicon nitride film on the deposition temperature (from reference [8]).

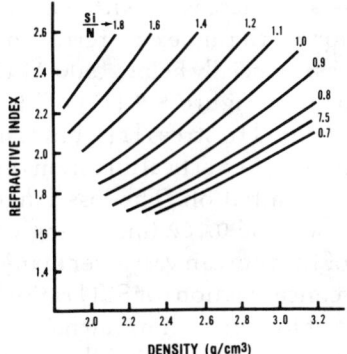

Figure 8.6 Typical Lorentz-Lorenz correlation curves for PECVD silicon nitride (from reference [17]).

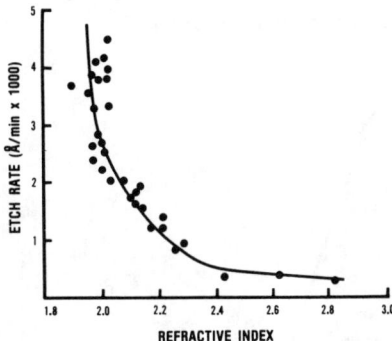

Figure 8.7 PECVD silicon nitride etch rate in 48% HF as a function of refractive index (from reference [8]).

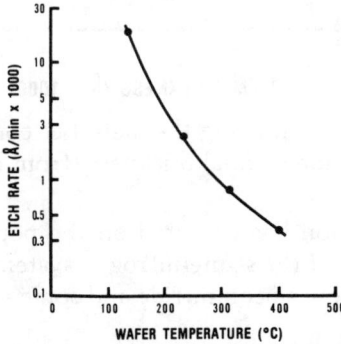

Figure 8.8 PECVD silicon nitride etch rate in 48% HF as a function of deposition temperature (from reference [8]).

The dielectric constant increases with silane concentration [11] (matching the behavior of the index of refraction). It increases with decreased film thickness, especially below 1500 Å (it is relatively constant above 1500 Å) as shown in Figure 8.9 [11]. This dependence on film thickness has been attributed to stress [11], but it may also be related to nonuniform composition in the vertical direction resulting from nonuniformities during growth initiation. The loss tangent at 1 kc has been measured between 0.0007 and 0.002 (increasing with silane concentration) [11]. Resistivity of the film can vary over many orders of magnitude as a function of silane concentration (or Si/N ratio) [11,15]. Low concentration silane (near stoichiometric) films can have a resistivity above 10^{17} ohm-cm, even up to near 10^{20} ohm-cm, while values below 10^{12} ohm-cm or less occur for high silane concentration (silicon rich films) [11,15]. Resistivity values as low as 10^{5} have been observed for highly silicon rich films [15]. Near-stoichiometric films show rapid increases in resistivity (and dielectric strength) with power [15].

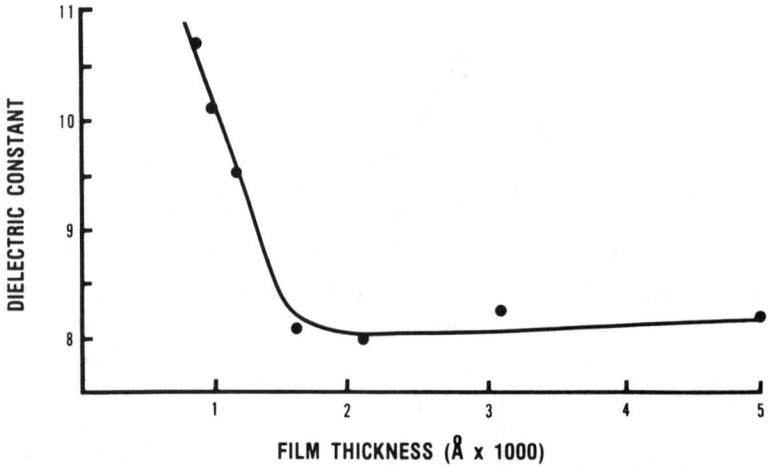

Figure 8.9 Typical data showing the dielectric constant of PECVD silicon nitride as a function of final thickness (from reference [11]).

The above discussion has centered on the popular silane/ammonia system with mention of the silane/nitrogen system. Other gas mixtures have been used to grow silicon nitride and some are listed in Table 8.3.

TABLE 8.3
Inorganic Films Made By Plasma Assisted Deposition

Film	Reactants	Reference
Aluminum nitride	$AlCl_3 + N_2$	[18]
Aluminum oxide	$AlCl_3 + O_2$	[19]
Boron nitride	$BBr_3 + NH_3 (H_2)$	[20]
	$B_2H_6 + NH_3$	[21]
Boron oxide	$B(OC_2H_5)_3 + O_2$	[22]
Carbon (amorphous)	C_4H_{10}	[23]
	C_2H_2	[24]
Gallium nitride	$Ga(CH_3)_3 + NH_3$	[25]
Germanium (amorphous)	GeH_4	[26]
Germanium carbide	$GeH_4 + C_2H_2$	[27]
Geramium oxide	$Ge(OC_2H_5)_4 + O_2$	[22]
Molybdenum; Nickel	Carbonyls	[28]
Silicon (amorphous)	SiH_4	[29]
Silicon carbide	$SiH_4 + C_2H_4$ or CH_4	[29]
Silicon carbonitride	$[Si(CH_3)_2 NH]_3$	[30]
Silicon nitride	$SiH_4 + N_2$	[31]
	$SiH_4 + NH_3$	[20,29]
	$SiBr_4 + N_2$	[32]
	$SiI_4 + N_2$	[33]
Silicon oxide	$SiH_4 + N_2O$	[20,29]
	$SiCl_4 + O_2$	[34]
	$SiH_4 + O_2 (N_2)$	[35]
	$Si(OC_2H_5)_4 + O_2$	[36]
Silicon oxynitride	combine above processes	
Tantalum oxide	not given	[20]
Titanium oxide	$TiCl_4 + O_2$	[20]
	$TiCl_4 + CO_2$	[28]
	Ti isopropylate + O_2	[22]

8.3.2 Silicon Oxide

Silicon dioxide (SiO_2) is used in GaAs processing for many of the same purposes as silicon nitride. Its low dielectric constant (compared to most other stable dielectrics including Si_3N_4) makes it a popular choice for use as an interlevel spacer to separate metal crossovers with minimum parasitic capacitance. Although it is common to refer to plasma deposited silicon oxide as "silicon dioxide", the same caveats on stoichiometry that applied to plasma silicon nitride also apply here: the Si/O ratio may depart from ½ and a certain amount of hydrogen will exist in the film. Nitrogen from either the input gasses or from background sources is very difficult to exclude completely. PECVD of silicon oxide is generally done using silane as the silicon source, and O_2, N_2O, CO, or other similar gas as the oxygen source. The desired reaction, for example using N_2O, would be

$$SiH_4 + 2N_2O \rightarrow SiO_2 + 2N_2 + 2H_2$$

Oxygen has a much greater affinity for reacting with silane than does nitrogen. Hence silicon oxide formation dominates over silicon nitride formation. The same type of radial flow planar reactors described for deposition of silicon nitride are also suitable for silicon oxide deposition, often without major changes in operating parameters other than gas composition and flow rates. Although the above process is the more popular one, other gas combinations such as the following have also been used (see Table 8.3).

$$SiCl_4 + O_2 \rightarrow SiO_2 + 2Cl_2$$

8.3.3 Other Materials

A number of other materials have been grown using PECVD. Some of these are of special interest to other fields such as optics. Also, some of these have been grown in small, research machines and reproducibility in production reactors may be uncertain. General descriptions are contained in two references [4,5] and Table 8.3 lists a number of these materials along with their reactant gasses and references. Some of these materials may be useful for transfer layers of multilevel resist processes (Chapter 6) or other patterning purposes. In any event, the compounds and methods shown in the table illustrate the breadth of possible films that can be grown using PECVD. Others are certainly possible and the process engineer should not be discouraged from attempting materials not listed.

REFERENCES

[1] R.F. Gould, editor, *Chemical Reactions in Electric Discharges*, Advances in Chemistry Series, No. 80. Washington, D.C.: Am. Chem. Soc. Publ., 1969 (especially the chapter by F. Kaufman, p. 29).

[2] F.K., McTaggart. *Plasma Chemistry in Electrical Discharges*. Amsterdam: Elsevier, 1967.

[3] J.R. Hollahan and A.T. Bell, editors, *Techniques and Applications of Plasma Chemistry*. New York: Wiley Interscience, 1974.

[4] M.J. Rand, *J. Vac. Sci. Technol.*, 16, 1979, p. 420.

[5] J.R. Hollahan and R.S. Rosler, in *Thin Film Processes*, J.L. Vossen and W. Kern, editors, New York: Academic Press, 1978, p. 335.

[6] J.W., Coburn, editors *Plasma Chemistry and Plasma Processing*, 2, 1982, p. 1.

[7] A.R. Reinberg, *Electrochem. Soc. Extend. Abstr.*, 74-1, 1974, p. 4; (U.S. Patent 3,757, 733 (1973)).

[8] R.S. Rosler, W.C. Benzing, and J. Baldo, *Solid State Technology*, 19 (6), 1976, p. 45.

[9] J.L. Vossen, *J. Electrochem. Soc.*, 126, 1979, p. 319.

[10] A.K. Gupta, D.P. Siu, K.T. Ip, and W.C. Petersen, *IEEE Trans. Electron Devices*, 30, 1983, p. 1850.

[11] R.C.G. Swann, R.R. Mehta, and T.P. Cauge, *J. Electrochem. Soc.*, 114, 1967, p. 713.

[12] W.A. Lanford and M.J. Rand, *J. Appl. Phys.*, 49, 1978, p. 2473.

[13] E.A. Taft, *J. Electrochem. Soc.*, 118, 1971, p. 1341.

[14] W. Kern and R.S. Rosler, *J. Vac. Sci. Technol.*, 14, 1977, p. 1082.

[15] A.K. Sinha and T.E. Smith, *J. Appl. Phys.*, 49, 1978, p. 2756.

[16] R. Gereth and W. Scherber, *J. Electrochem. Soc.*, 119, 1972, p. 1248.

[17] A.K. Sinha, *Electrochem. Soc. Extended Abstr.*, 76-2, 1976, p. 625.

[18] J. Bauer, et al., *Phys. Status Solidi (a)*, 39, 1977, p. 173.

[19] H. Katto and Y. Koga, *J. Electrochem. Soc.*, 118, 1971, p. 1619.

[20] J.H. Alexander, et al., in *Thin Film Dielectrics*, F. Vratny (editor) Princeton, N.J.: Electrochem. Soc., 1970, p. 186.

[21] C.J. Dell'Oca, et al., *Phys. Thin Films*, 6, 1971, p. 1.

[22] D.R. Secrist and J.D. Mackenzine, *Bull. Am. Ceram. Soc.*, 45, 1966, p. 784.

[23] L. Holland and S.M. Ojha, *Thin Solid Films*, 48, 1978, p. L21.

[24] D.S. Whitmell and R. Williamson, *Thin Solid films*, 35, 1976, p. 255.

[25] K.C. Wiemer, *Chem. Abstr.*, 80, 101347a.

[26] R.C. Chittick, *J. Noncryst. Solids*, 3, 1970, p. 255.

[27] D.A. Anderson and W.E. Spear, *Phil. Mag.*, 35, 1977, p. 1.

[28] H.F. Sterling, et al., *Vide*, 21, 1966, p. 80.
[29] H.F. Sterling and R.C.G. Swann, *Solid State Electron.*, 8, 1965, p. 653.
[30] A.M. Wrobel and M. Kryszewski, *Bull. Acad. Polon. Sci., Ser. Chim.*, 22, 1974, p. 471.
[31] Y. Kuwano, *Jpn. J. Appl. Phys.*, 7, 1968, p. 88; 8, 1969, p. 876.
[32] A. Androshuk, et al., U.S. Patent 3,424,661, 1969.
[33] M. Shiloh, et al., *J. Electrochem. Soc.*, 124, 1977, p. 295.
[34] D. Kuppers, et al., *J. Electrochem. Soc.*, 123, 1976, p. 1079.
[35] R. Kalnina, et al., *Chem. Abstr.*, 75, 124021a.
[36] S.W. Ing and W. Davern, *J. Electrochem. Soc.*, 111, 1964, p. 120; 112, 1965, p. 284.

CHAPTER 9

DRY ETCHING — PLASMA, RIE, RIBE, ION MILLING

9.1 INTRODUCTION

Dry etching techniques are those that use plasma driven chemical reactions and/or energetic ion beams to remove material. (Wet etching procedures use liquid etchants — Chapter 5.) Descriptions of plasma conditions and equipment considerations were given in sections 8.1 and 8.2 of Chapter 8 (Plasma Assisted Deposition). Most of that material is relevant to dry etching and should be considered a part of this introduction.

Dry etching has several advantages over wet etching. It can provide greater control at reduced cost. It offers substantial directionality and etch anisotropy because etching can proceed more rapidly in the vertical direction than in the horizonal. Some conditions result in lateral etch rates very close to zero, so undercutting of masking patterns can be greatly reduced. This is essential when etching geometries have lateral dimensions on the order of the material thickness. Therefore dry etching is capable of patterning smaller geometry features than wet etching. This dimensional control offers a substantial advantage over wet etching for fabrication of small features on the order of one micron. Etch conditions can also be adjusted to yield smoothly sloped edge profiles when needed for metal crossovers. Gas usage in dry etching is generally less than fluid use in wet etching; purity can be higher and contamination lower. The disadvantage of dry etching is the complex dependence of process results on process parameters (section 8.1).

Dry etching is particularly important to GaAs processing for one other reason. Unlike silicon, GaAs has no liquid etchants that can etch with little or no undercutting of the masking pattern. In fact, the amount of undercutting during liquid etching is generally about the same as the etch depth. Narrow, deep holes in GaAs (source via ground connections, for example) require dry etching techniques.

Dry etching encompasses many different techniques including *plasma etching*, *reactive ion etching* (RIE), *reactive ion beam etching* (RIBE), *sputter etching*, and *ion milling*. Unfortunately, there is no standardized nomenclature for these various techniques. Worse still, the distinction among several of them is not always clearly defined. Because of this, different authors have used terminology in varying ways. This is especially true in older references. In fact, there has been a conglomeration of terms encompassing almost every reasonable combination of the relevant adjectives and nouns. This chapter will adopt the definitions summarized in Table 9.1 and described below. These definitions are consistent with present trends. Nevertheless, the reader is again cautioned that many existing references do not conform to this nomenclature. As defined here, the various techniques are differentiated on the basis of etching mechanism, pressure range, equipment configuration, and anisotropy or directionality. The term *plasma assisted etching* is often used as a synonym for dry etching. In that sense it is a generic term and is different from the term *plasma etching* which describes a specific technique (see below). Each of the dry etching techniques will be described briefly in this introduction; all but sputter etching are treated in more detail in subsequent sections. General reviews of most dry etching techniques are found in references [1-3].

Plasma etching generally refers to any process in which the plasma generates reactive species that then serve to chemically etch material in immediate proximity to the plasma. The etching may be completely chemical and therefore tend to be isotropic, or the chemical reactions on the substrate may be driven or enhanced by the kinetic energy of the incoming ions. This situation will be referred to as *kinetically assisted chemical reaction*. This type of plasma etching can be highly directional and is generally performed in a radial flow planar reactor having parallel electrodes of equal area. The bottom electrode is grounded and holds the slices. Pressure is generally in the range of 0.1 to 5 Torr.

Reactive ion etching (RIE) is similar to plasma etching in that reactive species generated from a plasma are used to etch the material. Hence the distinction from plasma etching is somewhat nebulous. But RIE is gaining a separate identity of its own based on the following distiguishing characteristics. RIE utilizes only kinetically assisted chemical etching and aspires to high directionality; isotropic etching is not attempted. To enhance directionality, RIE is generally performed in a reactor different from the type used in plasma etching. The electrode which holds the slices is significantly smaller in area than the second electrode which usually is the chamber. This configuration results in most of the applied voltage being dropped at the electrode which holds the slices. This voltage is generally on the order of a few hundred volts. The smaller electrode is powered and the other electrode is ground. The process is performed at low pressure, typically between 0.01 and 0.1 Torr. RIE is used to etch holes in GaAs.

Reactive ion beam etching (RIBE) is similar to RIE except that the slices are separated from the plasma by a grid that accelerates ions (created in the plasma) toward the slice. The ion energy is generally higher than that in RIE, perhaps over 1 kV, and hence the kinetic assisted aspect of the etching process is emphasized. Physical etch mechanisms may also be involved. In this sense, RIBE can be thought of as an intermediate case between RIE and ion milling. Pressure is lower than in RIE.

Sputter etching is a purely mechanical phenomenon in which energetic ions from the plasma strike the substrate and physically blast atoms away. There is no attempt at chemical reactions; inert gasses are generally used. Although sputtering is a common technique used to apply material to slices (by sputtering the material from a target), this is not the technique being discussed here. Sputter etching in this chapter refers to using sputtering to remove material from the slice. The slice is placed near the plasma in a configuration similar to that of planar plasma reactors or RIE reactors. Sputter etching is not used very much in GaAs processing. The mechanical removal of material can be accomplished by ion milling which offers more flexibility, more control, and faster etching rates. For these reasons, sputter etching will not be considered further in this chapter. Further information can be found in references [2, 4, 5].

Ion milling is also a purely mechanical phenomenon in which energetic ions of inert gasses are used to erode the surface by bombardment. An ion mill differs radically from the equipment described previously (except, perhaps, the RIBE machines); it resembles a particle accelerator. The ions are generated, roughly collimated, and then impinged on a stage some distance away that has no part in the rf circuity. Ion milling is also unique in that the angle of incidence of the ions is selectable by adjusting the angle between the slice stage and the ion beam. The other procedures described above are limited to perpendicular incidence. Ion energy is also easily controlled by adjusting the accelerating voltage.

The above techniques use chemical and/or physical mechanisms to remove material. This is relevant to the issue of etch rate selectability. Physical etching mechanisms tend to be less selective; etch speed is not a strong function of the type of material etched. Chemical etching, however, can be extremely selective, etching one material rapidly and another hardly at all.

Dry etching has been a particularly active field of experimental research and many types of etching apparatus have been constructed. Some of these use techniques that do not fit cleanly into the above definitions. However, when reading papers describing such research, two cautions should be observed. First, there is a world of difference between obtaining a research result worthy of a paper and developing a process that is reliable, reproducible, and production-worthy. Second, fabrication of dry processing machinery and process development rapidly consume vast amounts of time and money. For these reasons, production techniques tend to utilize the more developed and popular methods and machines which are discussed in this chapter. There are enough problems with these popular techniques to discourage using new ones in production operations.

TABLE 9.1
Dry Etching Techniques

Characteristic	Plasma Etch	RIE	RIBE	Sputter Etch	Ion Mill
Mechanism	Chem	K.A. Chem	K.A. Chem/ Physical	Physical	Physical
Directionality	+/−	+	++	+	++
Pressure (Torr)	0.1-5	0.01-0.1	10^{-4}	0.001-0.1	$<10^{-4}$
Equipment Configuration	Barrel/ Planar $A_S = A_O$	Planar $A_O > A_S$	Planar with Grid	Planar with Target	Ion Beam Accelerator

K.A. = Kinectically Assisted
A_S = Area of Electrode Holding Slices
A_O = Area of Other Electrode

All of these techniques have been utilized in silicon processing and most of the existing references are written in that context. Nevertheless, most of the treatment discussed can apply to any substrate material. Specific materials such as silicon dioxide and silicon nitride are important in both fields. Papers on dry etching generally appear in the *Journal of the Electrochemical Society*, the *Journal of Vacuum Science and Technology*, and the journal *Plasma Chemistry and Plasma Processing*. The magazine *Solid State Technology* is also a good source of articles about dry etching.

9.2 PLASMA ETCHING

Plasma etching is commonly employed to remove materials such as resist, dielectrics, or metals. The plasma state can create highly reactive species of oxygen, fluorine, or chlorine that will readily attack many materials. Etching proceeds with three steps: absorption of the necessary species on the material surface, chemical reactions, and desorption of the products. Any of these may be the rate limiting step [1,7]. Etching that proceeds purely chemically will tend to be isotropic. Etching that requires ionic bombarding energy to aid the reaction (kinetically assisted chemical reaction) will tend to be anisotropic because the ion velocity is perpendicular to the surface. As indicated above, there is not a hard and fast distinction between plasma etching and reactive ion etching (RIE). Many of the considerations and process reviews contained in this section also apply to RIE. Topics specifically related to RIE will be addressed in section 9.3.

It is worth repeating a few generalizations made about plasma enhanced deposition (Chapter 8) because they apply equally to plasma etching: the number of chemical species and energy states in the plasma make exact process analysis extremely difficult; process results are dependent on almost every parameter possible; processes tend to be developed more by emperical means than by exact analysis; and sensitivity to equipment geometry makes it difficult to transfer a successful process to another machine without further process modification. Important etch characteristics such as absolute etch rate, relative etch rate ratios, and directionality are affected by most of the process parameters. The general complexity of the situation may be appreciated by considering dry etching of silicon (important in silicon processing), the most studied example of plasma etching. At least 43 different gas combinations have been reported for etching silicon [1] and the exact mechanisms are still being explored.

There are good general review articles which treat plasma etching [1,2]. Further information of general interest can be found in references [3,6-13].

Plasma chemistry can be complex. But generally, the incoming gasses supply a highly reactive species such as O, Cl, or F. For example, Freon (CF_4) is a source of fluorine under the plasma driven reaction

$$CF_4 \rightarrow CF_3 + F$$

Sometimes additional gasses can aid in this process. For example, oxygen reacts readily with CF_3 (in the plasma) to free even more fluorine and reduce recombination of radicals:

$$2CF_3 + O_2 \rightarrow 2CO_3 + 6F$$

or

$$2CF_3 + O_2 \rightarrow 2COF_2 + 2F$$

Hence, addition of oxygen can result in increased etch rates. Other oxidants such as CO_2, NO_2, NO, etc., will have similar effects. Note, however, that addition of oxygen can decrease the etch rate of materials which favor oxidation over halogenation reactions. Aluminum is one such material. Also, in some cases the CH_3 ion itself is quite active in etching, as is the case when etching SiO_2.

The reactive species, fluorine in this example, can then etch materials by the following representative reactions:

$$Si_3N_4 + 12F \rightarrow 3SiF_4 + 2N_2$$

$$SiO_2 + 4F \rightarrow SiF_4 + O_2$$

SiF_4 is a volatile compound and is pumped from the chamber. It should be emphasized that the above reactions simply indicate initial and final products. There may be intermediate reaction steps and the species, especially fluorine, may be ionized or in an excited electronic state. Hydrocarbons, such as resist, can be etched in oxygen plasmas. In this case the plasma generates atomic oxygen:

$$O_2 \rightarrow 2O$$

(again, the oxygen may be ionized or in an excited state) which then reacts with the hydrogen and carbon to form volatile product species such as H_2O, CO, and CO_2, which are pumped from the chamber.

One caution is worth noting. Dry etching commonly employs one or more halogens (especially F, Cl, and Br) or oxygen. These excited species can capture electrons from the plasma to form stable negative ions. Such a reaction reduces the conductivity of the plasma by replacing a highly mobile carrier (electron) with a heavier and less mobile ion. This situation can sometimes change the behavior of the discharge, cause instabilities, and lead to nonuniform etching [2]. These effects are especially prevalent in halocarbon gases containing Br atoms, two or more Cl atoms, or hydrogen and a halogen [2]. This situation can usually be alleviated by adding suitable diluent gasses and/or by adjusting power and pressure. The above is one example of the general fact that plasmas are notoriously susceptible to instabilities and again underscores the sensitivity of plasma processing to operating parameters.

Another general issue is selectivity or relative etch rate. Clearly, in patterned etching using a mask, the masking material should etch much more slowly than the material it is masking. Photoresist, a hydrocarbon, is much less sensitive to halogen etchants than are typical inorganic materials such as silicon dioxide or silicon nitride. Hence, photoresist is a suitable masking substance for etching common dielectrics. GaAs is relatively insensitive to plasma etching in CF_4, so dielectrics can generally be etched to the GaAs without etching the exposed slice. Oxygen plasmas do not etch dielectric films or GaAs (although oxidation of the GaAs will occur) so they are suitable for completely stripping hydrocarbons such as resist without attacking any exposed GaAs or dielectric film beneath.

The two general types of plasma machines or reactors, barrel and planar, were described in section 8.2. As noted there, the barrel or tube reactors are suitable for plasma etching materials when nonuniform etch rates are tolerable and no directionality is required. Slices in a barrel reactor are completely immersed in the plasma and are isolated from the rf electrodes. They assume a floating potential that is difficult to measure in a straightforward manner. But measurements have been made of the voltage difference between the slice and the plasma [14] and it was found to be low, generally less than a few tens of volts. Greater voltage differences were found between the plasma and the reactor walls. This low voltage greatly restricts the possiblity of anisotropic etching. The general configuration of barrel reactors results in poor uniformity over a slice. This is due to nonuniformity in both the rf field and the gas flow. Etch rates tend to increase from the center of a slice to the edges. Oxygen plasmas in barrel reactors are commonly used to remove photoresist or other organic materials in stripping or cleaning operations.

The planar type of reactor is more suitable for selective etching through masking patterns. That type of reactor was described in section 8.2, but will be considered in more detail here with respect to etching processes. Figure 9.1 illustrates the basic configuration usually used for planar plasma assisted etching. The two electrodes have the same area. As described in section 8.2, the plasma does not extend completely up to the electrodes. There is a zone, referred to as the plasma sheath, that separates the plasma from the electrode. The plasma is generally neutral (equal numbers of negative and positive species). The electrodes become charged by electrons which move from the plasma on the surfaces. The electrons have greater mobility than do positive ions and so the electrodes become negative with respect to the plasma. This results in an electric field across the plasma sheath, between the plasma and the electrodes. This field causes ions at the edge of the plasma to be accelerated across the plasma sheath. Because of the general geometry, they move perpendicular to the electrodes except near the outer radius where deviations can lead to corresponding distortions in etch profile. The plasma sheath is 0.1 to 10mm thick — it can easily be at the higher end of this range.

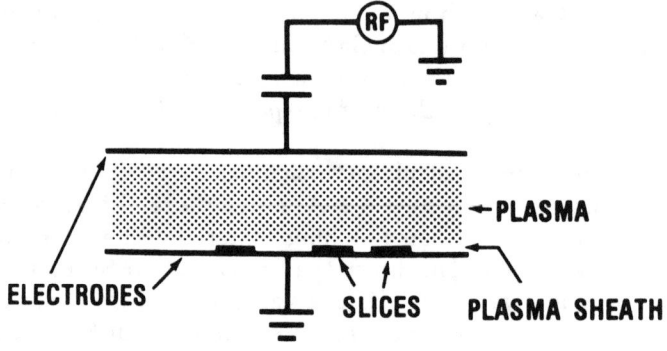

Figure 9.1 Configuration of parallel plate, planar plasma etching machine.

It is the perpendicular arrival of energetic ions that opens the possibility of anisotropic etching. As shown in Figure 9.2, any masking substance will prevent these ions from initially hitting the sidewalls of etching patterns. Therefore the bottom surface of the etched feature is subject to energetic ion bombardment while the sidewalls mainly see neutral species. The undercutting depends on the chemical etch rate of the neutral species. If etching can occur by purely chemical processes, it will tend to be isotropic. However, if additional energy from ionic bombardment is needed to help drive the total reaction, the process will be anisotropic (Figure 9.3). The precise role of energetic ions in stimulating etching is not understood for all reactions, but there is no doubt that they enhance

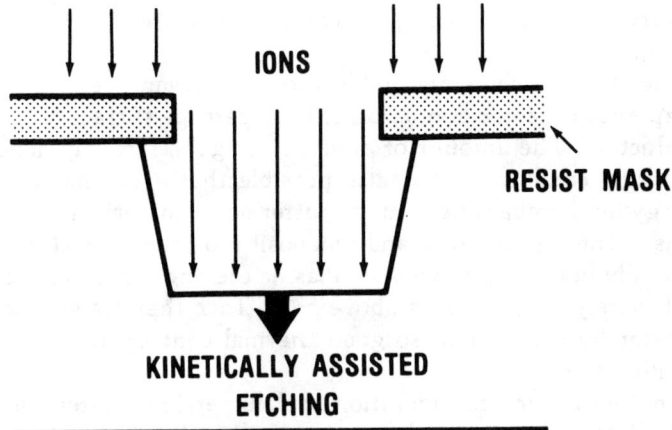

Figure 9.2 Anisotropic etching driven by kinetically assisted chemical reactions.

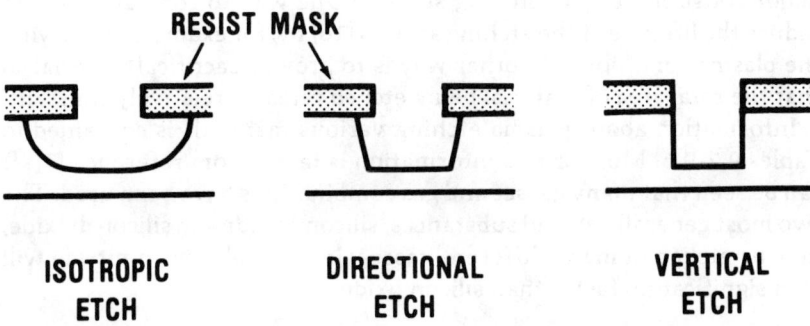

Figure 9.3 Range of directionality that may be obtained using dry etching techniques. These range from completely isotropic etching (completely chemical), through directional etching, to completely anisotropic etching (no undercutting).

etching by neutral species [15]. In practice, the directionality is sensitive to the operating parameters. Anisotropy is aided by lower pressure and/or increased power. Some materials, such as Si_3N_4, exhibit a transition from vertical to isotropic etching as pressure increases at constant voltage.

The plasma etching method has some disadvantages. As described in Chapter 8, the ionic bombardment can damage surface layers of semiconductors. The amount of damage is, again, strongly dependent on operating conditions. It is also possible that even the relatively low energy ion bombardment may sputter some material from other portions of the slice or stage and redeposit it on the slice, although this is generally not a major problem. Plasma etching can also generate heat. Fortunately, at pressures above ~0.1 Torr there is substantial heat transfer by convection, so good thermal contact to the stage is not usually important.

Another general consideration of planar etching systems is the "loading" effect; etch rates can decrease as the amount of material to be etched increases [1]. This makes the process sensitive to the number of slices present in the reactor. Worse still, the effect can lead to greatly increased lateral etching rates (undercutting) near the end of a run when most of the material has been removed. This loading effect is not completely understood. But, generally, it indicates that the etching process is the major consumer of the etching species. One way to combat this is to reduce the lifetime of the etching species [16] by, for example, modifying the plasma conditions. Another way is to provide sacrifical material so that the total area of material being etched remains relatively constant.

Information about plasma etching various materials is contained in Tables 9.2-9.6. Most of this information is taken from reference [1]. It can be seen that many gasses and gas combinations have been used. The two most generally etched substances, silicon nitride and silicon dioxide, may be etched in many different gasses. In general, silicon nitride will etch significantly faster than silicon oxide.

9.3 REACTIVE ION ETCHING

The distinctions between plasma etching and reactive ion etching (RIE) were discussed in section 9.1. As mentioned earlier, these two techniques are very similar and a firm distinction between them is difficult to make. But they each have sufficient distinct characteristics to treat RIE as a separate topic. Yet, clearly, many of the issues considered in plasma etching apply equally to RIE so a firm understanding of plasma etching is needed to consider RIE.

RIE is principally distinguished from plasma etching by its emphasis on

TABLE 9.2
Plasma Etching Si_3N_4
(From reference [1])

Gases	References
CF_4	17-21
$CF_4 + O_2$	2,6,22-32
C_2F_6	33,34
$C_2F_6 + C_2H_4$	35
C_3F_8	36
c-C_4F_8	35
CHF_3	37
CCl_4	38
CF_3Br	39
$SiF_4 + O_2$	27
$SiCl_4$	38

TABLE 9.3
Plasma Etching Of Si_2O_2 And Glass
(From reference [1])

Gases	References
CF_4	17-19,35,39-49
$CF_4 + O_2$	2,6,25-31,50-60
$CF_4 + H_2$	36,54,61-64
$CF_4 + Cl_2$	65
$CF_4 + C_2H_4$	35,66,67
$CF_4 + C_2F_4$	61
C_2F_6	18,27,33,34,68-70
$C_2F_6 + Cl_2$	51,71
$C_2F_6 + C_2H_4$	35
$C_2F_6 + CF_3Cl$	51,56,57
C_3F_8	17,18,20,23,36,69
c-C_4F_8	35,69
C_5F_{12}	69
CHF_3	8,36,37,46,69,72-76
CF_3Cl	51
CCl_4	27,38,77,78
C_2F_5Cl	50
$C_2F_3Cl_3$	79
$CCl_2F_2 + O_2$	79
Cl_2	80

(*Continued on next page*)

TABLE 9.3
(continued)

F_2	81,82
CF_3Br	39,83
SF_6	73,84
$SF_6 + O_2$	85
$SF_6 + N_2$	8
$SiF_4 + O_2$	27
$SiCl_4$	38

directionality, leading to changes in pressure and equipment configuration. RIE uses the directional ion bombardment as a fundamental part of its etching procedure. It is a highly anisotropic, directional etching mechanism in which the vertical etch rate far exceeds the lateral etch rate. The general equipment configuration is illustrated in Figure 9.4. A flat electrode holds the slices in the same manner as in plasma etching, but the area of the other electrode is much larger; often the entire chamber is used as the second electrode. The electrode that holds the slices is usually the rf powered electrode so that the chamber may be rf ground. The large ratio of electrode areas causes most of the voltage drop between electrodes to appear across the plasma sheath at the smaller electrode. As was discussed in section 8.2, the plasma is generally neutral and the voltage drop occurs across the plasma sheath. It is divided between the two sheaths. The expression usually quoted for this division is

$$\frac{V_1}{V_2} = \left(\frac{A_2}{A_1}\right)^4$$

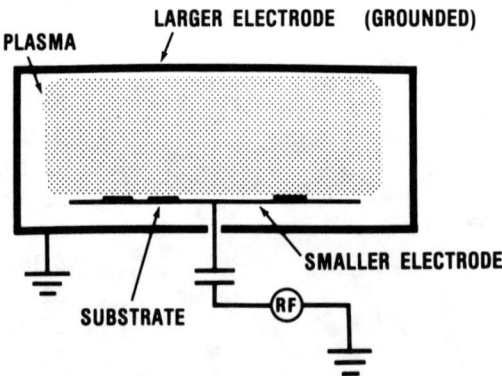

Figure 9.4 General configuration of a reactive ion etching (RIE) machine.

TABLE 9.4

Etching Of Resists, Carbon, And Other Organics
(From reference [1])

Gases	References
O_2	68,82,86-91
CF_4	19,37,39,44,46,92
$CF_4 + O_2$	2,6,28,51,57,91,93-96
$CF_4 + H_2$	62
$CF_4 + C_2H_4$	66,35
$C_2F_6 + Cl_2$	51,71
$C_2F_6 + CF_3Cl$	51,56,57
CHF_3	37,46,72
CF_3Cl	51
CF_2Cl_2	79
$C_2Cl_3F_3$	79
CCl_4	27,38,78,97
CF_3Br	39
$SiCl_4$	38

where A_i and V_i are the area and voltage drop associated with the *i*th electrode. But the derivation of this well quoted expression is clear only at low pressure, <1.0 mTorr. At higher pressures the exact value of the exponent is questionable [121], but the qualitative trend remains. A large asymmetry in electrode areas will place most of the voltage drop across the plasma sheath at the smaller electrode. This enhances ion energy and hence etch directionality at the electrode holding the slices. Conversely, the voltage drop across the other electrode (chamber) is minimal so there is no problem of sputtering material from the chamber.

Directionality is also enhanced in RIE systems by operating at lower pressures (0.01 to 0.1 Torr) than are generally used in plasma etching (0.1 to 5 Torr). In fact, it may be difficult to operate parallel plate reactors much below 0.1 Torr because of arcing between closely spaced plates. The lower pressure reduces scattering within the gas and hence enhances directionality. A disadvantage of RIE is that convective cooling, which is effective above 0.1 Torr, is reduced and the slices may need to be etched on a cooled stage to restrict temperature rise during etching.

The enhanced directionality associated with RIE makes it less sensitive to the loading effect (discussed in the previous section) than is plasma etching [10]. It should be noted that the directionality is derived from the kinetically assisted chemical reaction. Actual sputtering of material is almost nil [10].

TABLE 9.5
Etching Of Other Materials
(From reference [1])

Material	Gases	References
Al*	CCl_4	23,27,38,77,78,80,98-108
	$CCl_4 + O_2$	109
	BCl_3	100,108
	$BCl_3 + Cl_2$	98
	Cl_2	77,98
	$SiCl_4$	38
Au	$CClF_3$	28
	$C_2Cl_2F_4$	6
Cr	Cl_2	89
	$Cl_2 + O_2$	6,110-112
	CCl_4	111
	$CCl_4 +$ Air	111-114
	$CCl_4 + O_2$	112
CrO_x	CCl_4	111
	$CCl_4 +$ Air	112,114,115
	$CCl_4 + O_2$	112
Mo	CF_4	116
	$CF_4 + O_2$	6,23,28,116
	CF_3Br	39
	CCl_2F_2	79
	$C_2Cl_3F_3$	79
Ta	CF_4	43
	$CF_4 + O_2$	6
Ti	CF_4	45
	$CF_4 + O_2$	6
	C_2F_6	33,35
	$C_2F_6 + C_2H_4$	35
	$CF_3Br + O_2$	117
TiO_2	C_2F_6	33,35
W	CF_4	37,121
	$CF_4 + O_2$	6,23,25,28
	CHF_3	37

*Aluminum etching is more important in silicon processing than in GaAs processing. More information about aluminum etching (including doped Al) may be found in reference [1].

Because of the relatively new status of RIE as a topic distinct from plasma etching in general, there are no extensive review articles on RIE alone. But obviously, information about plasma etching is relevant to RIE so Tables 9.2-9.6 can be consulted regarding RIE processes. Several references address RIE of various materials including GaAs [98, 118-120]. RIE is used in GaAs processing to etch via holes from the back of the slice to the front metal. The resulting opening can be metalized and plated and then serve as a ground connection to the backside (Chapter 16). Small holes cannot be etched using wet etchants because of severe undercutting. Ion milling is generally too slow. Hence, RIE (or reactive ion beam etching — next section) is a major tool in the fabrication of monolithic GaAs devices and integrated circuits. The highly directional etching qualities of RIE also make it a major candidate for use in multi-level resist processes (Chapter 6).

9.4 REACTIVE ION BEAM ETCHING

Reactive ion beam etching (RIBE) is a further extension of reactive ion etching. In this technique, a grid separates the plasma from the slice. An accelerating voltage applied to the grid extracts ions from the plasma and directs them toward the slice. The general equipment configuration is illustrated in Figure 9.5. RIBE is distinguished from RIE by still higher voltages and lower pressures. Voltages may be between a few hundred to 2000 V, but are generally between 500-1000 V. The voltage can be

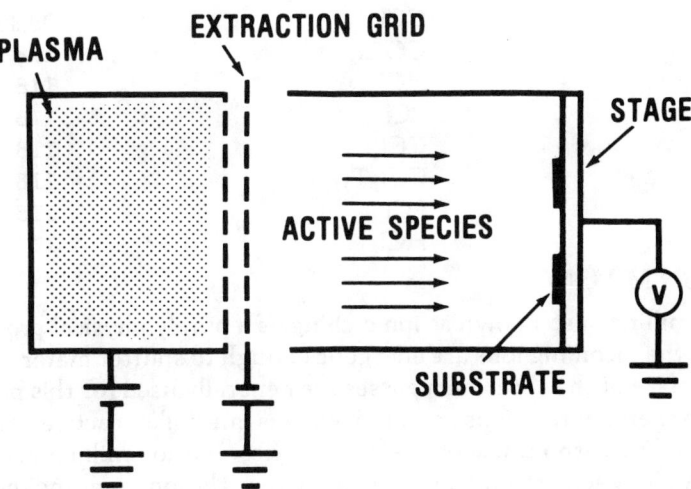

Figure 9.5 General configuration of a reactive ion beam etching (RIBE) machine.

controlled accurately because it is applied to the grid. Pressure is usually in the range of 10^{-4} Torr. Under these conditions, unlike RIE, some physical sputtering can play a role in the etching. In fact, RIBE may be considered an intermediate step between RIE and ion milling. To some extent the chemical and physical etching mechanisms can be independently controlled. This method is more flexible than RIE in the sense that any non-volatile products that result from the chemical reactions (and remain on the slice) can be removed by the physical sputtering aspect of the process. Thermal considerations may require the use of a cooled stage.

As with RIE, much of the information about gas combinations discussed with respect to plasma etching applies here also. Reference [122] is a good introduction to RIBE. Further information is contained in reference [123].

TABLE 9.6

Etching GaAs and Ga_2O_3
(From reference [1])

Material	Gases	References
GaAs	Cl_2	98,118
	$Cl_2 + BCl_3$	98
	CCl_2F_2	98,118,119
	$CCl_2F_2 + O_2$	98,120
	CCl_4	98,118
	$COCl_2$	118
	HCl	118
	PCl_3	118
Ga_2O_3	CCl_4	118
	CCl_2F_2	118
	PCl_3	118
	HCl	118

9.5 ION MILLING

Ion milling, also known as ion etching, is a purely physical process in which the incoming ions are energetic enough to sputter material from the surface of the slice. Inert gasses are generally used for this purpose and no chemical reactions are intended. It is much like reactive ion beam etching, but without use of chemical species. But ion milling machines usually allow adjustment of the angle between the ion beam and the slice. Electron sources may be present to neutralize the beam for etching insulating materials. The stage will generally have provisions for cooling and will rotate to aid in uniformity. Voltage is similar to that used in

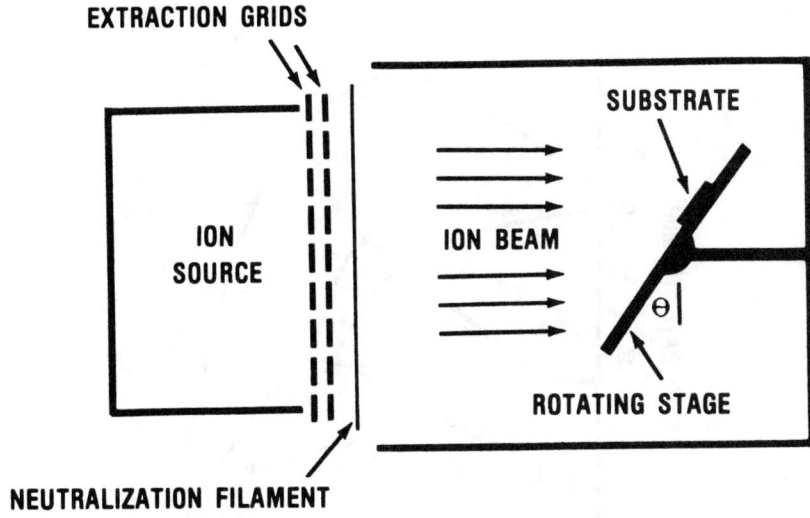

Figure 9.6 General configuration of an ion mill (or ion etching machine).

reactive ion beam etching, generally 500 to 1000 V. Pressure is also the same, in the 10^{-4} Torr range. Reviews of ion milling techniques may be found in several references [124-129]. Reference [124] describes equipment configuration and construction. The basic concept is illustrated in Figure 9.6. The significant advantage to ion milling is the complete absence of undercutting. However, as will be discussed below, there are a number of disadvantages also.

Table 9.7 lists typical ion milling rates for several materials [2,125]. These results are presented to indicate typical rates and to illustrate trends in relative etching rates. They also represent unpatterned material. If two materials are adjacent on the same slice, for example if a masking pattern is present, interactions can radically alter the relative etch rates. Relative etch rates will also be a function of ion beam angle, as shown below. Therefore, as always, the process engineer should optimize his own process using such information as guidelines.

Ion milling etch rate is a function of the angle between the beam and the slice. Because the slice is generally rotated during ion milling, variations in azimuth average out. Typical results are indicated in Figure 9.7 [126]. Examination of this data, and that of Table 9.7, indicates a general

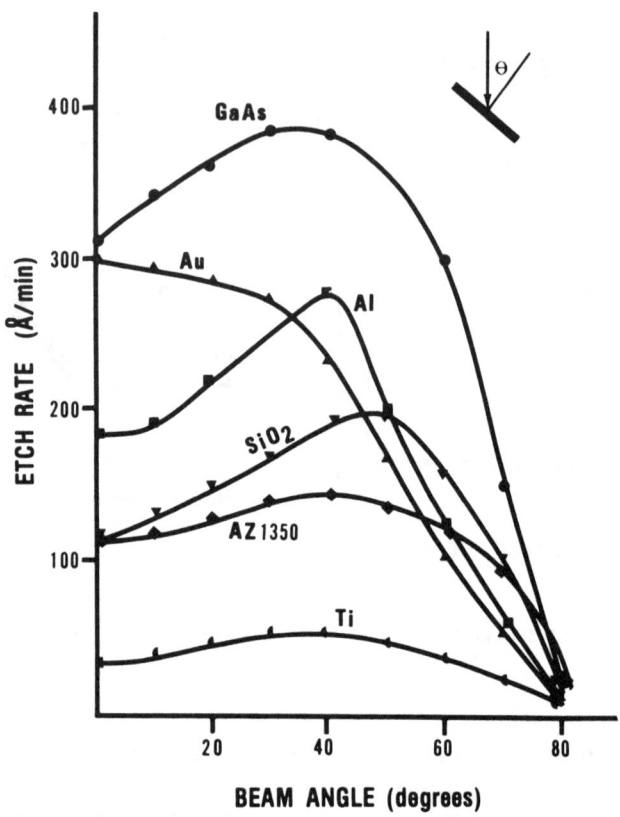

Figure 9.7 Ion milling rate of several materials as a function of beam angle (from reference [126]).

trend: photoresist etches much slower than many common metals used in fabrication, but faster than a few (titanium, for example). Dielectrics etch at roughly the same rate as resist. Unfortunately, GaAs etches fastest of all. This generally limits ion milling to applications in which the GaAs substrate is not exposed. Otherwise, it is extremely difficult (probably impossible) to cease etching before significant GaAs material is removed. Of course, there is no problem if GaAs is intended to be etched.

Ion milling is rather slow compared to plasma etching or reactive ion etching. It also generates a good amount of heat. This is a major consideration if resist is used as a mask. Overheated resist can be extremely difficult to remove. Note that although it is a rapidly etching material, it would take over one and one-half hours to etch through 0.001 inch of GaAs. That is assuming thermal considerations would allow continuous etching for that duration. This is why other techniques such as RIE are used to fabricate via holes in GaAs ICs.

The physical nature of the ion milling process leads to some unique features that can be disadvantages. These features follow from two general conditions. First, material that is sputtered from the surface can be redeposited on any surface in the line-of-sight. Second, scattering effects make near-vertical edges etch faster than other geometries. The results of these two effects are shown in Figures 9.8 and 9.9. Figure 9.8 shows typical results obtained using three types of masking patterns. The thick pattern with vertical edges suffers severe redeposition along the edges. The thick pattern with curved edges suffers very little of this effect, but results in sloped edges of the etched material. The thin pattern most closely replicates the pattern, having only a slight bevel on the upper edge of the etched material. Figure 9.9 illustrates the intermediate states that can occur during etching of material with vertical walled masks.

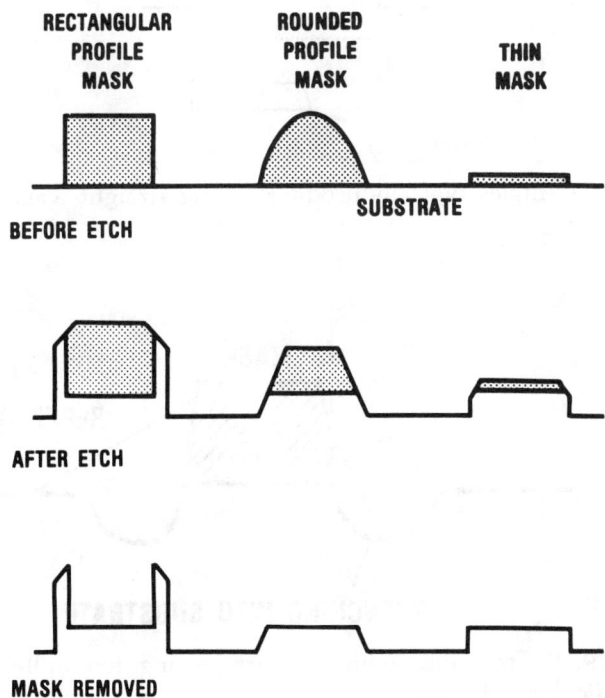

Figure 9.8 Typical results using different masking configurations for ion milling.

A second general ion milling effect that can occur is "trenching" at the bottom of the pattern, as shown in Figure 9.10. Again, both redeposition and scattering from the vertical walls can contribute to this effect. All these effects are functions of the etching angle which is usually measured from the vertical. As a very general statement, etch angles below about 10-15 degrees suffer most from redeposition and such effects.

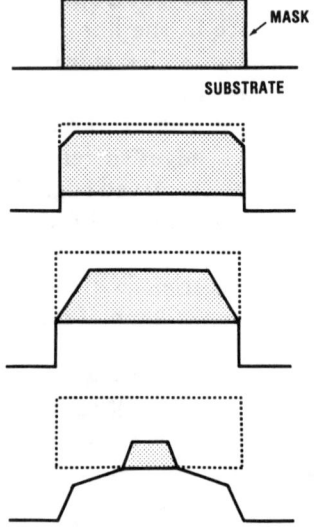

Figure 9.9 Angles typically produced using straight-walled masks and ion milling.

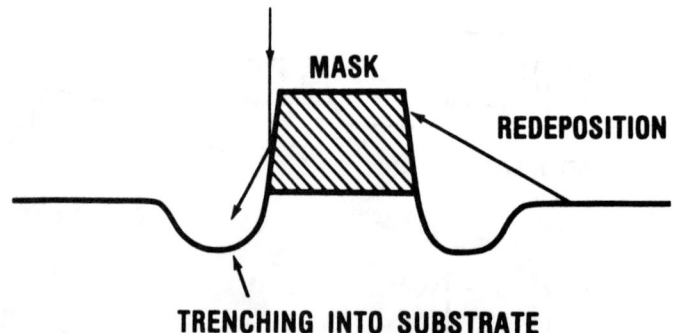

Figure 9.10 Trenching which occurs during ion milling for near-perpendicular incidence.

TABLE 9.7
Representative Ion Milling Rates For Various Materials
(From references [2] or [7])

	Rate (Å/min)	Ion Energy (eV)
Metals:		
Aluminum	300–700	500
	450–750	1000
Gold	1050–1500	500
	1600–2150	1000
Tungsten	180	500
Tantalum	150–330	500
Titanium	200	500
	200	1000
Molybdenum	230	500
	400	1000
Copper	450	500
Chromium	200–400	1000
Zirconium	320	1000
Silver	2000	1000
Manganese	270	1000
Vanadium	220	1000
Permalloy	330–450	500
Iron	320	1000
Niobium	300	1000
Dielectrics:		
SiO_2	280–420	500
	380–670	1000
Al_2O_3	83	500
	130	1000
$LiNbO_3$	640	1000
Semiconductors:		
Silicon	215–500	500
	360–750	1000
GaAs	650	500
	2600	1000
Resist Materials:		
AZ1350 (Shipley)	240–420	500
	600	1000
Riston 14 (duPont)	250	500
KTFR (Kodak)	390	1000
PMMA	840	1000

REFERENCES

[1] J.W. Coburn, *Plasma Chemistry and Plasma Proc.*, 2, 1982, p. 1.

[2] C.M. Melliar-Smith and C.J. Mogab, in *Thin Film Processes*. J.L. Vossen and W. Kern, editors. New York: Academic Press, 1978, p. 497.

[3] P.D. Parry and A.F. Rodde, *Solid State Technol.*, 22(4), 1979, p. 125.

[4] J.L. Vossen and J.J. O'Neill, *RCA Rev.* 29, 1968, p. 149.

[5] G.N. Jackson, *Thin Solid Films*, 5, 1970, p. 209.

[6] R.G. Poulsen, *J. Vac. Sci. Technol.*, 14, 1977, p. 266.

[7] J.W. Coburn and H.F. Winters, *J. Vac. Sci. Technol.*, 16, 1979, p. 391.

[8] H.W. Lehmann and R. Widmer, *J. Vac. Sci. Technol.*, 17, 1980, p. 1177.

[9] C.J. Mogab, *Inst. Phys. Conf. Ser.*, 53, 1980, p. 37.

[10] L.M. Ephrath, *J. Electrochem. Soc.*, 129, 1982, p. 62C.

[11] H.A. Clark, *Solid State Technol.*, 19(6), 1976, p. 51.

[12] R. Kumar, C. Ladas, and G. Hudson, *Solid State Technol.*, 19(10), 1976, p. 54.

[13] G.C. Schwartz, L.B. Zielinski, and T. Schopen, in *Etching*, M.J. Rand and H.G. Hughes, editors. Princeton, N.J.: Electrochem. Soc., 1976, p. 122.

[14] J.L Vossen, *J. Electrochem. Soc.*, 126, 1979, p. 319.

[15] J.W. Coburn and H.F. Winters, *J. Appl. Phys.*, 50, 1979, p. 3189.

[16] C.J. Mogab, *J. Electrochem. Soc.*, 124, 1977, p. 1262.

[17] T. Enomoto, *Solid State Technol.*, 23(4), 1980, p. 117.

[18] B.A. Raby, *J. Vac. Sci. Technol.*, 15, 1978, p. 205.

[19] J.A. Bondur, *J. Electrochem. Soc.*, 126, 1979, p. 226.

[20] T.C. Penn, *IEEE Trans. Electron Devices*, 26, 1979, p. 640.

[21] S.T. Griffin and J.T. Verdeyen, *IEEE Trans. Electron Devices*, 27, 1980, p. 602.

[22] T. Enomoto, M. Penda, A. Yasuoka, and H. Nakata, *Jpn. J. App. Phys.*, 18, 1979, p. 155.

[23] P.J. Marcoux and P.D. Foo, *Solid State Technol.*, 24(4), 1981, p. 115.

[24] K. Hirobe and T. Tsuchimoto, *J. Electrochem. Soc.*, 127, 1980, p. 234.

[25] R. Kumar, C. Ladas, and G. Hudson, *Solid State Technol.*, 19(10), 1976 p. 54.

[26] A. Jacob, *Solid State Technol.*, 21(4), 1978, p. 95.

[27] R.L. Bersin, *Solid State Technol.*, 21(4), 1978, p. 117.

[28] R.L. Bersin, *Solid State Technol.*, 19(5), 1976, p. 31.

[29] S. Chung, *Solid State Technol.*, 21(4), 1978, p. 114.

[30] H.P. Kleinknecht and H. Meier, *J. Electrochem. Soc.*, 125, 1978, p. 798.

[31] R.L. Maddox and H.L. Parker, *Solid State Technol.*, 21(4), 1978, p. 107.

[32] L.A. Coldren, K. Inga, B.I. Miller, and J.A. Rentschler, *Appl. Phys. Lett.*, 37, 1980, p. 681.

[33] S. Matsuo, *Jpn. J. Appl. Phys.*, 17, 1978, p. 235.

[34] M. Oshima, *Jpn. J. Appl. Phys.*, 20, 1981, p. 683.

[35] S. Matsuo, *J. Vac. Sci. Technol.*, 17, 1980. p. 587.

[36] R.A.H. Heinecke, *Solid State Electron.*, 18, 1975, p. 1146.

[37] H.W. Lehmann and R. Widmer, *J. Vac. Sci. Technol.*, 15, 1978, p. 319.

[38] M. Sato and H. Nakamura, *J. Vac. Sci. Technol.*, 20, 1982, p. 186.

[39] S. Matsuo, *Appl. Phys. Lett.*, 36, 1980, p. 768.

[40] H. Toyoda, H. Itakura, and H. Komiya, *Jpn. J. Appl. Phys.*, 20, 1981, p. 667.

[41] B.N. Chapman, T.A. Hansen, and V.J. Minkiewiez, *J. Appl. Phys.*, 51, 1980, p. 3608.

[42] D.J. DiMaria, L.M. Ephrath, and D.R. Young, *Jpn. J. Appl. Phys.*, 50, 1979, p. 4015.

[43] H.H. Busta, R.E. Lajos, and D.A. Kiewit, *Solid State Technol.*, 22(2), 1979, p. 61.

[44] J.A. Bondur and H.A. Clark, *Solid State Technol.*, 23(4), 1980, p. 122.

[45] T. Harada, K. Gamo, and S. Namba, *Jpn. J. Appl. Phys.*, 20, 1981, p. 259.

[46] H.W. Lehmann and R. Widmer, *Appl. Phys. Lett.*, 32, 1978, p. 163; *Appl. Phys. Lett.*, 33, 1978, p. 367.

[47] V.J. Minkiewiez and B.N. Chapman, *Appl. Phys. Lett.*, 34, 1979, p. 192.

[48] G.C. Schwartz, L.B. Rothman, and T.J. Schopen, *J. Electrochem. Soc.*, 126, 1979, p. 464.

[49] J. L. Mauer and S.J. Logan, *J. Vac. Sci. Technol.*, 16, 1979, p. 404.

[50] J. Hayes and T. Pandhumsoporn, *Solid State Technol.*, 23(11), 1980, p. 71.

[51] A.C. Adams, and C.D. Capio, *J. Electrochem. Soc.*, 128, 1981, p. 366.

[52] C.J. Mogab, A.C. Adams, and D.L. Flamm, *J. Appl. Phys.*, 49, 1978, p. 3796.

[53] D.L. Flamm *Solid State Technol.*, 22(4), 1979, p. 109.

[54] L.M. Ephrath and D. J. DiMaria, *Solid State Technol.*, 24(4), 1981, p. 182.

[55] R. d'Agostino, F. Cramarossa, S. DeBenedictus, and G. Ferraro, *J. Appl. Phys.*, 52, 1981, p. 1259.

[56] A.C. Adams, *Solid State Technol.*, 24(4), 1981, p. 178.

[57] A.C. Adams, and C.D. Capio, *J. Electrochem. Soc.*, 128, 1981, p. 423.

[58] A. Jacob, *Solid State Technol.*, 19(9), 1976, p. 70.

[59] A. Jacob, *Solid State Technol.*, 20(6), 1977, p. 31.

[60] K. Jinno, H. Kinoshita, and Y. Matsumoto, *J. Electrochem. Soc.*, 124, 1977, p. 1258.

[61] J.W. Coburn and E. Kay, *IBM. J. Res. Dev.*, 23, 1979, p. 33.

[62] L.M. Ephrath, *J. Electrochem. Soc.*, 126, 1979, p. 1419.

[63] J.W. Coburn *J. Appl. Phys.*, 50, 1979, p. 5210.

[64] H. Itakura, H. Komiya, and H. Toyada, *Jpn. J. Appl. Phys.*, 19, 1980, p. 1429.

[65] M. Shibagaki and Y. Horiike, *Jpn. J. Appl. Phys.*, 19, 1980, p. 1579.

[66] S. Matsuo and Y. Takehara, *Jpn. J. Appl. Phys.*, 16, 1977, p. 175.

[67] M. Oshima, *Jpn. J. Appl. Phys.*, 17, 1978, p. 579.

[68] S. Matsuo, Y. Takehara, and A. Ozawa, *Jpn. J. Appl. Phys.*, 17, 1978, p. 2017.

[69] R.A.H. Heinecke, *Solid State Electron.*, 19, 1976, p. 1039.

[70] M. Oshima, *Surf. Sci.*, 86, 1979, p. 858.

[71] C.J. Mogab, and H.J. Levinstein, *J. Vac. Sci. Technol.*, 17, 1980, p. 721.

[72] H. Toyoda, H. Komiya, and H. Itakura, *J. Electron Mater.*, 9, 1980, p. 569.

[73] H. Toyoda, M. Tobinaga, and H. Komiya, *Jpn. J. Appl. Phys.*, 20, 1981, p. 681.

[74] G.D. Boyd, L.A. Coldren, and F. G. Storz, *Appl. Phys. Lett.*, 36, 1980, p. 583.

[75] D.C. Flanders, H.I. Smith, H.W. Lehmann, R. Widmer, and D.C. Shaver, *App. Phys. Lett.*, 32, 1978, p. 112.

[76] K. Knop, H.W. Lehmann, and R. Widmer, *J. App. Phys.*, 50, 1979, p. 3841.

[77] P.M. Schaible, W.C. Metzger, and J.P. Anderson, *J. Vac. Sci. Technol.*, 15, 1978, p. 334.

[78] R.C. Booth, and C.J. Heslop, *Thin Solid Films*, 65, 1980, p. 111.

[79] N. Hosokawa, R. Matsuzaki, and T. Asamaki, *Jpn. J. Appl. Phys. Suppl.*, 2, P.t 1, 1974, p. 435.

[80] G.C. Schwartz and P.M. Schaible, *Solid State Technol.*, 23(11), 1980, p. 85.

[81] D.L. Flamm, C.J. Mogab, and E.R. Sklaver, *J. App. Phys.*, 50, 1979, p. 6211.

[82] M.J. Vasile, *J. App. Phys.*, 51, 1980, p. 2510.

[83] D.L. Flamm, P.L. Cowan, and J.A. Golovchenko, *J. Vac. Sci. Technol.*, 17, 1980, p. 341.

[84] K.M. Eisele, *J. Electrochem. Soc.*, 128, 1981, p. 123.

[85] R. d'Agostino and D.L. Flamm, *J. App. Phys.*, 52, 1981, p. 162.

[86] L. Holland *J. Vac. Sci. Technol.*, 14, 1977, p. 5.

[87] A. Rizk and L. Holland. *Vacuum*, 27, 1977, p. 601.

[88] W.R. Harshbarger, R.A. Porter, and P. Norton, *J. Electron Mater.*, 7, 1978, p. 429.

[89] H. Abe, *Jpn. J. Appl. Phys. Suppl.*, 14, 1975, p. 287.

[90] B.B. Strafford and G.J. Gorin, *Solid State Technol.*, 20(9), 1977, p. 51.

[91] G.N. Taylor and T.M. Wolf, *Poly. Sci. Eng.*, 20, 1980, p. 1087.

[92] H.W. Lehmann, K. Frick, R. Widmer, J.L. Vossen, and E. James, *Thin Solid Films*, 52, 1978, p. 231.

[93] P.D. Parry and A.F. Rodde, *Solid State Technol.*, 22(4), 1979, p. 125.

[94] K. Jinno, *Jpn. J. Appl. Phys.*, 17, 1978, p. 1283.

[95] H. Akiya, K. Saito, and K. Kobayashi, *Jpn. J. Appl. Phys.*, 20, 1981, p. 647.

[96] K. Harada, *J. Electrochem. Soc.*, 127, 1980, p. 491.

[97] S. Inamura, *J. Electrochem. Soc.*, 126, 1979, p. 1628.

[98] V.M. Donnelly and D.L. Flamm *Solid State Technol.*, 24(4), 1981, p. 161.

[99] B.N. Chapman and M. Nowak, *Semicond. Int.*, 3(10), 1980, p. 139.

[100] K. Tokunaga, F.C. Redeker, D.A. Danner, and D.W. Hess, *J. Electrochem. Soc.*, 128, 1981, p. 851.

[101] W.Y. Lee and J.M. Eldridge, *J. App. Phys.*, 52, 1981 p. 2994.

[102] P.M. Schaible and G.C. Schwartz, *J. Vac. Sci. Technol.*, 16, 1979, p. 377.

[103] B.J. Curtis and H.J. Brunner, *J. Electrochem. Soc.*, 125, 1978, p. 829.

[104] B.J. Curtis, *Solid State Technol.*, 23(4), 1980, 129.

[105] K. Ukai, and K. Hanazawa, *J. Vac. Sci. Technol.*, 16, 1979, p. 385.

[106] K. Ukai and K. Hanazawa, *J. Vac. Sci. Technol.*, 15, 1978, p. 338.

[107] M. Oda and K. Hirata, *Jpn. J. Appl. Phys.*,119, 1980, p. L405.

[108] D.W. Hess, *Solid State Technol.*, 24(4), 1981, p. 189.

[109] K. Tokunaga and D.W. Hess, *J. Electrochem. Soc.*, 127, 1980, p. 928.

[110] M.L. Hitchman and V. Eichenberger, *J. Vac. Sci. Technol.*, 17, 1980, p. 1378.

[111] H. Abe, K. Nishioka, S. Tamura, and A. Nishimoto *Jpn. J. Appl. Phys. Suppl.*, 15, 1976, p. 25.

[112] H. Nakata, K. Nishioka, and H. Abe, *J. Vac. Sci. Technol.*, 17, 1980, p. 1351.

[113] T. Yamazaki, Y. Suzuki, and H. Nakata, *J. Vac. Sci. Technol.*, 17, 1980, p. 1348.

[114] T. Yamazaki, Y. Suzuki, J. Uno, and H. Nakata, *Jpn. J. Appl. Phys.*, 19, 1980, p. 1371.

[115] T. Yamazaki, Y. Suzuki, J. Uno and H. Nakata, *J. Electrochem. Soc.*, 126, 1979, p. 1794.

[116] S. Takahashi, F. Murai, and H. Kodera, *IEEE Trans. Electron. Devices*, 25, 1978, p. 1213.

[117] C.J. Mogab and T. A. Shankoff, *J. Electrochem. Soc.*, 124, 1977, p. 1766.

[118] G. Smolinsky, R.P. Chang, and T.M. Mayer, *J. Vac. Sci. Technol.*, 18, 1981, p. 12.

[119] R.E. Klinger and J.E. Greene, *Appl. Phys. Lett.*, 38, 1981, p. 620.

[120] E.L. Hu and R.E. Howard, *Appl. Phys. Lett.*, 37, 1980, p. 1022.

[121] B. Chapman, *Glow Discharge Processes,* , New York: Wiley, 1980.

[122] D. F. Downey, W.R. Bottoms, P.R. Hanley, *Solid State Technol.*, February, 1981, p. 121.

[123] D.M. Brown, B.A. Heath, T. Coutumas, and G. R. Thompson, *Appl. Phys. Lett.*, 37, 1980, p. 159.

[124] E. G. Spencer and P.H. Schmidt, *J. Vac. Sci. Technol.*, 8, 1971, p. S52.

[125] C.M. Melliar-Smith, *J. Vac. Sci. Technol.*, 13, 1976, p. 1008.

[126] R.E. Lee, *J. Vac. Sci. Technol.*, 16, 1979, p. 164.

[127] P.G. Gloersen, *J. Vac. Sci. Technol.*, 12, 1975, p. 28.

[128] H. Dimigen and H. Luthji, *Philips Tech. Rev.*, 35, 1975, p. 199.

[129] J.M.E. Harper, J.J. Cuomo, P.A. Leary, G.M. Summa, H.R. Kaufman, and F.J. Bresnock, *J. Electrochem. Soc.*, 128, 1981, p. 1077.

PART III.
SPECIFIC PROCESS STEPS

CHAPTER 10

DEVICE ISOLATION

Device isolation is often the first major step in a process flow. Section 10.1 reviews the major objectives of device isolation. The remaining sections describe the three principal means used to achieve isolation: mesa etching, ion bombardment, and selective implantation.

10.1 PURPOSE

Isolation means restricting the electrically conductive portion of the slice to specific parts of its surface area so that electrical current is restricted from flowing to other areas. This electrically conductive portion is called the "active" part of the slice. Almost all devices or monolithic circuits fabricated on GaAs will require some form of isolation early in the process flow. For discrete devices, this usually means that the contact pads are on "inactive," electrically insulating material. Monolithic circuits will have all the passive elements (transmission lines, capacitors, pads) on inactive material.

GaAs has a major advantage over silicon with respect to device isolation; GaAs substrates can be made semi-insulating. As was described in Chapter 2, these substrates are, colloquially speaking, insulating. This means that only the surface of the GaAs slice need be considered in achieving isolation. Of course, these semi-insulating substrates are not perfectly insulating and their residual conductivity must sometimes be considered in device or circuit design.

Isolation serves a number of purposes. In active devices it restricts the current flow to the desired path (under the gate in FETs). It electrically isolates separate devices from each other. It reduces parasitic capacitance and resistances. It provides a sufficiently insulating surface for construction of capacitors and transmission lines.

Some of the purposes of isolation are illustrated in Figure 10.1. As shown in the figure, for the case of FET, the active portion of the slice is located so that source and drain metal contact this conductive region; the source-to-drain current is forced to flow under the gate metal. Any current flow between source and drain that does not pass under the gate represents a parasitic resistance that will degrade the rf performance of the device.

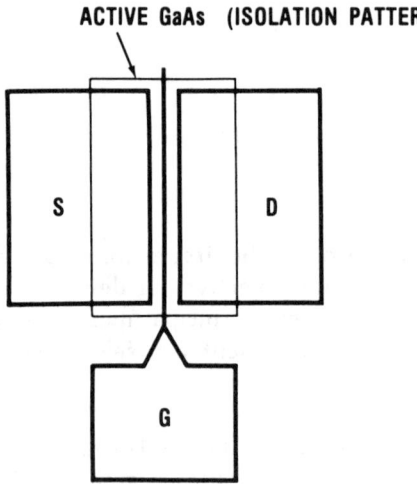

Figure 10.1 Use of device isolation to restrict current flow and isolate devices.

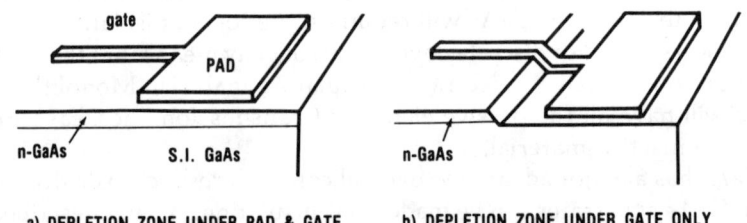

Figure 10.2 Use of isolation to reduce parasitic capacitance: (a) pad on active material has large associated capacitance (b) pad on semi-insulating material has minimal associated capacitance.

Use of isolation to reduce parasitic capacitance is illustrated in Figure 10.2. The gate stripe forms a Schottky barrier and has a depletion region beneath it; a capacitance is associated with this depletion region (Chapters 2 and 12). The amount of capacitance is a function of the metal area and of the doping in the semiconductor. If the gate pad were placed on active material, it would have an associated capacitance much larger than that of the gate stripe. Although the capacitance associated with the gate stripe is essential to the FET operation (Chapter 3), the capacitance associated with the pad is parasitic. As indicated by the equivalent circuit in Figure 10.3, most of the input rf signal would appear across the undesired, parasitic gate pad capacitance. The FET would appear perfectly good when examined on a 60 Hz curve tracer, but would not exhibit gain at microwave frequencies. Placing the pad on semi-insulating material greatly reduces, but does not completely eliminate, the parasitic capacitance.

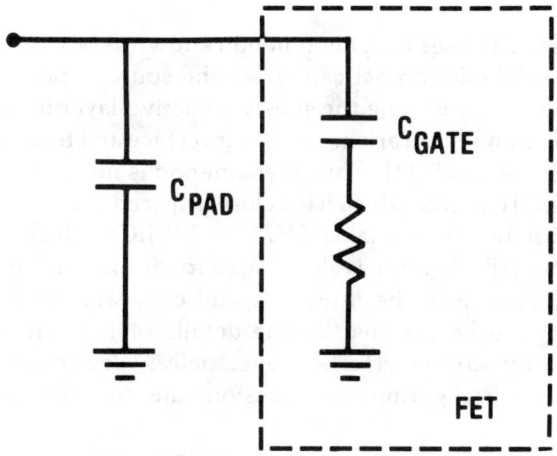

Figure 10.3 Equivalent input circuit of a FET showing the undesirable effect of pad capacitance.

Transmission lines (inductors) on monolithic circuits need to be fabricated on nonconductive material. The losses (the Q) associated with these transmission lines can be important to proper rf performance. The losses associated with transmission lines on GaAs are either dielectric or ohmic. Little can be done to affect dielectric losses (see Chapter 14). Fortunately, they are not excessive in GaAs; the dielectric loss tangent for GaAs is approximately 2.5×10^{-4} at 10 GHz [1]. Ohmic losses are a property of substrate resistivity. An approximate expression for the Q is [1]

$$Q = \omega \rho \epsilon$$

where ω is the angular frequency, ρ is the resistivity of the substrate, and ϵ is the permittivity. Higher resistivity in the substrate results in improved Q of transmission lines. (Of course, the Q of transmission lines is also a function of other parameters, including conductor losses; see Chapter 14.)

Capacitors on GaAs MMICs generally have their bottom plates fabricated on the semi-insulating substrate. Obviously any undesired ohmic conduction between these bottom plates and other portions of the slice is detrimental.

Isolation also addresses the phenomenon known as *backside gating*. Negative bias on an ohmic contact can affect the source-drain currents of nearby FET devices by biasing the substrate/active-layer interface. This expands a depletion region arising at that interface and tends to gate the device from the backside [2]. This phenomenon is more detrimental to monolithic digital circuits, which have closely spaced devices under varying bias conditions, than analog FETs or MMICs which have well-separated devices. Backgating is also related to the material properties of the semiconductor near the interface, and can vary from crystal to crystal. It can also be effected by the details of implant and anneal processes, and by surface effects. Nevertheless, device isolation processes can reduce backgating and this is one measurement of isolation effectiveness.

10.2 ISOLATION BY ETCHING — SLICE ORIENTATION

Isolation by etching consists of etching away portions of the electrically active surface layer of the slice, leaving mesas of the active layer in the desired locations as illustrated in Figure 10.4. The thickness of material removed is usually less than 0.5 μm. The inactive (etched) surface remaining is either a buffer layer (if one was present) or the GaAs substrate itself. Photolithographic techniques (Chapter 6) are used to define an etch mask in the appropriate pattern. Usually the photoresist itself is used for this purpose. Then the slice is etched to remove material in the unmasked areas. This is generally accomplished by wet etching. The undercutting characteristic of wet etching (Chapter 5) will make the mesa a few tenths of a micron smaller than the mask pattern dimension.

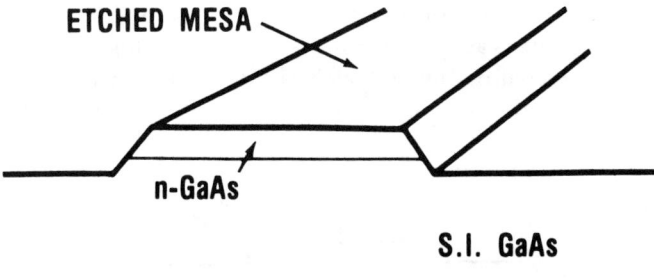

Figure 10.4 An etched mesa created by etching away other portions of the surface of the slice.

Mesa etching is the simplest means of providing isolation and for this reason it is widely used. But there are disadvantages to this technique and a trend exists to use alternative procedures. One disadvantage is the lack of planarity after mesa formation. Even though the non-planarity is only a few tenths of a micron, it is enough to cause difficulties in subsequent resist application, pattern exposure, and metal step coverage. Resist thickness will not be uniform near a mesa step, resulting in patterning variations (Chapter 6). The mesa step also means that the mask (in contact photolithography) cannot be in good contact with material just off the mesa edge; this leads to further patterning variations in these areas. The presence of a step can cause difficulties in metal step coverage. Finally, the isolation produced is not as good as can be achieved using ion bombardment.

None of these disadvantages is seriously restrictive. In fact, mesa etching has been the overwhelming method of choice throughout the development of GaAs devices. Many modern, high performance production devices employ this technique. The point is that alternative methods are rapidly gaining favor by eliminating these disadvantages, but at the cost of increased fabrication complexity.

The shape of the mesa edge is important. Metal patterns must generally "step over" this edge. If the edge profile is too steep or undercut, this will be difficult. A gradual slope of the type illustrated in Figure 10.5(a) is ideal. This type of slope may be obtained using sulfuric acid/hydrogen peroxide type etches. As was described in Chapter 5, most GaAs etches are non-isotropic. This is certainly true for (100) oriented slices (Chapter 2), the almost universal choice for GaAs device processing — this type of slice will be assumed in all following discussions. The etchant that produces the slope shown in Figure 10.5(a) will produce the edge profile shown in Figure 10.5(b) on the side of the mesa, 90° away. Only the former edge profile is suitable for metal step-over. It is therefore necessary to know which direction will yield the desirable, obtuse angle before beginning mesa fabrication. Attaining this information is known as orienting the slice.

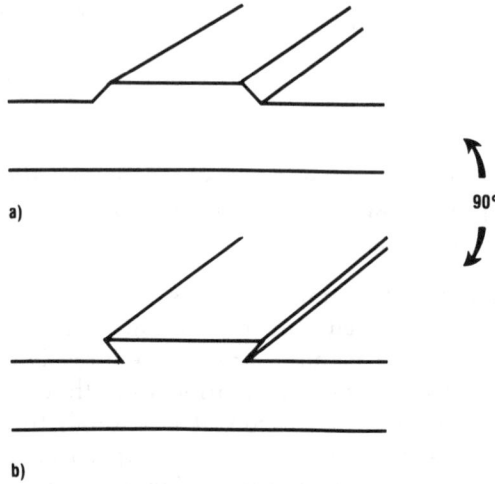

Figure 10.5 Mesa edge profiles produced on (100) GaAs surfaces using most wet etchants. The edges are parallel to the cleavage directions.

Orienting the slice was especially important during the early years of GaAs development in the laboratory. Slices were often grown and supplied without the strict standards prevalent in production so they might have had to be oriented before beginning processing. This problem should not exist in a production environment. The crystal grinding and slice generation procedures should yield round slices with flats, the location of the flats implying the orientation. Nevertheless, it is sometimes useful to check the orientation to confirm the flat location is correct. Also, experiments or other work may occasionally be done using non-standard slices or pieces of slices. For these reasons it is useful to understand the most common methods used to orient slices. All involve etching.

An obvious method for orienting slices is to pattern a slice and etch it; then examine the slopes using a high power microscope. In fact, the pattern can be as simple as a drop of resist placed on the slice and dried. This is a very easy method. But unless the surface of the slice is polished, the etch depth must be substantial to determine the slope. The front, polished side of a slice is clearly not a desirable place to perform this test. The back of the slice may be rough. An easier method that is very reliable is to forgo patterning. A drop of an anisotropic etchant (e.g., $H_2SO_4/H_2O_2/H_2O$) is placed on the slice and left to etch to a very moderate depth. The slice is then rinsed, dried, and examined under a high power microscope. The pattern should appear similar to that shown in the photograph contained in Figure 10.6. Phase contrast microscopy has been used to enhance the pattern shown in the photomicrograph. But the pattern is readily discernable without using phase contrast. There is a clear elongation of the roughness in one direction. This direction is perpendicular to the obtuse slope direction (see Figure 10.6). This method is best performed on the back of the slice so that the front remains unmarred. However, it is crucial to realize that the obtuse slope direction as determined on the back of the slice will be 90° away from the obtuse slope direction on the front of the slice. This is shown in Figure 10.7.

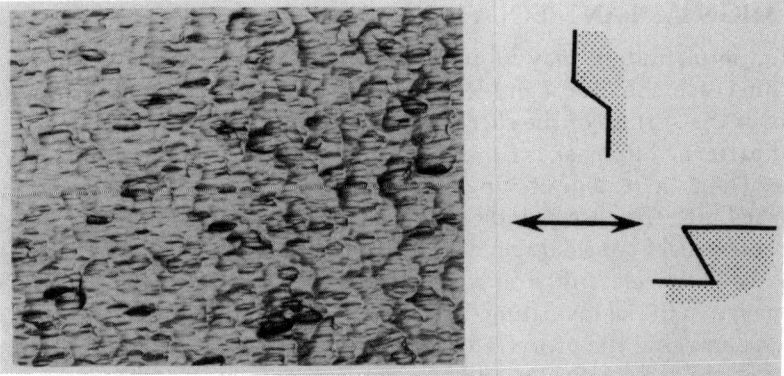

Figure 10.6 Photomicrograph of elongated patterns produced in (100) GaAs using anisotropic etchants. As indicated, the elongated direction is perpendicular to the "smooth step-over" direction of an etched mesa.

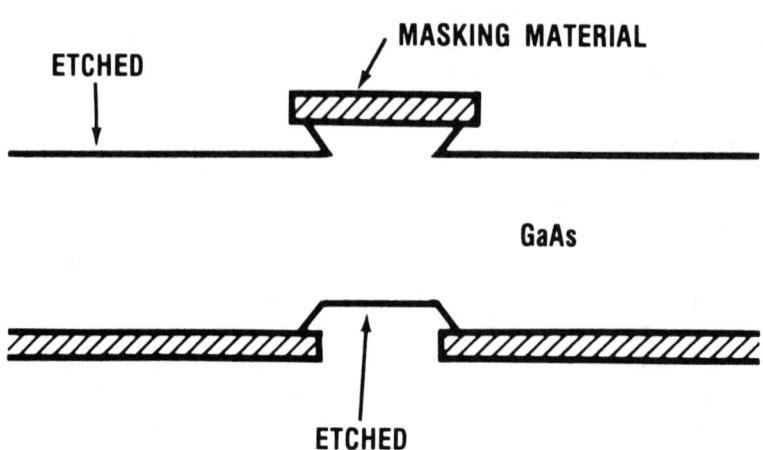

Figure 10.7 The obtuse (smooth step-over) slope determined from the back of the slice is perpendicular to the obtuse slope created on the front of the slice.

10.3 ION IMPLANT ISOLATION

Ion implantation may be used for device isolation. As in the mesa etching case, the slice begins with the active layer extending continuously over the surface of the slice. Resist (and/or other materials) is applied and patterned to mask off the areas that are desired to remain conductive. Then, instead of etching away the exposed material, ions are implanted into the slice as indicated in Figure 10.8. The ion implantation process causes considerable damage to the crystal lattice. This damage creates many electron trapping centers (Chapter 2) and renders the damaged material insulating. The masking substance prevents the ions from damaging the protected areas. Ions commonly used for this purpose are hydrogen, boron, and oxygen. Common masking materials are thick photoresist, silicon nitride, silicon dioxide, or combinations thereof. The crystal damage can only be removed by a high temperature anneal (as is done when forming active layers by ion implantation).

It should be noted that in conversation, process engineers sometimes use the term "ion implanted" in two different contexts. The first refers to slices in which the active layer is formed by ion implantation followed by an activating anneal. The second refers to slices in which device isolation is achieved using ion implantation (without a subsequent high temperature anneal). This can be confusing. Sometimes the term "ion bombardment" is used to distinguish the latter context. The term "implant isolation" may be the best to use since there can be no confusion concerning its meaning.

Another confusing habit of process engineers is to refer to the active (isolated) part of the slice as the mesa even if implant isolation was used and there really is no physical mesa present. This can obviously create confusion and should be discouraged.

Implant isolation offers several advantages over mesa etching. The slice remains planar, so it has none of the problems associated with mesa edges which were described in the previous section. Also, implant isolation provides better isolation, as assessed by leakage current measurements and backgating, than does mesa etching. This better isolation exists because the induced damage of ion implantation results in resistivities greater than those of the (crystalline) semi-insulating substrate itself. This point is particularly relevant to fabrication of high-Q transmission lines on monolithic devices. As described above, the Q is a function of slice resistivity. This is one reason that implant isolation is sometimes used to augment mesa etching. In this joint procedure, etching is used to remove part or all of the active, surface layer. Then the same masking pattern remains on the slice during ion implantion which completes the isolation. The etching part of the procedure leaves a visible pattern for subsequent alignments.

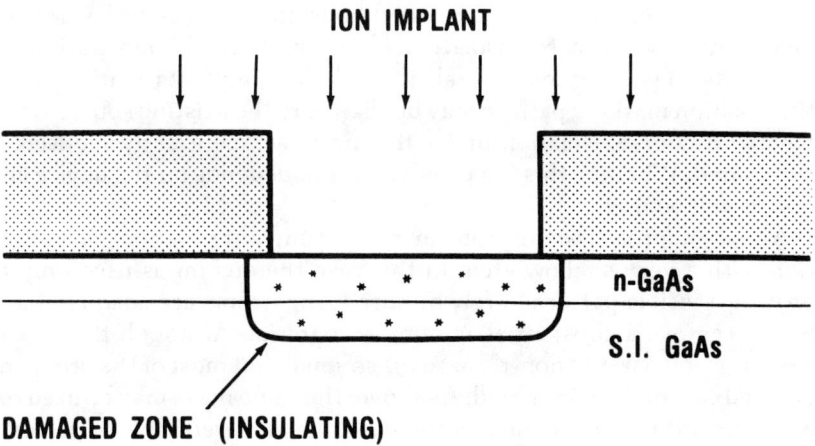

Figure 10.8 Implant isolation achieved using unannealed implants to damage the crystal lattice.

This joint procedure is especially appropriate for slices which have a highly doped, N⁺ contact layer at the surface. Usually such a layer is quite thin (approximately 1000-2000 Å) and is used to enhance ohmic contact, reduce parasitic resistances, and improve reliability (section 3.2). Such layers are often doped near 1×10^{18} cm^{-3}. Ion implant isolation may have difficulty compensating for such high conductivity, especially given the near-Gaussian distribution of ion implant damage as a function of depth (most damage occurs below the surface). It is therefore useful to etch off the N⁺ material and then use ion implantation to render the underlying material resistive.

Implant isolation alone has one major disadvantage. If the isolation step is the first major process step, the desired planarity means that there is no visible pattern to use in subsequent alignments. There are several ways to avoid this problem. One obvious way is to fabricate alignment marks (perhaps by etching) on the slice before the isolation step. Then these alignment marks can be used to align both the isolation pattern and a subsequent metalization pattern. The disadvantage of this approach is the necessity for a separate process step to fabricate the alignment markers. A second approach is to use a dielectric as part of the implant mask. The dielectric pattern can remain on the slice and serve as an alignment reference in a subsequent metalization step. Before metalization, the portions of the dielectric that are exposed by the resist pattern can be dry etched to expose the slice, and metalization can proceed by dielectric-assisted lift off (Chapter 13).

Another procedure would be to change the process order so that metalization occurs before isolation. For example, in FET fabrication the source-drain pattern could be fabricated before implant isolation. Then the isolation masking pattern may be aligned to the existing source-drain metal. Of course, isolation under the metal will be greatly reduced or eliminated. Whether this is a problem or not depends on the device and how it will be used.

An easy approach to the problem is to use implant isolation in conjunction with a very shallow etch. In this case the etching is used only to provide a visible pattern (a few hundred Angstroms are adequate) and not to remove a substantial amount of material. Although the slice is again non-planar, the non-planarity is so small that most of the problems detailed earlier are eliminated. As above, the same mask may be used for both implanting and etching. Note that it may be preferable to perform the etching step after the implant step because the uppermost material will not be as damaged by the Gaussian distribution of the implant as is the underlying material.

Several species of ions have been used for ion implant isolation, including protons [1-8], boron [9], and oxygen [10]. Significant damage and hence, isolation can extend much deeper than the peak of the implant profile [2,11]. Protons have received substantial study [1,2]. Protons are usually implanted at doses of 10^{14}-10^{15} cm^{-2} and energies of 50-140 keV. Low dosage implants result in damage sufficient to remove three conduction electrons for every proton implanted [2]. But isolation does not increase linearly with dosage. Optimum doses may be two orders of magnitude greater than would be predicted from low dosage results [2]. Interestingly, isolation (as assessed by leakage current between ohmics) does not monotonically increase with dosage. Very high doses, around 10^{15}, have been reported to result in isolation less than that achieved with half that dose [2,5]. The reasons for this are unclear but may be related to defect-level banding [5] or enhanced hopping conduction mechanisms [8]. Greater energies generate deeper damage, but also require thicker or more dense masking patterns. Although optimum doses and implant energies have been suggested for proton isolation [2], the optimum will depend on the thickness and carrier concentration of the active levels present in a particular slice. As always, the process engineer should perform experiments to determine the optimum dose for his own application.

Proton isolation, and presumably other ions as well, generate substantial damage at the surface of the semiconductor, resulting in decreased sensitivity of devices to surface conditions [2]. Compared to selective implantation (next section), proton implant isolation has resulted in better isolation as assessed by leakage current and backgating [2]. The proton implants seem thermally stable for temperatures up to about 400°C [2,3,7], which is above the temperature an AuGeNi ohmic contact should be subjected to. Multiple implants result in stability at 500°C [6,8,12].

Less detailed information is available concerning boron or oxygen implants, although both are used by some institutions. Dosage for these heavier ions can be an order of magnitude less than that required for protons (in the 10^{13} cm^{-12} range rather than the 10^{14} range). In summary, all three species seem to provide adequate isolation for most GaAs devices applications. An exact assessment of the relative merits of these, or other, species has yet to be made.

10.4 SELECTIVE IMPLANTATION

Selective implantation consists of forming the active layer by ion implanting only into the areas of the slice that are desired to be conductive. Then the usual activating anneal follows. The remaining portions of the substrate never become electrically conductive. This procedure has two disadvantages. First, a masking step is required to protect most of the slice from the implanted ions (usually the entire slice is implanted and no masking step is required). Second, the isolation provided by the substrate is not as great as can be obtained using implant isolation (section 10.3). But this technique may be useful for specific applications. For example, if different types of implants are needed on the same MMIC (perhaps for both low noise and power FETs), then the masking must be performed anyway and does not constitute an extra step. Also, even though the isolation is not as good as can be achieved by an unannealed implant, it certainly is not bad and may be entirely adequate for most applications.

REFERENCES

[1] R. Esfandiari, M. Feng, H. Kanber, *IEEE Electron Device Lett.*, 4, 1983, p. 29.

[2] D.C. D'Avanzo, *IEEE Trans. Electron Devices*, 29, 1982, p. 1051.

[3] A.G. Foyt, W.T. Lindley, C.M. Wolfe, and J.P. Donnelly, *Solid-State Electron.*, 12, 1969, p. 209.

[4] J.C. Dyment, J.C. North, and L.A. D'Asaro, *J. App. Phys.*, 44, 1973, p. 207.

[5] B.R. Pruniaux, J.C. North, and G.L. Miller, *Proc. 2nd Inter. Conf. Ion Implantation in Semiconductors*, New York: Springer-Verlag, 1971, p. 212.

[6] J.D. Speight, P. O'Sullivan, P.A. Leigh, N. McIntyre, K. Cooper, and S. O'Hara, *Proc. Inst. Phys. Conf.*, ser. no. 33a, 1977, p. 275.

[7] T. Sakurai, Y. Bamba, and T. Furuya, *Fujitsu Sci. & Tech. J.*, June 1975, p. 71.

[8] J.P. Donnelly and F.J. Loenberger, *Solid-State Electron.*, 20, 1977, p. 183.

[9] D. Boccon-Gibod, M. Gavant, M. Rocchi, and M. Cathelin, *Research Abstracts of the IEEE GaAs IC Symp.*, paper 7, 1980.

[10] J.L. Vorhaus, W. Fabian, P.B. Ny, and Y. Tajima, *IEEE Trans. Electron Devices*, 28, 1981, p. 204.

[11] K.C. Lee, J. Silcox, and C.A. Lee, *Appl. Phys. Lett.*, 43, 1983, p. 488.

[12] K. Steeples, G. Dearnley, and A.M. Stoneham, *Appl. Phys. Lett.*, 36, 1980, p. 981.

CHAPTER 11

OHMIC CONTACTS

11.1 OVERVIEW AND DEFINITIONS

The purpose of an ohmic contact on a semiconductor is to allow electrical current to flow into or out of the semiconductor. The contact should have a linear I-V characteristic, be stable over time and temperature, and contribute as little parasitic resistance as possible. This seemingly simple task represents an extraordinary amount of solid-state physics and experimental development because simply placing a metal in contact with a wide bandgap III-V semiconductor such as GaAs generally results in a rectifying contact (a diode) rather than an ohmic contact. Achieving a stable, low resistance ohmic contact has been as much pragmatic engineering as science. This problem has generated a large amount of research over two decades (general reviews are contained in references [1-6]). A complete theoretical understanding has still not been achieved.

Because ohmic contacts are so important to GaAs devices, substantial experimental effort has been directed toward developing practical metalization and processing procedures. Even with the best presently known procedures, the contact resistance in a modern GaAs FET may be half of the total parasitic source resistance. These parasitic resistances limit performance. Noise figure is particularly sensitive to such resistances. Of all the metalizations investigated, none has proven superior to gold germanium based systems for contacting n-type GaAs over reasonable doping ranges (1×10^{16} to 5×10^{17} cm^{-3}). This particular metalization will therefore be discussed in detail.

Section 11.2 reviews the theoretical considerations involved in ohmic contact formation from the point of view of semiconductor physics. Section 11.3 discusses fabrication and physical characteristics of ohmic contacts on GaAs. The gold germanium system is emphasized. Section 11.4 discusses methods of measuring contact resistance and associated parameters and discusses the models relating these parameters.

11.1.1 Definitions

As the term "contact resistance" implies, there is a resistance associated with an ohmic metal-to-semiconductor junction. This is illustrated in Figure 11.1 which shows a slab of semiconductor material having a cross-sectional area A and length L. An ohmic contact covers each end of the slab. The resistance of the semiconductor material from end to end is simply

$$\rho \frac{L}{A}$$

where ρ is the resistivity of the semiconductor. However, there is also a contact resistance, R_c, associated with each ohmic contact. Hence the total measured resistance would be

$$R = 2R_c + \rho \frac{L}{A}$$

(Note that any actual resistance measurement would also include the resistance of the external circuitry and the probes used to contact the metal pads. This external resistance will be considered in more detail in section 11.4. Until then, it is assumed that this "probe resistance" can be determined accurately and subtracted from the measurement.)

The fundamental entity characterizing the resistance of the contact is the specific contact resistance, r_c, which is the contact resistance of a unit area for current flow perpendicular to the contact. It has units of ohms-cm^2. For the geometry of Figure 11.1, r_c is simply

$$r_c = R_c A.$$

Many modern GaAs devices, such as FETs, generally have a planar configuration, so the situation is slightly more complicated, as illustrated in Figure 11.2(a). In this case the ohmic contact is formed on a thin, conductive layer having a semi-insulating layer beneath it (this layer is "thin" with respect to the lateral extent of the contact metal; the semi-insulating layer is regarded as being perfectly insulating). Further, the

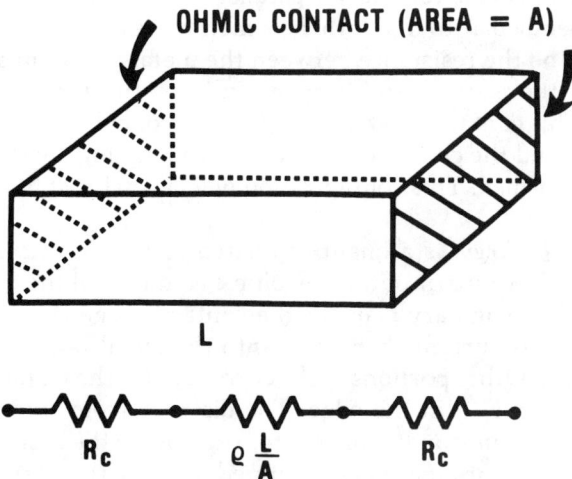

Figure 11.1 A slab of material with ohmic contacts on the two ends exhibits a resistance composed of the end-to-end resistance of the material, plus the two contact resistances.

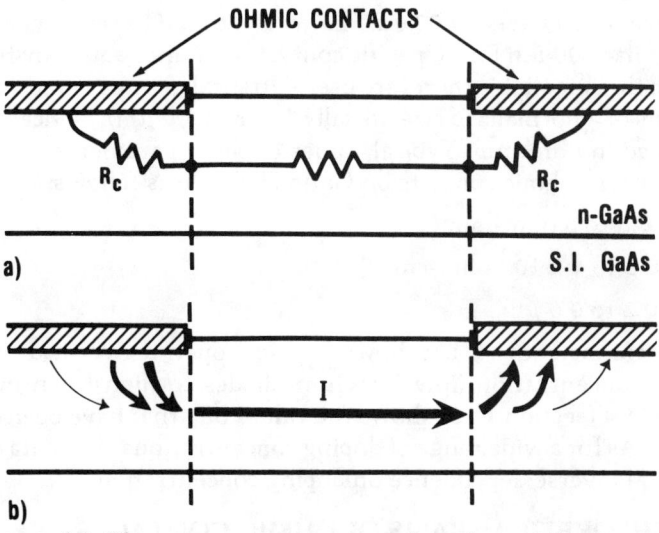

Figure 11.2 (a) The contact resistance of a planar ohmic contact is defined to be the resistance between the metalization and an imaginary plane at the edge of, and perpendicular to, the ohmic metalization. (b) Most of the current flows into the contact relatively near the edge. The magnitude of current remaining in the semiconductor decreases exponentially from the contact edge.

final direction of current flow is parallel to the plane of the metal rather than perpendicular to it. For this geometry, the contact resistance, R_c, is defined to be the resistance between the metal and an imaginary plane that is at the edge of and perpendicular to the metalization (Figure 11.2(a)). This contact resistance is a function of both the specific contact resistance and the characteristics of the conducting layer on which the contact is formed. This configuration will be analyzed in more detail in section 11.4

The *transfer length* is a quantity related to the distance required for current to flow into or out of the ohmic contact. All the current passes through the imaginary plane at the contact's edge (Figure 11.2(b)). A portion of this current then flows into the metalization very near the contact edge. Other portions of the current enter the metal contact some further distance from the edge. The distance from the edge that the current in the semiconductor falls to, $1/e$, (e being the base of the natural logarithm) of its original value is defined to be the transfer length, L_t. For modern GaAs FETs, the transfer length may be only several tenths of a micron. Hence, an ohmic contact that is 10 μm or more in width may be considered semi-infinite for modeling purposes.

Contact resistance, R_c, is measured in ohms, but is obviously a function of device size. That is, a 600 μm FET will have half the contact resistance of a similar 300 μm FET (specific contact resistance and transfer length remain the same). It is therefore useful to speak of the contact resistance in terms of a normalized size, usually 1 mm (1000 μm). Hence R_c is often described in ohm-mm. Typical values of these parameters for modern AuGe planar ohmic contacts on GaAs FETs are as follows:

R_c = 0.5 to 5 ohm-mm

r_c = 0.8 to 4 × 10^{-6} ohm-cm^2

L_t = 0.2 to 0.6 μm

Ohmic contacts to n-GaAs, however, must operate over wide ranges of doping concentration (low for Gunn diodes to high for n$^+$ material). Figure 11.4 (section 11.3) shows the values of r_c that have been obtained on n-GaAs for a wide range of doping concentrations. The data indicates a general inverse dependence on doping concentration.

11.2 THEORETICAL BASIS OF OHMIC CONTACTS

A detailed discussion of the physics of metal-semiconductor junctions is beyond the scope of this book and is not required to understand other portions of the chapter. Nevertheless, it is helpful to have a basic understanding of the considerations involved. This section reviews these considerations. More complete descriptions of the relevant physics may be found in references [6] and [7].

Ohmic Contacts 229

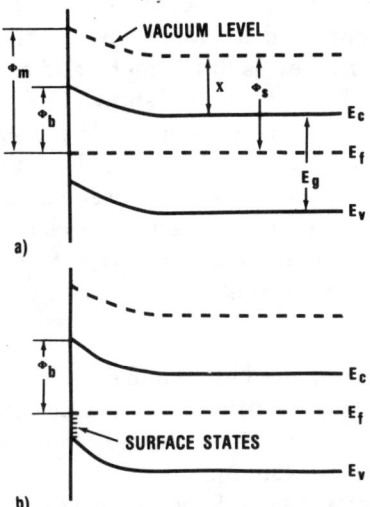

Figure 11.3 Metal-semiconductor junction: (a) idealized energy band diagram showing the barrier height, ϕ_b, as the difference between the work function of the metal, ϕ_m, and the electron affinity, χ; (b) more realistic band diagram, showing that surface states tend to pin the Fermi level and determine the barrier height.

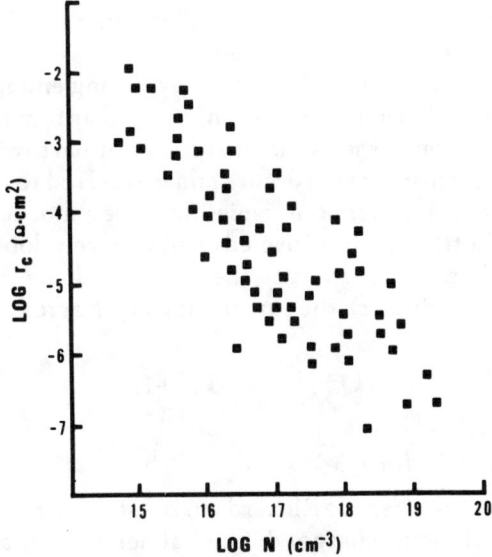

Figure 11.4 Experimental determinations of contact resistance as a function of n-GaAs doping concentration (after reference [41]).

When a metal is placed in intimate contact with a III-V semiconductor such as GaAs, the valence and conduction bands of the semiconductor "bend" to make the Fermi levels in the metal and semiconductor equal (thermal equilibrium). This situation is shown in Figure 11.3(a) for the case of n-GaAs with no bias voltage applied. In the simplest approximation, the resulting potential barrier depends only on the work function of the metal, ϕ_m, the work function of the semiconductor, ϕ_s, and the electron affinity, χ. The potential barrier to electron flow from the semiconductor to the metal is the quantity $\phi_m - \phi_s$. The potential barrier to electron flow from the metal is

$$\phi_b = \phi_m - \chi$$

where ϕ_b is the barrier height (Figure 11.3(a)). Such a metal-semiconductor junction is rectifying and is called a Schottky barrier, named after Schottky who investigated the theories of metal-semiconductor barriers in 1938 [8].

If this was an accurate model, Schottky barrier heights would be a strong function of the metal chosen and would range between about 0.07 eV to 0.57 eV for common metals [6]. However, these barrier heights can be experimentally measured as described in Chapter 12 and, as shown in Table 11.1, all are close to 0.8 eV. This discrepancy is considered to be caused by the large density of surface states present on GaAs [7,9-11]. It is the surface state density that tends to set the barrier height as indicated in Figure 11.3(b) and this is generally a function of the semiconductor only and is independent of the metal.

Electrons may traverse the barrier by having enough thermal energy to pass over it (thermionic emission) or by quantum mechanical tunneling (field emission) if the barrier is sufficiently narrow [12,13]. A mixture of these mechanisms often occurs and is referred to as thermionic-field emission [14, 15]. Thermionic emission is the major current flow mechanism for Schottky barriers formed on moderately doped GaAs. If V is the forward voltage applied across the Schottky diode (between the metal and the semiconductor), the current density, J, across the barrier is given by [7]:

$$\begin{aligned} J &= A^{**} T^2 \exp(-q\phi_b/kT) [\exp(qV/kT) -1] \\ &= J_s [\exp(qV/kT) -1] \\ &= J_s \exp(qV/kt) \text{ for } V \gg kT/q \end{aligned} \quad (11.1)$$

where A^{**} is the effective Richardson constant, T the absolute temperature, q the electron charge, ϕ_b the barrier height, and k Boltzmann's constant. Actual measurements of current density usually fit the expression

$$J = J_s \exp(qV/nkT) \tag{11.2}$$

where n is called the ideality factor and represents departures from an ideal Schottky junction ($n=1$). Actual Schottky barriers on moderately doped n-GaAs usually have n values of 1.0 to 1.25, the value increasing for higher doping in the semiconductor. A Schottky diode operating by thermionic emission is obviously not a good ohmic contact.

It is the second mechanism, tunneling or field emission, that makes present-day ohmic contacts possible. As the doping in the semiconductor increases, the width of the potential barrier decreases [6,7] and tunneling can begin to occur [12]. The current density for field emission has the form [7]:

$$J \sim e^{-q\phi_b/E_{oo}}$$

where

$$E_{oo} = \frac{g\hbar}{2} \sqrt{\frac{N}{\epsilon m^*}}$$

and $\hbar$ is Plank's constant divided by 2π, ϵ is the permittivity, N is the doping concentration, and m^* is the effective electron mass. Note that the tunneling current increases as the square root of doping concentration (because barrier width decreases as the square root of doping concentration).

In the intermediate case, thermionic-field emission, electrons have enough thermal energy to reach the upper, narrower portion of the barrier where tunneling may occur. In all of these regimes (thermionic, thermionic-field, or field emission) a potential difference must exist across the barrier to cause current flow. This is the fundamental origin of contact resistance.

The principal strategy employed to achieve an ohmic contact is to dope the surface of the semiconductor sufficiently high enough to assure that the dominant conduction mechanism is field emission (tunneling). In fact, the presence of a highly doped layer between the metalization and the lower doped semiconductor seems to be a necessary condition for achieving the best quality ohmic contacts [16,17] and the resistance of the n^+ to n junction may be a limiting factor in lowering contact resistance [18]. The required value of doping for n-GaAs is on the order of 1×10^{19} [3]. The major procedure used to achieve these levels is to employ metals that, upon alloy, can dope the surface layer of the semiconductor very highly. Germanium is one such element and its role in ohmic contact formation will be discussed in section 11.3.

The above discussion centered on the use of highly doped surface layers to allow ohmic conduction. This is the major approach used in most GaAs devices. However, other mechanisms have been proposed to achieve suitable contacts. These assume a graded transition of either crystallinity or crystal type between the surface of the semiconductor and the bulk beneath it. In the case of a gradual transition from amorphous (surface) to crystalline (bulk) structure, current flow can proceed by electron hopping between mobility gap states [19-22].

In the case of graded crystal type (heterojunctions), the barrier height can be reduced to negligible values by choosing an appropriate semiconductor (such as InAs) for the surface and forming a gradual compositional change to GaAs as a function of depth [23-26]. However, there are many potential problems to this procedure such as achieving good lattice matching. In fact, neither of the latter two mechanisms has found much use in GaAs processing.

11.3 FABRICATION AND STRUCTURE OF OHMIC CONTACTS

A complete understanding of ohmic behavior in metal-semiconductor contacts is yet to be achieved. Even though the underlying physics (summarized in the previous section) seem well established, exactly what occurs during actual processes used to fabricate ohmic contacts is much less clear. The importance of ohmic contacts for fabricating devices has inspired a great many experimental and empirical approaches to the problem. Table 11.2 lists many of the metalizations that have been investigated for contacting both n-type and p-type GaAs. For n-GaAs,

TABLE 11.1
Experimental Barrier Heights, eV, at 300 K

Metal	Barrier Ht	Reference
Al	0.80	[1]
Ag	0.88	[1]
Au	0.86	[27]
Bi	0.89	[28]
Cr	0.77	[29]
Cu	0.82	[1]
In	0.82	[28]
Ni	0.83	[30]
Pt	0.86	[1]
Sb	0.86	[28]
Ti	0.82	[29]
W	0.64	[31]
Au-Ga	0.71-.75	[27]
Au-Ge	0.27-.35	[32]
Au-Sn	0.72	[33]
Pt-Ni	0.95	[34]
Ti-Ag	0.80	[35]

TABLE 11.2
Metal Systems Used to Form Ohmic Contacts on GaAs

n-Type GaAs	Metal System	Reference
Ag-In	75% Ag, 25% In; sinter 500°C	[53]
Ag-In-Ge	90% Ag, 5% In, 5% Ge; alloy 600°C	[54,55]
Ag-Sn	98% Ag, 2% Sn; alloy 550–650°C	[37]
	33% Ag, 67% Sn; alloy	[56]
Au-Ge	12% Ge (alloy); alloy 450°C	[57,58]
	Au overlayer	[59]
Au-Ge-Ni	12% Ge (alloy), Ni; alloy 480°C	[56,60,61]
Au-Ge-In		[62]
Au-Si	alloy 425°C	[63]
Au-Sn	20% Sn; alloy 450°C	[3,64]
Au-Sn-Ni-Au	alloy 300°C	[65]
Au-Te	10% Te; laser alloy	[66]
	2% Te; alloy 500°C	[3]
In	300°C melt	[67]
In-Al	alloy 320°C	[68]
In-Au	90% In; alloy 550°C	[69–71]
In-Ni	Ni plated to In	[71]
Pd-Ge	sinter 500°C for two hours	[72]
Sn-Ni	Ni plated to Sn; alloyed	[71]
Sn-Sb	4% Sb; alloy 300-350°C	[73]
p-Type GaAs		
Ag-In	25% In; alloy 500°C	[53]
Ag-In-Zn	80% Ag, 10% In, 10% Zn; alloy 600°C	[54,60]
Ag-Zn	90% Ag, 10% Zn; alloy 450°C	[74]
Au-Be	1% Be;	[75]
Au-Zn		[76,77]
In-Zn	(to S.I. GaAs)	[78,79]

none have been found superior to gold germanium eutectics for reasonable doping ranges (1×10^{16} to 5×10^{17}). This section will therefore discuss the AuGe metalization at some length. Other metalizations and procedures will be considered to a lesser extent. The discussions will include alloying procedures and the resulting physical morphology.

11.3.1 Ohmic Metallization

If the surface layer of the GaAs is doped highly enough, almost any metal placed in intimate contact with the surface will result in an ohmic contact without having to be alloyed. This has certain advantages: the slice does not have to be exposed to elevated temperatures; possible difficulties arising from alloying (irregular metal edges, poor surface morphology, etc.) are eliminated; a complete process step can be skipped. Unfortunately, the necessary doping level is at least 1×10^{19} [3] and these levels are not easily achieved by most growth or doping techniques including vapor phase epitaxy, liquid phase epitaxy, and ion implantation (generally regarded as the most cost effective option for production). Molecular beam epitaxy (MBE) is capable of providing highly doped layers and can do so with excellent control of both doping level and layer thickness. However, this technology is still in its infancy and is by far the most expensive growth technology. Although, this might change in the future.

Other, more esoteric, procedures are under investigation in the laboratory. These include use of ion implantation in conjunction with electron beam or laser annealing. Some of these efforts have produced sufficiently high enough doping levels to allow TiPtAu metalization to form ohmic (tunneling) contacts without alloy [36]. Nevertheless, many difficulties remain and it is unlikely that these procedures will mature in the foreseeable future.

The most common approach to fabricating ohmic contacts on GaAs is to apply an appropriate metalization to the slice (in the desired pattern) and then alloy the metal into the GaAs. During the alloy and cooling period a component of the metal enters into the GaAs and highly dopes the surface layer. Candidate species for this function are Si, Ge, Sn, Se, and Te for n-type GaAs and Zn, Cd, Be, and Mg for p-type GaAs [6]. The following discussion will center on gold germanium contacts to moderately doped n-GaAs because this is the most common procedure and is representative of the essential elements of other metalizations.

Gold germanium is usually applied with an overlay of another metal such as nickel or silver. Nickel is the most common choice. The germanium is responsible for doping the GaAs during alloy. The exact function of the nickel is still somewhat debatable. It originally was included to act as a wetting agent and prevent the AuGe metal from "balling up" during alloy [37-40]. This balling up is likely to occur if AuGe alone is used and results in contacts exhibiting poor contact resistance, poor morphology, and very irregular edges. Recent evidence also suggests that nickel aids in the diffusion of the Ge into the GaAs. This will be considered in more detail below.

Gold germanium is applied in proportions that represent an eutectic alloy (88% Au, 12% Ge by weight). The eutectic melts at 360°C. The metal may be applied to the slice in many ways. The most common is to employ a precompounded alloy of gold and germanium and evaporate it onto the slice. If this procedure is used, the entire charge of metal must be evaporated to dryness to assure correct stochiometry on the slice. If this is not done, the resulting Au/Ge ratio is likely to be incorrect and difficult to reproduce. The nickel layer may be evaporated over the AuGe. Approximately 280 Å of nickel for every 1000 Å of AuGe is a common choice for use in evaporation. The materials can be applied separately by any method including e-beam evaporation, thermal evaporation, or even sputtering (although, as explained in Chapter 6, sputtering is generally not compatible with liftoff procedures). The exact total thickness of AuGeNi applied is not critical. However, 1000 Å to 2500 Å is a useful range; contact resistance will suffer if the thickness is reduced below 1000 Å. The order of application of the metals also does not seem critical. The nickel, for example, could be evaporated first, followed by the AuGe.

Ohmic contacts on n-GaAs are required to contact a wide range of doping levels, which depend on device type and function. Figure 11.4 shows the values of specific contact resistances, r_c, that have been reported on n-GaAs [41]. Generally the data follow an inverse doping law. This will be considered further in the discussion of alloy in section 11.3.2.

It is also possible, and often desirable, to apply a thick gold overcoating before alloy. Such an overcoat has several potential advantages. Alloyed AuGeNi metalization has poor sheet resistance (generally in the range of 1 to 2 ohms/square) and usually requires that an overcoat of gold be added in a subsequent step. If the extra gold can be put in place during the ohmic fabrication procedure, a step can sometimes be eliminated. Further, it is difficult to probe test devices having alloyed AuGeNi pads and obtain accurate and reproducible results. The high sheet resistance of the metalization makes such measurements sensitive to the exact placement of the probe on the pad. Also, the resistance of the probe-to-metal contact is overly sensitive to probe pressure. An overlaying layer of gold greatly alleviates these problems and enhances the ability to make accurate measurements immediately after ohmic contact formation. The extra gold results in improved surface morphology and does not generally degrade contact resistance (it may even improve it). These issues will be considered again in the discussion on alloying.

If e-beam lithography is used to define gates, the necessary e-beam alignment marks are usually formed as part of the ohmic contact pattern (for proper e-beam alignment to the source-drain pattern). The extra gold layer usually results in markers having improved contrast and edge definition (alloyed AuGeNi on GaAs has poor contrast for e-beam acquisition). Such improvements can result in faster acquisition of the markers and more accurate placement of the e-beam pattern.

Surface cleanliness is essential for reliable, reproducible ohmic contacts. A final cleanup may be employed after the pattern is defined in resist and immediately before the slice is placed in the evaporator. These cleanups may include a light etch and/or other procedures to remove any oxide layer present.

11.3.2 Alloy

Before the AuGeNi (or AuGeNiAu) can be used as an ohmic contact the surface of the slice must be heated until the metalization alloys into the GaAs. This is a very complex process which is still not completely understood. What follows is a qualitative description of what is generally believed to occur.

As temperature rises, the AuGe alloy begins to melt and gallium diffuses into the metal. A lesser amount of arsenic evolves. Germanium diffuses into the slice and acts as a dopant. Germanium is an amphoteric dopant of GaAs and thus it is desired that it come to rest on the gallium sites of the crystal in quantities sufficient to dope the GaAs close to 2×10^{19}. The diffusivity of germanium in GaAs is strongly affected by other impurities [42,43]. Nickel is one of these impurities — apparently, it enhances diffusion as well as serving its traditional role as a wetting agent which prevents "balling up" of the metal. Damage near the surface caused by defects or strains in the crystal structure can also affect the rate of diffusion.

The desired placement of germanium on the gallium site is aided by diffusion of gallium into the gold. Gallium has even been found to puddle on the surface of gold overlays [44]. The diffusion of gallium into the gold frees gallium sites in the crystal for occupation by the germanium. Contact resistivities have been found to improve as the thickness of the gold capping layer increases [44]. Hence, the use of a top, capping layer of gold can improve contact resistance. However, it is also possible that too much gold could getter more gallium than there is germanium available to replace it. If this occurs, the resulting gallium vacancies can cause a high resistance region [39,44]. The use of a barrier layer, such as Ti-W, placed between the AuGeNi and the extra gold has been suggested to prevent this effect [36,44]. Obviously, the exact situation is complex. In practice, an extra gold layer does not seem to degrade contact resistance for gold thicknesses in the range of 0.3 to 0.5 μm.

The heating of the metalization may be accomplished by several methods including strip heaters, ovens, flash lamps, and lasers. Using heaters has the disadvantage that results depend on the thermal contact between the slice and the heater, and may not be reproducible for larger slices. Flash lamps and lasers have some advantages in rapid temperature rise. The use of lamps seems especially promising. Laser alloy for GaAs is still largely experimental. Ovens (furnaces) are the most popular present choice for production alloying. High quality, multi-zone furnaces are readily available because of their use in silicon technology. They are also capable of accepting several slices in one processing run. The alloy operation in these ovens is usually performed in the presence of an inert gas such as helium or argon, or forming gas (15% hydrogen, 85% nitrogen). The alloy temperature is generally near 450°C.

There is an optimum time and temperature for alloying the metalization. Yet there is a rather wide variation in the temperatures and times used by various institutions. Part of this variation is caused by equipment differences or differing methods of measuring temperature. The temperature of the surface of the slice is very difficult to ascertain. Usually the temperature that is monitored is that of the strip heater stage, the interior of the oven, or the slice holder in the oven. Even if a thermocouple is placed directly against the slice, questions of thermal transfer make its accuracy questionable. In ovens, the optimum temperature (and time) will also depend on oven size and configuration (due to radiation effects), gas flow rate, and numerous other parameters. For these reasons, and also because of other minor variations in the process, each user should optimize his own procedure by experiment. Even apparently identical systems may need to be optimized individually. Optimizations may have to be repeated following any change in procedure. For example, replacing a quartz slice holder with another holder having less mass can result in a clear change in the optimum temperature. The optimum will also depend on the heating and cooling conditions.

The straightforward way to optimize alloy parameters is to run a series of experiments in which these parameters are varied and contact resistance is measured using appropriate test structures. Section 11.4 contains detailed descriptions of the procedures and test patterns necessary for such measurements. As described in that section, accurate contact resistance measurements are a difficult and time-consuming task. Sometimes accuracy is relaxed to obtain rapid results. For process optimization, however, the highest accuracy possible is desired. Figure 11.5 shows typical data that might be taken to optimize alloy temperature.

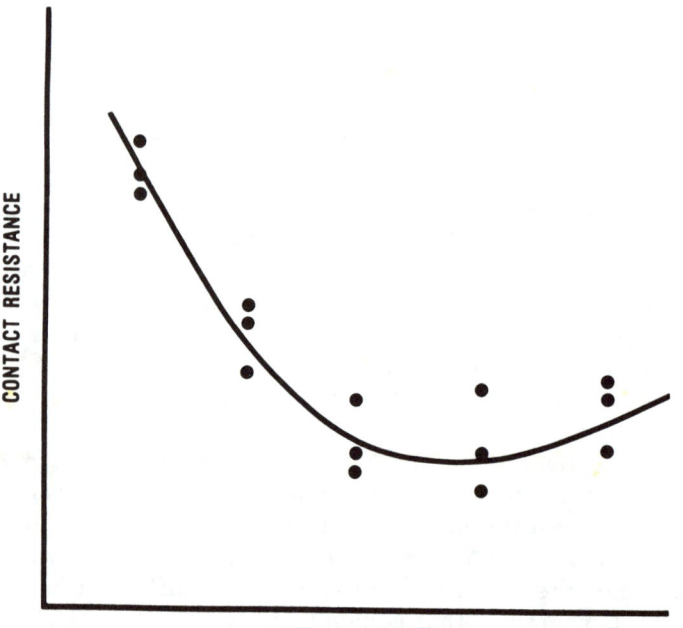

Figure 11.5 Typical data that might result from an alloy optimization experiment. Contact resistance is plotted as a function of alloy temperature, other parameters remaining fixed.

If an overcoat of "thick" gold is used over the AuGeNi, several precautions are necessary. If the gold overcoat is too thin, the extra gold will take part in the alloy to an extent that will render the surface morphology poor, thereby removing one of its primary advantages. This minimum thickness is approximately 0.3 μm. Smaller thicknesses may be used with good results if a capping layer such as silicon nitride or silicon dioxide is present on the slice during alloy. This requires an extra process step, but can preserve surface morphology. But difficulties can be encountered using such an encapsulation. Arsenic evolution during alloy can cause blistering or bubbling of the capping layer and defeat its purpose locally. This effect is dependent on the parameters of the encapsulating film.

Practical considerations may also place an upper limit on the thickness of the extra gold layer. If the ohmic metalization is too thick, it may be difficult to achieve proper resist coating across narrow source-drain channels of FETs, as is required for subsequent gate definition.

If AuGeNi, without the extra gold overlay, is used to form the ohmic contact, the alloyed metal will have a very distinctive appearance when viewed using either an optical or scanning electron microscope (SEM). The metal will appear splotchy, having dark and light spots when viewed in an SEM. In fact, it seems such an appearance is necessary for the lowest resistance contacts. A contact that is either "over-alloyed" or "under-alloyed" will not only have a poorer contact resistance, but also will have a different appearance. Examples of these appearances are illustrated in Figure 11.6. Of course, if the extra overcoat of gold is present, the surface will not show this effect.

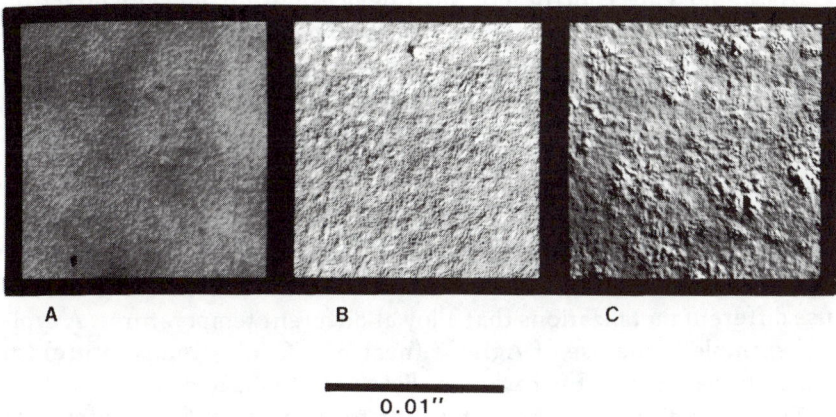

Figure 11.6 Alloyed AuGeNi contacts, showing typical patterns if (a) underalloyed; (b) alloyed correctly; (c) overalloyed. The pictures were taken using phase contrast microscopy.

The surface nonuniformity also extends into the GaAs. The physical structure and metallurgy of alloyed AuGeNi, as well as other contacts, is very complex. A great many analytical techniques have been used to examine this structure [6]. The resulting data show that metalization extends into the slice in irregular depressions or spikes that can extend even as deep as one micron [45]. They certainly extend several thousand Angstroms into the slice [41]. This is easily seen by angle lapping a slice having an alloyed metal on its surface and examining it under the microscope. These depths are substantial considering that active layer thickness on modern GaAs devices often are in the range of 0.1 to 0.4 μm. The general inverse dependence of specific contact resistance on doping concentration (Figure 11.4) has been attributed to these alloy inhomogeneties [41]. In this model, germanium-rich depressions carry most of the current from the metal contact into the GaAs. Spreading resistance results in inverse doping dependence [41].

This general roughness becomes more critical for thinner active layers or for novel geometries such as vertical FETs (section 3.4). In the latter case the alloy spikes cannot be allowed to extend into the active region of GaAs under gate control. For such devices, an unalloyed contact on highly doped material (produced, for example by molecular beam epitaxy) would have strong appeal.

Although AuGe based systems are presently the most generally applicable choice for forming ohmic contacts to moderately doped n-GaAs, it is sometimes necessary to form ohmic contacts at two separate process steps. For example, two terminal devices are often fabricated on conductive substrates and may require ohmics on both the front and back of the slice. If the same metalization is used in both steps, the second alloy may degrade the first, already allowed, ohmic. For this reason, it is useful to use different metalizations that alloy at different temperatures. A typical example is the use of AgInGe (near 600°C alloy temperature) for fabricating ohmic on the back of a slices used to make diodes.

The above discussion has centered on the popular AuGe based system for contacting n-type GaAs. As indicated in Table 11.1, other metalizations are used to fabricate ohmic contacts on p-type material. AuZn based systems work well on p-type material and can exhibit specific contact resistances within an order of magnitude of the best values obtained using AuGe on n-type material. Most p-type layers of device interest are heavily doped. This alleviates the problem of making suitable ohmic contact.

11.3.3 Reliability

Ohmic contacts are obviously an important element affecting device reliability [38, 39]. Several investigations have shown that although contact degradation does occur, it is not a serious threat to device reliability [46-49]. These studies are often performed by subjecting the contact to high temperature for a given time. Application of electrical bias can accelerate the degradation [50].

Other studies have shown that AuGePt ohmic contacts degrade rapidly if a TiAu overlay is present [51,52]. AuGeNi (with and without a TiAu overlay) and AuGePt alone did not show degradation under the test conditions. Similar results were obtained when TiPtAu or TiWAu overlays were used instead of TiAu. The cause of this phenomenon is not completely understood. But Pt does not seem to be a good replacement for Ni in the AuGe system.

11.4 MEASUREMENT OF CONTACT RESISTANCE PARAMETERS

Contact resistance and other associated parameters can be measured using appropriate test structures fabricated on the slice. These measurements are important for in-process monitoring and also for assessing the effects of design or process variations. It is relatively straightforward to ascertain contact resistance given a suitable test pattern. It is more difficult to obtain accurate values of specific contact resistance. To extract such parameters from experimental measurements requires the assumption of a model establishing the relationships among the various parameters.

11.4.1 Models

The model described here addresses the planar type of contact illustrated in Figure 11.2. This is the most common type of ohmic contact. Test patterns and techniques for slices having conductive substrates (rather than semi-insulating substrates) or for slices that cannot use isloation to restrict current flow are more complicated [6,80,81]. Nevertheless, the following considerations apply also to them.

The general class of model used for analyzing planar FET contacts (Figure 11.2) is illustrated in Figure 11.7. The model assumes that an electrically conductive semiconductor layer, situated above an insulating substrate and below an ohmic metalization, exists. This conductive layer is assumed to be thin with respect to the lateral width of the metalization (usually assumed to be semi-infinite). A resistance (the specific contact resistance) is associated with current flow between the conductive layer and the metalization. Such models have been extremely useful both for understanding and analyzing contact resistance. Nevertheless, it must be emphasized that such models do not represent reality for alloyed contacts. As described in the previous section, alloy spikes may extend significant distances into the conductive layer or even completely through it. Irregularity is the rule. Hence, the model may seem inappropriate. Nevertheless, the relationships derived from the model correspond well with other experimental determinations.

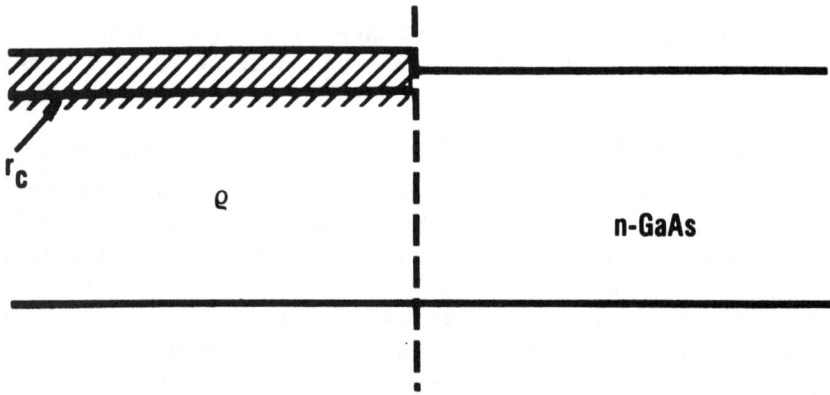

Figure 11.7 Physical model for treating planar ohmic contacts. The material under the contact is characterized by a bulk resistivity, ρ; the transition between the bulk material and the ohmic contact is characterized by the specific contact resistance, r_c.

It is instructive to consider several specific versions of the general model of Figure 11.7. These are illustrated in Figure 11.8. All are slightly different approaches to the same problem and it is reassuring that all yield the same result. Each will be considered briefly.

11.4.2 Resistor Network Model

The resistor network model shown in Figure 11.8(a) is a particularly simple implementation of the general model. The horizontal resistors represent the resistance of the electrically conductive semiconductor layer and have a resistance of

$$r_1 = \frac{R_{sc}}{W} dx$$

where R_{sc} is the sheet resistance of the conductive layer in ohms per square, W is the width of the contact, and dx is an incremental length. The extra subscript c is appended to the sheet resistance to emphasize that it is the sheet resistance of material under the contact. The vertical resistors represent the transition to the metalization and have a resistance of

$$r_2 = \frac{r_c}{W} dx$$

where r_c is the specific contact resistance. Applying usual network theory to this semi-infinite array results in the equation

$$R_c^2 + r_1 R_c + r_1 r_2 = 0$$

Solving this equation and letting dx approach zero results in

$$R_c = \sqrt{r_1 r_2}$$

or

$$R_c = \sqrt{\frac{r_c R_{sc}}{W}} \tag{11.3}$$

and

$$r_c = \frac{R_c^2 W^2}{R_{sc}} \tag{11.4}$$

11.4.3 Differential Model

Another approach to the situation is illustrated in Figure 11.8(b). In this case differential quantities are handled analytically. The current, I, that flows across any differential length is

$$I = \frac{dV}{dR} \tag{11.5}$$

where

$$R = \frac{R_{sc} dx}{W} \tag{11.6}$$

and dV is the voltage across the differential length, dx. The current, dI, that leaves the semiconductor and enters the metalization across a length dx placed a distance x from the contact edge is given by

$$dI = \frac{-V}{(r_c/Wdx)} = \frac{-W\,V\,dx}{r_c} \tag{11.7}$$

Combining equations (11.5)-(11.7) yields

$$\frac{r_c}{R_{sc}} \frac{d^2V}{dx^2} - V(x) = 0$$

which has the solution

$$V(x) = V_o \exp(-x/L_t), \quad \text{where } L_t^2 = r_c/R_{sc} \tag{11.8}$$

and

$$I(x) = I_o \exp(-x/L_t), \quad \text{where } I_o = V_o W L_t/r_c$$

The contact resistance is simply V_o/I_o, or

$$R_c = \sqrt{\frac{r_c R_{sc}}{W}} \tag{11.9}$$

Note that this is the same answer for R_c obtained above (equation 11.3) using the resistor network model. However, this model also easily yields an analytic expression for the transfer length (equation (11.8)).

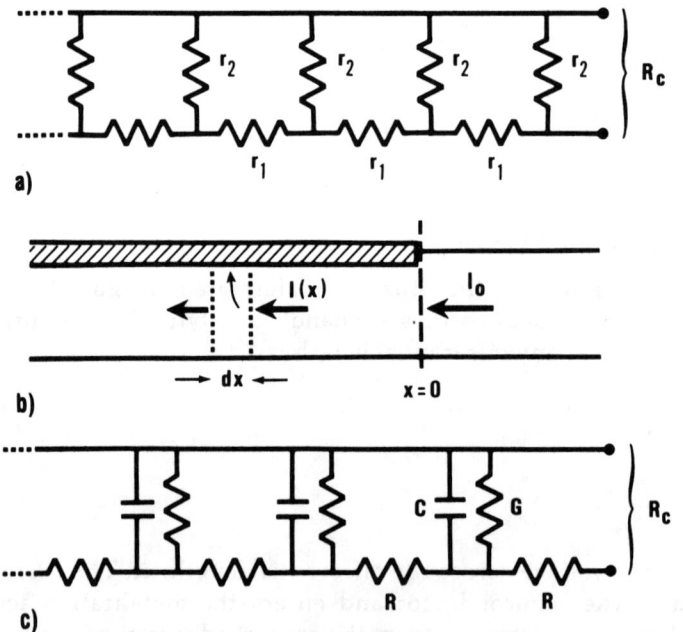

Figure 11.8 Three approaches used to model the ohmic contact of Figure 11.7 (a) a network of resistors, (b) a sequence of differential currents and voltages, (c) a transmission line. See text for details.

11.4.4 Transmission Line Model

A different and seemingly more accurate approach for microwave currents is to treat the contact as a transmission line [80] as indicated in Figure 11.8(c). The transmission line parameters R and G (Figure 11.8) are given by

$$R = \frac{R_{sc}}{W}$$

and

$$G = W\left(\frac{1}{r_c} + j\omega C\right)$$

where C is the capacitance per unit area, ω the angular frequency, and j the square root of negative one. No inductance is included because it is small compared to the series resistance, even at high frequencies [80]. The voltage and current distributions along the contact are given by the well-known transmission line equations

$$V(x) = V_1 \cosh(\gamma x) - I_1 Z \sinh(\gamma x) \qquad (11.10a)$$

$$I(x) = I_1 \cosh(\gamma x) - V_1 Z \sinh(\gamma x) \qquad (11.10b)$$

where Z is the characteristic impedance of the transmission line and γ is the propagation constant. These are defined by

$$Z = \sqrt{\frac{R}{G}} = \frac{\sqrt{R_{sc} r_c}}{W} \cdot \frac{1}{\sqrt{1 + j\omega C r_c}}$$

$$\gamma = \sqrt{RG} = \sqrt{\frac{R_{sc}}{r_c}} \cdot \sqrt{1 + j\omega C r_c}$$

If the analysis is restricted to $\omega=0$ and the contact is assumed infinite in length ($Z = R_c$), then

$$Z = R_c = \frac{r_c R_{sc}}{W} \qquad (11.11)$$

and

$$\gamma = 1/L_t = \sqrt{\frac{R_c}{r_s}}$$

where γ is real for $\omega=0$ and obviously is the inverse of the transfer length, L_t. Hence we again obtain the same result as above.

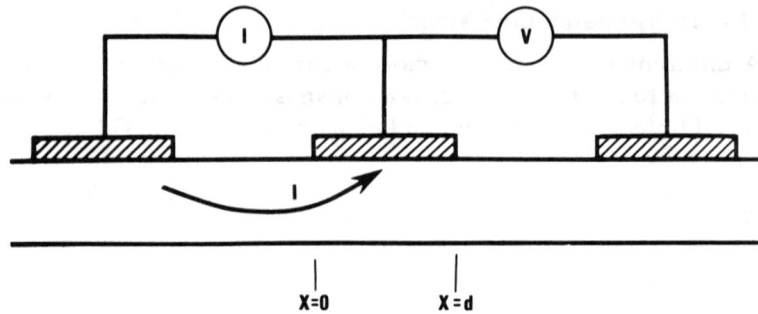

Figure 11.9 The "end resistance" measurement using a perfect current source and voltmeter, corresponding to the fundamental definition of this quantity. See text.

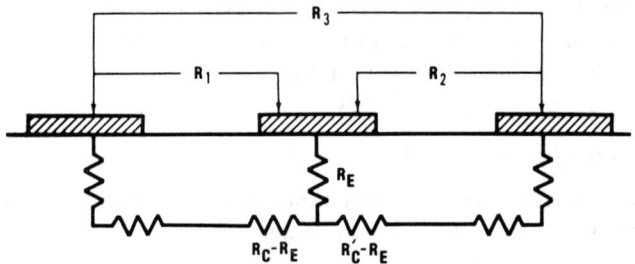

Figure 11.10 Equivalent resistor network representing the end effect and the contact resistance. Several simple measurements will yield the end resistance.

11.4.5 Finite Length Contacts

It will be useful to consider the case in which the contact is not assumed to be semi-infinite. It will be seen that, for contacts to moderately-doped GaAs, the contacts must be very short to exhibit any observable effect. We will use the transmission line model and continue to take $\omega=0$. In this case the contact resistance is (using equations (11.10a) - (11.10b)):

$$R_c = \frac{V_{(o)}}{I_{(o)}} = \frac{V_1}{I_1} \tag{11.12}$$

where $x = 0$ is at the front edge of the contact. Note that for a semi-infinite contact, $I(x)$ approaches zero as x gets larger and hence by equation (11.10b), $I_1 = V_1 Z$, and equation (11.12) gives $R_c = Z$ as was assumed in equation (11.11). If the contact is not semi-infinite, but extends only to $x = d$, then $I(d) = 0$ and using equations (11.10) and (11.12) gives

$$R_c = Z \coth\left(\frac{d}{L_t}\right) = \frac{\sqrt{R_s r_c}}{W} \coth\left(\frac{d}{L_t}\right) \qquad (11.13)$$

Clearly, if d is only a few times L_t, the coth term is very near one.

As part of the discussion of finite length contacts, it is useful to consider a quantity called the *end resistance*, R_E. The fundamental definition of this quantity, as indicated in Figure 11.9, is the voltage at the end of the contact ($x = d$) divided by the incoming current at the front of the contact ($x = 0$), under the condition $I(d) = 0$ [80,82,83]:

$$R_E = \frac{V(d)}{I(0)} \text{ with } I(d) = 0 \qquad (11.14)$$

(The term end resistance has sometimes been used in a slightly different manner [84]). This quantity could be measured as indicated in Figure 11.9, using a high impedance voltage meter between pads two and three so that no current flows. Hence the voltage that is measured is $V(d)$. At first glance, the end resistance seems awkward to measure and of little use. But consider the resistor network shown in Figure 11.10. Note that this equivalent resistor network satisfies both the definition of contact resistance and end resistance; appropriate measurements of contact resistance would yield R_c and appropriate measurements of the end resistance would yield R_E. Examination of the figure reveals a straightforward method of measuring R_E [83]. The resistances R_1, R_2, and R_3 are measured, then the end resistance is

$$R_E = (R_1 + R_2 - R_3)/2 \qquad (11.15)$$

A more meaningful expression for end resistance can be obtained using equations (11.10) in equation (11.14):

$$R_E = \frac{Z}{\sinh \frac{d}{L_t}} \qquad (11.16)$$

or, in combination with equation (11.13),

$$\frac{R_c}{R_E} = \cosh\left(\frac{d}{L_t}\right) \qquad (11.17)$$

Note that R_E rapidly approaches zero as d becomes several times greater than L_t. Hence, R_E is a measure of the ability of the contact to carry the current without "crowding" effects (which depends on the length of the contact). $R_E \approx 0$ is the desirable case.

11.4.6 Summary of Equations

The relationships derived in the previous sections are summarized here:

$$R_c = \frac{\sqrt{R_{sc} r_c}}{W} \cdot \coth\left(\frac{d}{L_t}\right) = \frac{R_{sc} L_t}{W} \cdot \coth\left(\frac{d}{L_t}\right) \quad (11.18)$$

$$= \frac{\sqrt{R_{sc} r_c}}{W} = \frac{R_{sc} L_t}{W} \quad \text{for } L_t \ll d \quad (11.19)$$

$$r_c = R_{sc} L_t^2 \quad (11.20)$$

$$\frac{R_c}{R_E} = \cosh\left(\frac{d}{L_t}\right) \quad (11.21)$$

where equation (11.20) has been used to obtain the furthest-right part of equations (11.18) and (11.19).

Note that for reasonable values of r_c and R_s, the transfer length, L_t, is several tenths of a micron. Therefore any contact several microns in length essentially extracts all of the current and may be treated as semi-infinite. Also, the above models assumed that the sheet resistance of the metal itself is small compared to any of the other resistances. If this is not the case, more complex expressions are needed to include this effect [85]. If an overcoat of gold is present (AuGeNiAu for example) there should be no need to include such effects.

11.4.7 Measurements

It must be emphasized again that there are really two values of sheet resistance that are of interest: the sheet resistance of the GaAs material between the ohmic contacts and the sheet resistance of the material under the ohmic contacts. Because of the alloy process, the two quantities will not be the same. The sheet resistance between the contacts will be simply designated R_s. The sheet resistance of the material under the contact will continue to be designated R_{sc}.

The basic technique used to measure contact resistance of planar ohmic contacts employs a test pattern composed of differently spaced ohmic contacts as illustrated in Figure 11.11. Ohmic contacts are formed on a semiconductor surface and separated by a distance, L. The contacts have a width, W, and the pattern is isolated to restrict current to flow only across the distance, L. The resistance between two such contacts consists of the two contact resistance plus the resistance of the semiconductor layer between the two contacts. The latter resistance should depend only on the sheet resistance, the width of the contact, and the distance between the contacts. Hence the total resistance is

$$R = 2R_c + \frac{R_s}{W} L \qquad (11.22)$$

$$= \frac{2R_{sc}L_t}{W} + \frac{R_s}{W} L \qquad (11.23)$$

(It is assumed that the external resistance of the measuring system, the "probe resistance," is accurately known and subtracted from the experimental measurement. This probe resistance can be determined by placing both probes close together on the same test pad and measuring the resulting resistance. More details on probe resistance problems can be found in Chapter 18.) Hence, assuming sheet resistance is constant, a plot of measured resistance as a function of spacing, L, will yield a straight line as shown in Figure 11.12. The slope of the line gives the value R_s/W and the intercept with the R axis gives the value $2R_c$. The intercept with the L axis, $-L_x$, is related to the transfer length, L_t

$$L_x = \frac{2R_c W}{R_s} = \frac{2R_{sc}L_t}{R_s} \qquad (11.24)$$

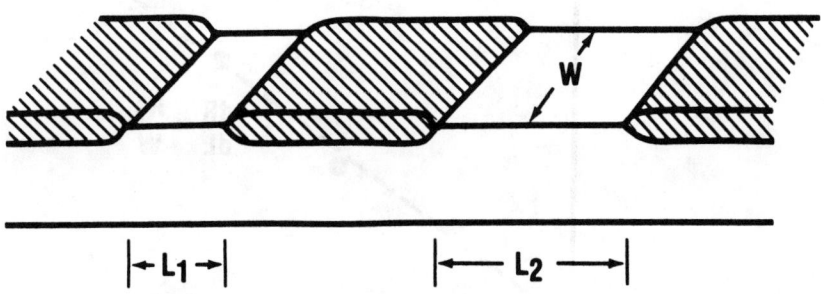

Figure 11.11 Basic pattern used to experimentally determine contact resistance parameters. Ohmic contacts are separated by increasing distances.

where equation (11.19) has been employed. Thus, the L-axis intercept would give the value of L_t if the sheet resistance under the contact were the same as the sheet resistance between the contacts. Note that use of equation (11.22) gives unequivocal values for R_s and R_c; these are the experimentally determined variables. Additional data is needed to accurately determine the quantities r_c, R_{sc}, and L_t. Technically the end resistance, R_E, can supply the necessary information if the length, d, of the contact is known. Equations (11.20), (11.21), and (11.24) can then be used to determine the remaining quantities. But the end resistance measurement is useful if d is not greatly larger than L_t. For FET type devices, this would be on the order of a micron, an almost prohibitively small contact. As a practical matter, it is sometimes useful to (incorrectly) adopt the assumption that $R_s = R_{sc}$. It is then straightforward to compute r_c and L_t based on this assumption. These difficulties, coupled with experimental error (see below), make it exceedingly difficult using this method to determine r_c accurately for state-of-the-art AuGeNi ohmic contacts on moderately doped material. Claims of specific contact resistances less than about 1×10^{-6} ohm-cm^2 should be examined critically. Fortunately, R_c is the quantity most relevant to device performance and it can be determined accurately.

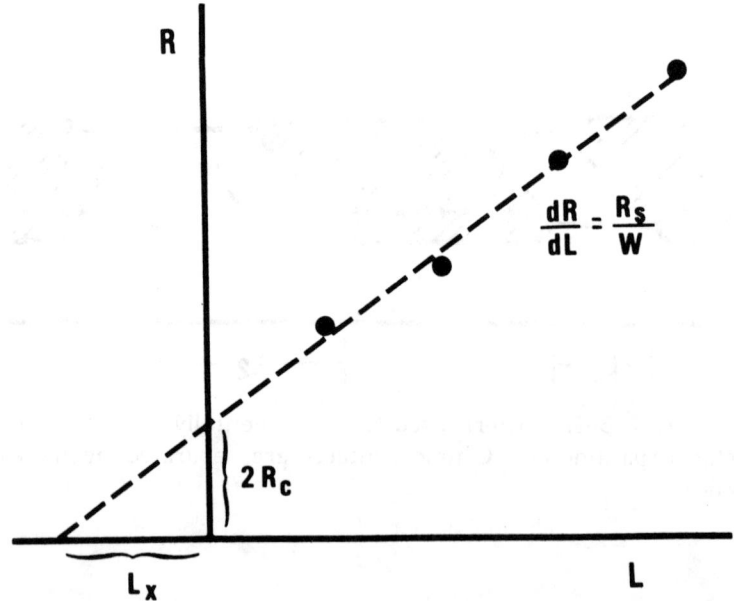

Figure 11.12 Plot of measured resistance as a function of contact separation (using the test pattern of Figure 11.11) yields sheet resistance, contact resistance, and other parameters (see text).

Obviously only two spacings are needed to use this technique (Figure 11.11). But experimental error is often large enough that several spacings are more useful and decrease the uncertainties. A least-squares straight line is fit to the data and any point well off the line should be discarded. Usually such deviations can be correlated to irregular contact edges, materials defects, etc.

The measurement technique described above is sensitive to measurement error. This is easily seen graphically. Consider Figure 11.12 and note that a systematic error in probe resistance (which is subtracted from each measurement) translates the line vertically and results in an error in R_c equal to half the error in probe resistance. A systematic error in contact spacing translates the line horizontally and also introduces error. Because the y-intercept is usually small compared to the measured resistances, minor errors in resistance or spacing can translate into significant percentage errors in contact resistance. Obviously great care must be taken to assure accurate data. Sometimes care can be reduced to obtain speed. But if the best accuracy is desired, as for process optimization, a number of precautions are necessary. These include the following:

1) Minor variations in contact spacing can affect accuracy, so every spacing on every test pattern used should be individually measured.

2) The test pattern should be completely isolated so that current flow is only possible between pads. Ill-planned mask designs can result in high resistance alternate current paths arising, for example, from a general grid of mesa material.

3) A gold or conductive overlay metal should be present on the test pads extending up to or very near to the contact edge. Alloyed AuGeNi (without any overlay) is too resistive and probe resistance will depend on exact probe placement and probe pressure; it will be virtually impossible to reproduce accurately.

4) The test patterns should be examined individually at high magnification and patterns having irregular edges or obvious defects should not be used.

These precautions are time consuming. If a large percentage error (50% to 100%) can be tolerated, all of these precautions may be relaxed. For example, in routine in-process evaluation of contact resistance there will exist a maximum permissible value. If measured values usually fall below this specification, even with experimental error, extra uncertainty may be acceptable to gain speed. Assuming a fixed set of contact spacing (rather than measuring each one) greatly eases the measurement difficulty.

Figure 11.13 shows several test patterns that have been used to measure contact resistance. The pattern in Figure 11.13(a) is the type alluded to above and consists of a number of pads spaced at unequal distances. The test pattern is isolated (Chapter 10) so that current flow only occurs in the space between pads. The spacing should not be so small that irregular edges cause a problem. Typical spacings for material doped for FET devices are 5, 10, 15, 20, and 25 μm. It is useful to include this type of test pattern on all mask sets used to fabricate devices so that contact resistance can be monitored. The patterns contained in Figures 11.13(b) and 11.13(c) are sometimes used to evaluate material or contact resis-

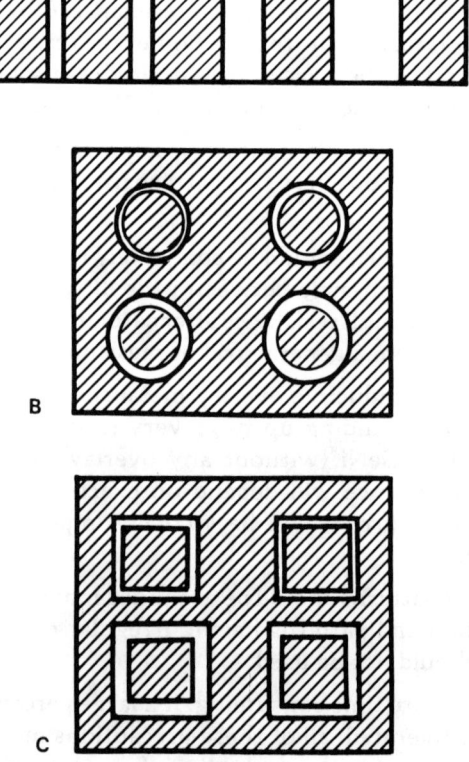

Figure 11.13 Several different possible patterns used to assess contact resistance: (a) linear array of the type shown in Figure 11.11; (b) circular contacts; (b) square contacts.

tance when it is desired to perform only one process step (ohmic fabrication). The device isolation step may be omitted because there is no leakage path for current flow. The geometry, however, makes data analysis slightly more complicated. For circular test patterns (Figure 11.13(b)), if r_1 is the inner radius, r_2 the outer radius, and d the difference $(d = r_2 - r_1)$ then the relationship analagous to equation (11.23) is [85]

$$R = \frac{1}{2\pi} \left[\frac{R_{sc} L_t}{r_1} \frac{I_o\left(\frac{r_1}{L_t}\right)}{I_1\left(\frac{r_1}{L_t}\right)} + \frac{R_{sc} L_t}{r_2} \frac{K_o\left(\frac{r_2}{L_t}\right)}{K_1\left(\frac{r_2}{L_t}\right)} + R_s \ln\left(\frac{r_2}{r_1}\right) \right]$$

where I_o, I_1, K_o, and K_1 are the modified Bessel functions. If the radii are greater than L_t by a factor of four or more, then I_o/I_1 and K_o/K_1 approach unity and the above equation becomes [85]

$$R = \frac{1}{2\pi} \left[R_{sc} L_t \left(\frac{1}{r_2} + \frac{1}{r_2 - d} \right) + R_s \ln\left(\frac{r_2}{r_2 - d}\right) \right] \quad (11.25)$$

Note that if $R \gg d$, then the above equation reduces to equation (11.23) using $2\pi r_2 = W$.

REFERENCES

[1] A.G. Milnes and D.L. Feucht, *Heterojunctions and Metal-Semiconductor Junctions*. New York: Academic Press, 1972.

[2] B.L. Sharma and R.K. Purohit, *Semiconductor Heterojunctions*. Oxford: Pergamon, 1974.

[3] V.L. Rideout, *Solid-State Electronics*, 18, 1975, p. 541.

[4] B. Schwartz, editor, *Ohmic Contacts to Semiconductors*. New York: Electrochem. Soc., 1969.

[5] W.D. Edwards, W.A. Hartman, and A.B. Torrens, *Solid-State Electronics*, 15, 1972, p. 387.

[6] B.L. Sharma, in *Semiconductors and Semimetals*, vol. 15, 1981, p. 1.

[7] S.M. Sze, *Physics of Semiconductor Devices*. New York: Wiley, 1981.

[8] W. Schottky, *Naturwissenshaften*, 26, 1938, p. 843.

[9] J. Bardeen, *Physics Review*, 71, 1947, p. 717.

[10] A.M. Cowley and S.M. Sze, *J. Appl. Phys.*, 36, 1965, p. 3212.

[11] S. Kurtin, T.C. McGill, and C.A. Mead, *Physics Review Lett.*, 22, 1969, p. 1433.

[12] F.A. Padovani, in *Semiconductors and Semimetals*, Vol. 7A, R.K. Willardson and A.C. Beer, editors. New York: Academic Press, 1971, p. 7.

[13] C.B. Duke, *J. Vac. Sci. Technol.*, 7, 1970, p. 22.

[14] C.R. Crowell and V.L. Rideout, *Solid-State Electronics*, 12, 1969, p. 89.

[15] C.R. Crowell and V.L. Rideout, *Appl. Phys. Lett.*, 14, 1969, p. 85.

[16] T. Sebestyen, H. Hartnagel, and L.H. Herron, *Proc. Int. Symp. Gallium Arsenide Relat. Compd.*, 5th Deaville, 1974, Inst. Phys. Conf. ser. 24, 1975, p. 77.

[17] T. Sebestyen, H. Hartnagel, and L.H. Herron, *IEEE Trans. Electron. Devices*, 22, 1975, p. 77.

[18] S.C. Gupta, B.L. Sharma, and A.K. Sreedhar, *Solid-State Electronics*, 14, 1971, p. 427.

[19] T. Sebestyen, in *Amorphous Semiconductors '76*. I. Kosa Somogyi, editor. Budapest: Akademiai Kiado, 1977, p. 321.

[20] A.K. Johnscher and R.M. Hill, *Phys. Thin Films*, 8, 1975, p. 196.

[21] H.Y. Wey, *Physics Rev. B*, 13, 1976, p. 3495.

[22] K.E. Peterson and D. Addler, *IEEE Trans. Electron. Devices*, 23, 1976, p. 471.

[23] R.M. Raymond and R.E. Hayes, *J. Appl. Phys.*, 48, 1977, p. 1359.

[24] J.F. Womac and R.H. Rediker, *J. Appl. Phys.*, 43, 1972, p. 4129.

[25] D.T. Cheung, S.Y. Chiang, and G. L. Pearson, *Solid-State Electronics*, 18, 1975, p. 263.

[26] W.G. Oldham and A.G. Milnes, *Solid-State Electronics*, 6, 1963, p. 121.

[27] S. Guha, B.M. Arora, and V.P. Salvi, *Solid-State Electronics*, 20, 1977, p. 431.

[28] M.S. Tyagi, *Jpn. J. Appl. Phys.*, 16, Suppl. 1, 1977, p. 333.

[29] Y. Sato, M. Uchida, K. Shimada, M. Ida, and T. Imai, *Rev. Electr. Commun. Lab.*, 18, 1970, p. 638.

[30] R. Hackman and P. Harrop, *IEEE Trans. Electron. Devices*, 19, 1972, p. 1231.

[31] K.J. Linden, *Solid-State Electronics*, 19, 1976, p. 843.

[32] W.J. Moroney and Y. Anand, *Proc. Int. Symp. Gallium Arsenide Relat. Compd.*, 3rd, London, 1971, p. 259.

[33] L. Buene, T. Finstad, K. Rimstad, O. Lonsjo, and T. Olsen, *Thin Solid Films*, 34, 1976, p. 149.

[34] M. Ogawa, D. Shinoda, N. Kawanura, T. Nozaki, and S. Asanabe, *Proc. Int. Symp. Gallium Arsenide Relat. Compd.*, 3rd, London, 1971, p. 268.

[35] J.C. Irvin and N.C. Vanderwal, in *Microwave Semiconductor Devices and Their Circuit Applications*, H.A. Wilson, editor. New York: McGraw-Hill, 1969, p. 340.

[36] M.N. Yoder, *Solid-State Electronics*, 23, 1980, p. 117.

[37] S. Knight and C. Paola, in *Ohmic Contacts to Semiconductors*, B. Schwartz, editor. New York: Electrochem. Soc., 1969, p. 102.

[38] A.M. Andrews and N. Holonyak, *Solid-State Electronics*, 15, 1972, p. 601.

[39] G.Y. Robinson, *Solid-State Electronics*, 18, 1975, p. 331.

[40] K. Heime, U. Konig, E. Kohn, and A. Wartmann, *Solid-State Electronics*, 17, 1974, p. 835.

[41] N. Braslau, *J. Vac. Technol.*, 19, 1981 p. 803.

[42] W.T. Anderson, A. Christou, and J. Davey, *IEEE J. Solid-St. Circuits*, SC-13, 1978, p. 430.

[43] D. Sigurd, G. Ottaviani, V. Marrello, J.W. Mayer, and J.O. McCaldin, *J. Non-Cryst. Solids*, 12, 1975, p. 135.

[44] S.G. Bandy, R. Sankaran, and D.M. Collins, Annual Report No. 3, Contract N00014-76-0303, Varian Associates, 1979.

[45] D.C. Miller, *J. Electrochem. Soc.: Solid-State Science & Tech.*, 127, 1980, p. 467.

[46] H.M. Macksey, in *Inst. of Physics Conf. Series*, 33b, 1976, p. 254.

[47] M. Wittmer, R. Pretorios, J.W. Mayer, and M.A. Nicolet, *Solid-State Electronics*, 20, 1977, p. 433.

[48] D.A. Abbott and J.A. Turner, *IEEE Trans. Microwave Theory Tech.*, 24, 1976, p. 317.

[49] T. Irie, I. Nagasako, H. Kohzu, and K. Sekido, *IEEE Trans. Microwave Theory Tech.*, 24, 1976, p. 321.

[50] A. Christou and K. Sleger, *Proc. 6th Annual Biennial Cornell Electrical Engineering Conference*, vol. 56, 1976, p. 169.

[51] C.P. Lee, B.M. Welch, and W.P. Fleming, *Electron. Lett.*, 17, 1981, p. 407.

[52] A.K. Gupa, D.P. Siu, K.T. Ip, and W.C. Petersen, *IEEE Trans. Electron. Devices*, 30, 1983, p. 1850.

[53] H. Matino and M. Tokunaga, *J. Electrochem. Soc.*, 16, 1969, p. 709.
[54] R.H. Cox and H. Strack, *Solid-State Electronics*, 10, 1967, p. 1213.
[55] T. Sebestyen, H. Hartnagel, and L.H. Herron, *Electron. Lett.*, 10, 1974, p. 372.
[56] W.D. Edwards, W.A. Hartman, and A.B. Torrens, *Solid-State Electronics*, 15, 1972, p. 387.
[57] N. Yokoyama, S. Ohkawa, and H. Ishidawa, *Jpn. J. Appl. Phys.*, 14, 1975, p. 1071.
[58] M. Fukuta, K. Suyama, H. Suzuki, and H. Ishikawa, *IEEE Trans. Electron. Devices*, 23, 1976, p. 390.
[59] B.L. Sharma, P.L. Bharti, S.N. Mukerjee, and S. Mohan, *Indian J. Pure Appl. Phys.*, 16, 1978, p. 727.
[60] K. Chino and Y. Wada, *Jpn. J. Appl. Phys.*, 16, 1977, p. 1823.
[61] M. Otsubo, H. Kuambe, and H. Miki, *Solid-State Electronics*, 20, 1977, p. 617.
[62] A. Christou, *Solid-State Electronics*, 22, 1979, p. 141.
[63] L.K.J. Vandamme and R.P. Tijburg, *J. Appl. Phys.*, 47, 1976, p. 2056.
[64] R. Gulati, R.K. Purohit, and I. Chandra, *J. Inst. Telecommun. Eng.*, New Delhi, 15, 1969, p. 815.
[65] W.M. Kelly and G.T. Wrixon, *Electron. Lett.*, 14, 1978, p. 80.
[66] A.N. Pikhtin, V.A. Popov, and D.A. Yas'kov, *Sov. Phys.-Semicond.*, 3, 1970, p. 1383.
[67] C.R. Wronski, *RCA Review*, 30, 1969, p. 314.
[68] M.F. Healy and R.J. Mattauch, *IEEE Trans. Electron. Devices*, 23, 1976, p. 374.
[69] V.K. Handu and M.S. Tyagi, *J. Inst. Telecommun. Eng.*, New Delhi, 18, 1972, p. 527.
[70] C.R. Paola, *Solid-State Electronics*, 8, 1970, p. 1189.
[71] B.W. Hakki and S. Knight, *IEEE Trans. Electron. Devices*, 13, 1966, p. 94.
[72] A.K. Sinha, T.E. Smith, and H.J. Levinstein, *IEEE Trans. Electron. Devices*, 22, 1975, p. 218.
[73] J.R. Dale and R.G. Turner, *Solid-State Electronics*, 6, 1963, p. 388.
[74] O. Ishihara, K. Nishitana, H. Sawano, and S. Mitsui, *Jpn. J. Appl. Phys.*, 15, 1976, p. 1411.

[75] I. Ishida, S. Wako, and S. Ushio, *Thin Solid Films*, 39, 1976, p. 227.
[76] B.L. Sharma and S. N. Mukerjee, *Phys. Status Solidi A*, 29, 1975, p. k141.
[77] H.J. Gopen and A.Y.C. Yu, *Solid-State Electronics*, 14, 1971, p. 515.
[78] R.K. Purohit, *Phys. Status Solidi*, 24, 1967, p. k57.
[79] S.G. Suleimanov, *Sov. Phys.-Semicond.*, 11, 1977, p. 844.
[80] H.H. Berger, *Solid-State Electronics*, 15, 1972, p. 145.
[81] H. Cox and H. Strack, *Solid-State Electronics*, 10, 1967, p. 1213.
[82] H.H. Berger, *Dig. Tech. Pap. ISSCC*, 1969, p. 160.
[83] G.K. Reeves and H.B. Harrison, *IEEE Electron. Devices Lett.*, 3, 1982, p. 111.
[84] K. Lee, M. Shur, K.W. Lee, T. Vu, P. Roberts, and M. Helix, *IEEE Electron. Devices Lett.*, 5, 1984, p. 5.
[85] G.S. Marlow and M.B. Das, *Solid-State Electronics*, 25, 1982, p. 91.

CHAPTER 12

SHOTTKY BARRIERS
AND
GATE FORMATION

The Schottky barrier gate is one of the two most important elements of many GaAs devices, the other being an ohmic contact. This chapter, which describes Schottky barrier gates, is placed after the chapter on ohmic contacts (Chapter 11) to correspond to the most common processing order, in which gate fabrication follows ohmic contact formation. However, the reverse order is used for the fabrication of some digital FETs. Those cases require that Schottky gate characteristics be thermally stable during the subsequent ohmic contact anneal. Both approaches are described in section 12.3. Section 12.1 serves as an introduction to Schottky barrier characteristics and reviews the theoretical foundation. Section 12.2 describes the common techniques used to characterize Schottky barriers.

12.1 INTRODUCTION

A discussion of the basic physics involved in metal-semiconductor junctions was given in the initial part of section 11.2 in the chapter on ohmic contacts. That material is equally important as a foundation for this section and should be considered a part of this discussion. The major points are summarized here. When a metal is placed in intimate contact with a wide-bandgap semiconductor of moderate doping (less than 5×10^{17}), the resulting junction is rectifying: it is a diode. The current density in the forward direction may be expressed as (see equations (11.1) and (11.2)):

$$J = A^{**} T^2 \exp(-q\phi_b/kT) \exp(qV/nkT) \text{ for } V \gg kT/q \quad (12.1a)$$

$$= J_s \exp(qV/nkT) \quad (12.1b)$$

where J is the current desnsity, A^{**} is the effective Richardson constant, T is the absolute temperature, q is the electron charge, ϕ_b is the barrier height, k is Boltzmann's constant, V is the applied voltage, and n is the ideality factor. The ideality factor would be 1 for a perfect Schottky diode (current flow via thermionic emission — see section 11.2). In practice, n is generally between 1.0 and 1.25, with the value increasing with increased doping in the semiconductor. (This concludes the summary of material from section 11.2.)

When a metal is placed on a wide-bandgap semiconductor, a portion of the semiconductor beneath the metal becomes depleted of carriers. We will assume the material is n-type and so the carriers are electrons, but the reverse situation (p-type and holes) is equally valid. This region is called a *depletion zone* and is indicated in Figure 12.1(a). As was described briefly in section 3.2, the applied voltage can alter the depth that this region extends into the semiconductor. The physics will be considered in greater detail here.

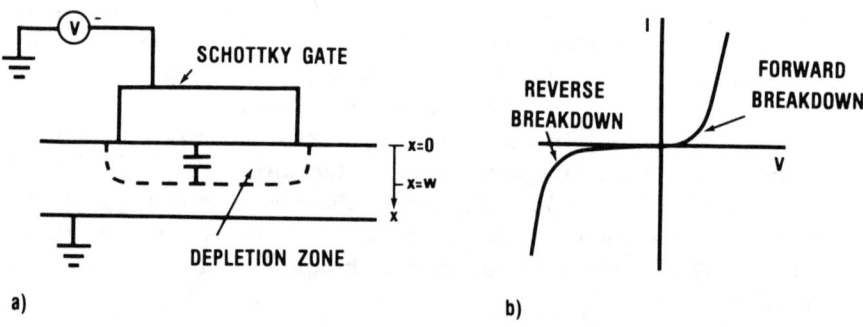

Figure 12.1 Schottky barrier diode: (a) Depletion zone beneath a Schottky barrier. The depth to which it extends into the semiconductor is a function of the applied voltage and the doping concentration. A capacitance is associated with this depletion zone. (b) The IV characteristics of a Schottky diode showing forward and reverse breakdown.

12.1.1 Depletion Region

Given the fact that a depletion region does exist, and assuming a sharp boundary, it is relatively easy to calculate the depth of the depletion region as a function of applied voltage. Other parameters may then be calculated. Note that in the bulk of the n-doped layer there is charge neutrality: the negative charge of the electron carriers is balanced by the positive charge of the ionized dopant atoms. However, electrons are restricted from being in the depletion region, and therefore a charge density exists in this region that is equal to the product of the doping density, N and the magnitude of the electron charge q.

The fundamental equation for electrostatics is Poisson's equation which for one dimension is

$$\frac{d^2 V}{dx^2} = - \frac{\rho(x)}{\epsilon} \tag{12.2}$$

where $V(x)$ is the potential, $\rho(x)$ is the charge density, and ϵ is the permittivity of the medium (i.e., the dielectric constant, k, times the permittivity of free space, ϵ_o; that is $\epsilon = k\epsilon_o$). Consider the situation indicated in Figure 12.1(a), in which a voltage V is applied across the depletion region which has a depth, w. The coordinate x represents depth into the semiconductor and $x = 0$ represents the surface. In this case the charge density is determined by the doping concentration. Poisson's equation becomes

$$\frac{d^2 V}{dx^2} = - \frac{qN}{\epsilon} \tag{12.3}$$

where N will be assumed constant for now. Integrating equation (12.3) and applying the boundary condition $dV/dx = 0$ at the boundary $x = w$ gives

$$\frac{dV}{dx} = \frac{qN(w - x)}{\epsilon}$$

Integrating once more and using the boundary condition $V(0) = 0$ gives

$$V(x) = \frac{qNw}{\epsilon}\left(x - \frac{x^2}{2w}\right)$$

and hence,

$$V(w) = V = \frac{qNw^2}{2\epsilon} \tag{12.4}$$

Re-arranging yields the standard expression for the Schottky barrier depletion width:

$$w^2 = \frac{2\epsilon}{qN}(V_{bi} - V_g) \tag{12.5}$$

where the voltage V has been separated into two components, the applied gate voltage, V_g and the built-in voltage, V_{bi}. As described in Chapter 11, the built-in voltage represents the fact that a depletion zone exists even if no external voltage is applied. As discussed in that chapter, the value of the built-in voltage is approximately 0.8 volts for metals placed on GaAs. Confusion sometimes occurs about the signs in equation (12.5) with respect to the derivation. The equation is usually given in the form of equation (12.5) in which the charge q is the magnitude of the electron charge (and so is a positive number) and the applied voltage, V_g, carriers its own sign. Thus, $V_g = 0$ results in the *zero bias* depletion region (corresponding to V_{bi}), $V_g < 0$ increases the depth of the depletion region, and $V_g > 0$ decreases the depth of the region. If V_g becomes too negative, reverse breakdown occurs (see below). If V_g becomes too positive (near the built-in voltage, V_{bi}), forward breakdown occurs. These features are illustrated in Figure 12.1(b), which shows the typical IV characteristic of a Schottky diode.

A more exact analysis would take into account the fact that the electrons (for the case of n-GaAs) are not uniformly distributed immediately below the depletion zone boundary at $x = w$, but have a distribution tail. This gives rise to an extra term of kT/q in equation (12.5):

$$w^2 = \frac{2\epsilon}{qN}\left(V_{bi} - V_g - \frac{kT}{q}\right) \tag{12.6}$$

At room temperature, $kT/q = 0.026$ V and so is often ignored. Equation (12.5) will generally be used in this chapter. But any expression can be modified for the more exact case by replacing V_{bi} with the quantity $(V_{bi} - kT/q)$.

Equation (12.5) is valid for uniform doping. But many GaAs devices have nonuniform doping profiles produced, for example, by ion implantation. If $N(x)$ represents the doping concentration as a function of depth, then remembering that the electric field, E, is related to the potential by $E = -dV/dx$,

$$-V = \int_0^w E(x)\,dx$$

which by partial integration yields

$$-V = xE(x)\Big|_0^w - \int_0^w xE'(x)\,dx$$

or

$$V = \frac{q}{\epsilon}\int_0^w xN(x)\,dx \tag{12.7}$$

This general equation reduces to equation (12.5) for $N(x) = N$.

12.1.2 Depletion Region Capacitance

The absence of carriers in the depletion region causes it to act as an insulator between the metal on the semiconductor surface and the conductive material beginning at the depletion zone edge (Figure 12.1(a)). Therefore, the depletion region represents a parallel plate capacitor (ignoring end effects) having a space w between the plates and a permittivity ϵ:

$$C = \frac{\epsilon A}{w} \tag{12.8}$$

where A is the metal plate area. This same result may be obtained more formally by noting that the fundamental definition of capacitance is $C = dQ/dV$ where Q represents the charge ($Q = qNwA$). Therefore,

$$C = \frac{dQ}{dV} = \frac{d(qNwA)}{d(qNw^2/2\epsilon)} = \frac{\epsilon A}{w}$$

or more rigorously,

$$C = \frac{dQ}{dV} = qNA\frac{dw}{dV} = \sqrt{\frac{q\epsilon N}{2V}} = \frac{\epsilon A}{w} \tag{12.9}$$

giving the same result as equation (12.8). Equations (12.5) and (12.8) are plotted in Figure 12.2 which shows the depletion depth as a function of (uniform) doping and the corresponding capacitance. The plots of depletion depth terminate at a voltage representing the breakdown voltage of bulk material; the depletion region cannot be extended further. Two qualifications apply to the above equations and to Figure 12.2. First, the capacitance represents parallel plate capacitance and end effects will produce greater capacitance in an actual structure. The end effects can be substantial if the Schottky barrier is very narrow, such as represented by a half-micron gate stripe. Second, the breakdown voltage indicated in the figure is for bulk material of uniform doping. This bulk breakdown voltage is plotted in Figure 12.3 as a function of doping concentration. Planar configurations typical of GaAs FETs exhibit breakdown voltages different than bulk (see section 12.1.3 below).

The behavior of the depletion zone capacitance as a function of voltage is relevant to several measurement techniques. Equations (12.9) and

(12.4) yield

$$\frac{1}{C^2} = \frac{2V}{q\epsilon NA^2} \quad (12.10a)$$

or

$$N = \frac{2}{q\epsilon A^2} \frac{-1}{d[1/C^2(x)]/dV} \quad (12.10b)$$

It can be verified, beginning with equation (12.7), that the above result also holds for nonuniform doping concentrations:

$$N(x) = \frac{2}{q\epsilon A^2} \frac{-1}{d[1/C^2(x)]/dV} \quad (12.10c)$$

This result is very important in that it allows doping profiles to be determined from a measurement of the CV profile of a Schottky diode.

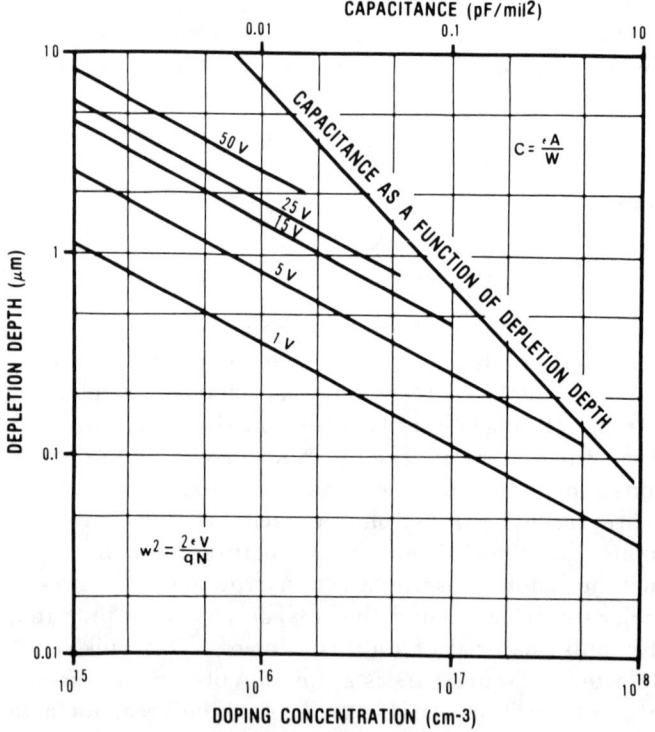

Figure 12.2 Depletion depth and associated capacitance as a function of (uniform) doping concentration and applied voltage. The plots stop at the bulk breakdown voltage of GaAs.

The basic procedure is to dc bias the Schottky barrier to produce a given depletion depth and capacitance. That capacitance is measured using an ac capacitance bridge. The amplitude of the ac test signal is chosen to be much less that the dc bias, and so does not appreciably alter the depletion depth or the capacitance. The presence of electron traps in the GaAs material (Chapter 2) presents a difficulty. These may fill and deplete in response to the ac signal and alter results. This phenomenon makes the measurement results sensitive to the ac frequency and can be a useful technique in investigating traps [1]. Most commercial instruments used for CV profiling purposes operate at 1 MHz. This is usually too rapid for the traps to follow, and valid results for profiling purposes can be obtained. The determination of doping profile by the CV method is considered again in Chapter 18.

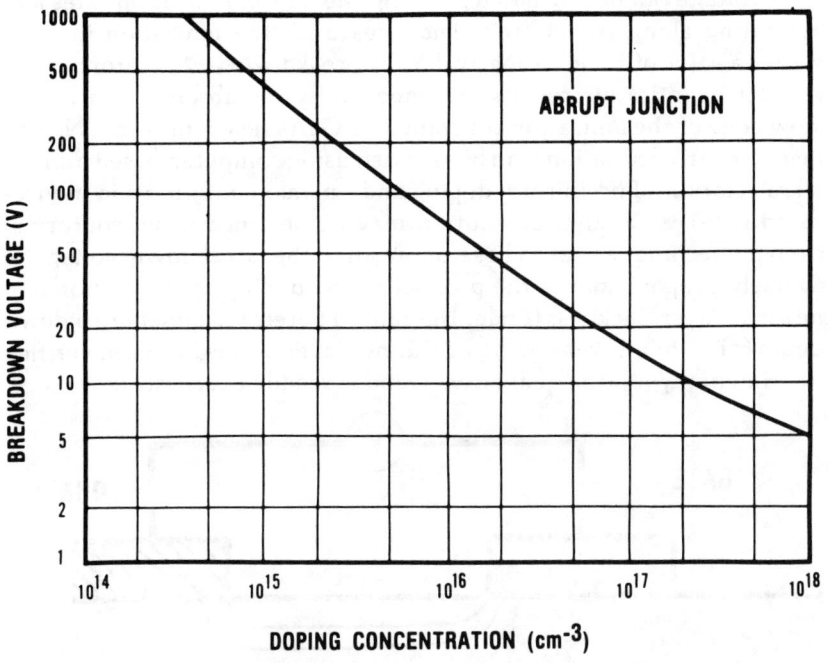

Figure 12.3 Bulk breakdown voltage of GaAs as a function of doping concentration (after reference [1]).

12.1.3 Reverse Breakdown Voltage

The reverse breakdown voltage is an important parameter of many Schottky barrier devices; it is one power limitation for GaAs FETs. The most important mechanism in junction breakdown is *avalanche multiplication* (or *impact ionization*). As the electric field in the semiconductor increases, the carriers gain sufficient energy to create electron-hole pairs by impact ionization. These carriers also gain energy in the field and create more electron-hole pairs. The process can avalanche, leading to breakdown and large current flow. Because breakdown is intimately related to the electric field, geometrical effects that influence the electric field configuration will also affect breakdown voltage. As illustrated in Figure 12.4, the electric field lines between the gate and the drain of an FET exhibit sharp curvature at some locations. In principle, the breakdown voltage can be calculated by using the impact ionization rates and integrating along the electric field lines until the ionization integral reaches a value of 1, indicating avalanche breakdown [1,2]. Unfortunately, exact quantitative results are impeded by the absence of accurate knowledge of the ionization constants for GaAs field ionization. Nevertheless, relative situations can be assessed using computer-based numerical calculations. For uniform doping and a planar configuration (nonrecessed gate), an analytical solution may be obtained using conformal mapping techniques [2]. These show that the breakdown voltage is inversely proportional to the product of the doping concentration and the active layer thickness (under the gate). Therefore, higher breakdown occurs for either lower doping or thinner layers. This has been verified experimentally [3]; the qualitative trend is readily observable.

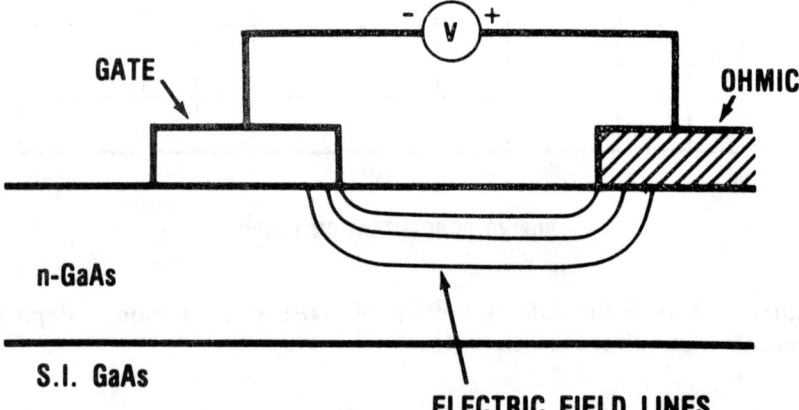

Figure 12.4 Electric field configuration in a planar FET results in breakdown voltages different from the bulk values.

Unfortunately, breakdown voltages on actual devices are also affected by other considerations, including surface effects. These make the breakdown voltage of actual devices sensitive to process details. Any phenomenon that affects the surface between the gate and the ohmic contact is a candidate for affecting breakdown also. It seems plausible that minor surface effects could modify the electric field configuration near the gate edge, thereby changing the breakdown voltage. These phenomena are not understood, and very little relevant material exists in the literature. It is well-known that deposition of dielectric films can affect several FET parameters including breakdown voltage (Chapter 13). Some complexities that relate to the measurement of breakdown voltage are considered in Chapter 18. Although there is almost nothing quantitative to relate here, the process engineer must be cognizant of the possibly detrimental effect of any procedure that affects the sensitive gate area.

12.2 MEASUREMENT OF SCHOTTKY BARRIER PARAMETERS

It is useful to have reasonably simple means to experimentally ascertain the barrier height and the ideality factor of Schottky barriers using either test structures or actual devices. The gate region is critical to FET operation and the parameters ϕ_b and n are often sensitive to process problems. For example, the author has noted a general correspondance between poor gate metal adhesion and high values of barrier height.

As was noted in the previous section, the voltage that appears in the Schottky barrier equations is a combination of the external gate voltage, V_g, the built-in voltage, V_{bi}, and the term kT/q. In fact, in some cases other terms must also appear [1]. An exact understanding of the detailed physics requires inclusion of all terms. But for device processing applications, an entirely practical and useful approach is to regard the built-in voltage as consisting of all voltage components except V_g, and not to worry excessively about the detailed complexities. This approach will be adopted below. Process uniformity and/or diagnosis of problems usually relate to changes in the measured parameters and not their precise value.

There are essentially four methods used to measure the barrier height of metal-semiconductor junctions: current-voltage, capacitance-voltage, activation energy, and photoelectric methods [1]. The first two methods are the simplest for general in-process characterization; all but the last one will be described below.

12.2.1 Current-Voltage Method

Equation 12.1 may be expressed as

$$\ln(J) = \ln(J_s) + \frac{qV}{nkT} \quad \text{for } V \gg kT/q \tag{12.11}$$

where

$$J_s = A^{**} T^2 \exp(-q\phi_b/kT)$$

Therefore, a plot of $\ln(J)$ as a function of V should result in a straight line having slope q/nkT and intercept $\ln(J_s)$ as shown in Figure 12.5. Note that for sufficiently small and sufficiently large currents the linear relationship does not hold, but is generally quite accurate over many orders of magnitude of current. Measurement of the slope allows calculation of the ideality factor, n:

$$\text{slope} = \frac{dV}{d(\ln J)} = \frac{q}{nkT}$$

Knowing the value of J_s (from the intercept point), the barrier height may be calculated as follows:

$$\phi_b = \frac{kT}{q} \ln\left(\frac{A^{**} T^2}{J_s}\right) \tag{12.12}$$

At first glance, this seems troublesome because the answer depends on knowledge of the constant A^{**} and the area of the Schottky contact (which is needed to calculate current density). For actual devices such as FETs, the gate length may be so small that gate area is difficult to measure accurately. However, because both these terms appear inside the logarithm, the barrier height ϕ_b is relatively insensitive to errors in those terms. In fact, at room temperature, a 100% error in A^{**} will cause an increase of only about 0.018 V in ϕ_b [1]. A suggested value of A^{**} is 120 A/cm^2/K^2 [1]. Hence, this technique is reasonably well-suited for use with discrete FETs despite their typically short gates.

The validity of equation (12.11) requires $V \gg kT/q$. At room temperature, this means $V \gg 0.026$ volts. This is approximately the location at which the plot in Figure 12.5 deviates from linearity.

12.2.2 Capacitance Voltage Method

This method of determining the barrier height applies if doping concentration is uniform. According to equation (12.10), a plot of $1/C^2$ as a function of V will yield a straight line (for uniform doping!), as shown in Figure 12.6. From the equation, it can be seen that such a line will have

$$\text{slope} = \frac{2}{q\epsilon N}$$

and

$$\text{voltage intercept} = V_{bi} \tag{12.13}$$

The disadvantage of the technique is that it applies only for uniform doping and does not provide the ideality factor. The advantage is that the area of Schottky diode does not have to be known (there are no current densities to calculate). Even so, the Schottky diode must be large enough to allow accurate capacitance measurements. This may eliminate discrete FETs as suitable test objects. If so, other test diodes placed on the slice for such purposes can be used instead.

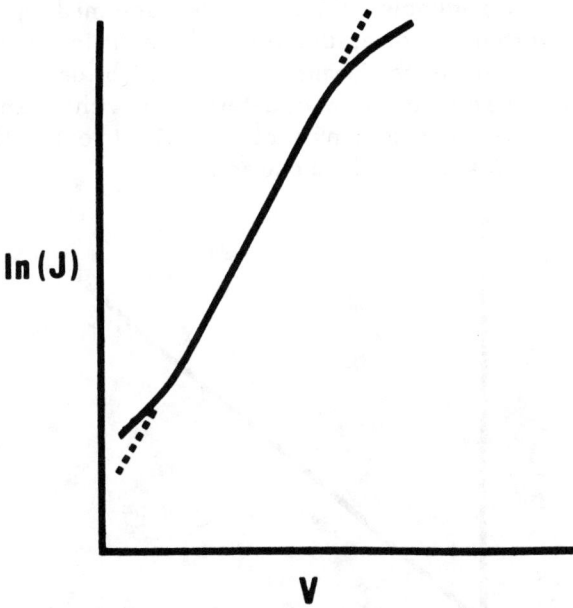

Figure 12.5 Current-voltage method to determine barrier height and ideality factor.

12.2.3 Activation Energy Method

This technique also requires no knowledge of the diode area, but does require measurements be made at several temperatures. This is usually not appropriate for in-process testing of slices, but certainly can be used if suitable temperature-controlled test apparatus is available. If A is the area of the diode, then the forward diode current I is $I = J_s/A$ and equation (12.11) can be written as

$$\ln(I/T^2) = \ln(AA^{**}) - \frac{q}{kT}\left(V_{bi} - \frac{V_f}{n}\right)$$

where the subscript f has been attached to the forward voltage to emphasize that it remains fixed in this measurement (the validity of equation (12.11) requires that $V_f \gg kT/q$). A plot of the quantity I/T as a function of $1/T$ (Figure 12.7) then gives a straight line having slope

$$\text{slope} = \frac{q}{kT}\left(V_{bi} - \frac{V_f}{n}\right) \tag{12.14}$$

(The quantity $q(V_{bi}-V_f/n)$ is the activation energy.) Unfortunately, this is one equation in two unknowns (V_{bi} and n). If n is assumed approximately equal to one (often a poor assumption), the built-in voltage can be calculated. Alternatively, the ideality factor, n, can be determined using the current voltage method (n can be determined) without knowing the diode area) and that value used in equation (12.14) to obtain the built-in voltage, all without knowing the diode area.

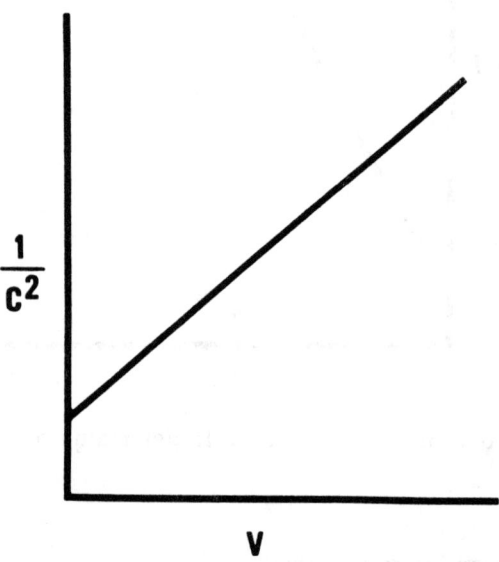

Figure 12.6 Capacitance-voltage method to determine barrier height.

12.3 GATE FABRICATION

This section describes gate fabrication techniques for GaAs FETs. The size and placement of the gate stripe is critical to FET performance in both power and low noise FETs (Chapter 3). Uniformity of device performance also requires reproducible size and placement of the gate.

Several basic methods of fabricating gates are addressed in the following subsections. One reminder: for historical reasons, the short dimension of the gate is called the length; the other, much longer dimension is called the width. Even though this convention can be confusing, it is universal and will be continued here.

12.3.1 Choice of Metalization

Although almost any metal placed on GaAs will yield a Schottky barrier (section 11.2), it must also exhibit two other characteristics: good adhesion and thermal stability. These two requirements eliminate many metals. Gallium is a rapid diffuser in many metals, especially above room temperature. Such diffusion encourages similar diffusion of the metal into the GaAs. In either case, the diode properties of the Schottky junction are degraded or destroyed. Gold is eliminated for both possible reasons: it exhibits poor adhesion to GaAs and is highly susceptable to diffusion. Metals meeting both criteria include aluminum, chromium, titanium, and molybdenum. Having selected a metal for the Schottky barrier, there remains the question of electrical conductivity. Digital FETs tend to have gate widths of 20 μm or less and can employ reasonably resistive gates. However, analog FETs have much wider gate fingers (75 to 300 μm) and gate resistance is an important performance factor.

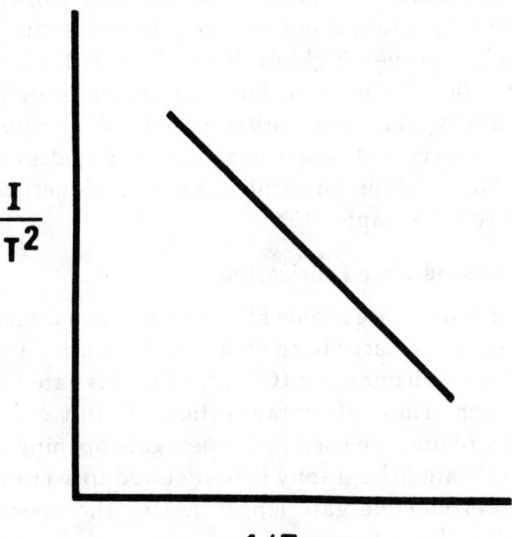

Figure 12.7 Activation energy method to determine barrier height.

Of acceptable Schottky metals, only aluminum has sufficient electrical conductivity to be used as the entire gate for analog devices. Other metals (chromium, titanium) are more resistive and require an overlay metal to enhance conductivity. This is universally gold because of its high electrical conductivity. But gold can diffuse through these resistive metals and initiate diffusion problems. Hence, barrier metals are used between the Schottky metal and the gold. This barrier metal is usually platinum, but molybdenum and paladium have also been used.

The general question of suitable metalizations for GaAs is considered in more detail in Chapter 13. Suffice it to say that most analog FET gates are either aluminum or TiPtAu (the first metal listed is nearest the substrate). Other compositions such as TiPdAu, TiMoAu, CrPdAu, and MoAl have also been used. (Some digital FETs use gate materials such as TiW because of special fabrication processes that require the gate to endure very high temperature. These are considered in section 12.3.3.). It might seem simplest to use aluminum rather than having to evaporate three successive metal layers, but aluminum can have problems interfacing with gold (Chapter 13). Also, composite (layered) metalizations are commonly used in GaAs FET processing (Chapter 13), and therefore, such use in gate fabrication is not uniquely troublesome. Further, the TiPtAu system has proven highly reliable. The Ti thickness is on the order of 1000 to 2000 Å, the Pt thickness on the order of 500 to 1000 Å, and the remainder is gold (on the order of 5000 Å). Platinum is a refractory metal and is very hot when evaporated. Rapid evaporation can damage resist. More details concerning metals and methods of application are considered in Chapter 13.

12.3.2 Basic Recessed Gate Fabrication

Most power and low noise analog FETs use a recessed gate geometry in which the gate stripe is placed in an etched slot to locate it slightly below the surface of the semiconductor (Chapter 3). This gate formation process is usually performed after fabrication of source-drain contacts. Lithographic techniques are used to define a gate opening in the source-drain spacing. E-beam lithography is best suited to accurate placement and uniform, reproducible gate length. After the opening has been defined in resist, a slot is fabricated in the exposed slice to "recess" the gate. This is almost always done by wet etching (some exceptions are considered below). The basic configuration is illustrated in Figure 12.8, using a resist profile typical of e-beam exposure. Note that in such cases it is the opening in the top of the resist that determines the size of the

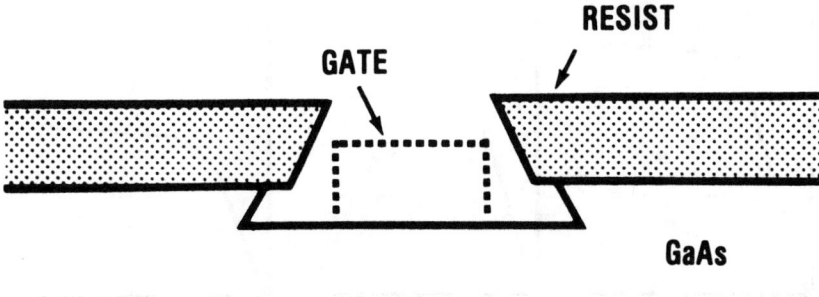

Figure 12.8 Basic recessed gate techniques.

gate; it is the opening at the bottom of the resist that (along with the amount of etch undercutting) determines the size of the etched slot. When using wet etching, the undercutting effect will always make the slot larger than the gate. Upon metal evaporation, the gate will be self-aligned to the center of the etched slot (Figure 12.8). This self-alignment is important. If the gate metal were to touch the side walls of the slot, an extra depletion zone would be formed at the side wall. Its associated capacitance would be a parasitic capacitance degrading the performance of the FET. The self-alignment feature prevents this, even if the slot is only 500 Å wider than the gate.

The depth to which the gate is recessed is a critical parameter in FET performance (Chapter 3). The method used to control the etch depth is to monitor the source-to-drain current during the etching process. The saturated current is reduced as the slot is etched into the slice (Figure 12.9). The slot is etched until the target recess current is reached. This requires alternate steps of etching and current measurement — it is not possible to monitor current while wet etching is proceeding. Etch depths are typically 800 Å to 2500 Å. The strength of the etchant should be adjusted (by dilution) to provide an etch rate that is sufficiently slow to allow good control over the recess process, being able to carefully approach the target current value without overshooting it. The wet etchant will leave a thin oxide on the GaAs (Chapters 4,5). This oxide can be removed using either an acid or a base or a commercial composition intended to remove oxides (Chapter 4).

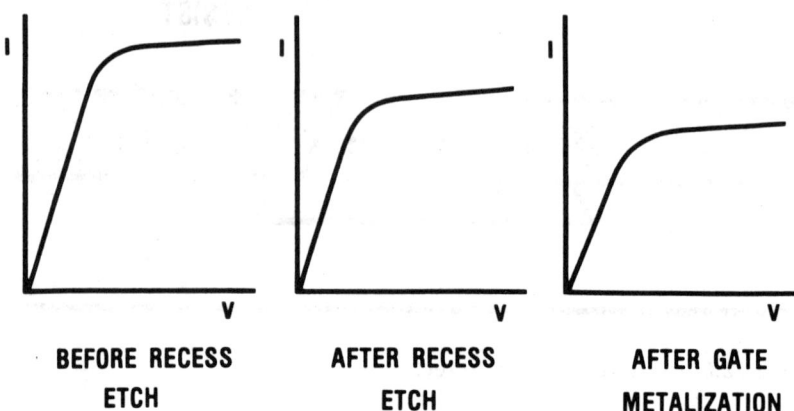

Figure 12.9 Saturation current before gate recess, after gate recess (but before metalization), and after gate metalization.

The current monitoring is best done on a special test FET provided on the slice for that purpose. Otherwise, sliding probes may damage an actual device. The gate exposure step should open appropriate areas in the resist to allow the probes to contact the source and drain of the test structure. (Under special circumstances, however, it is not difficult to force probes through resist to contact pads.) Usually, FETs are fabricated with the goal of achieving a given I_{dss} value after the gate is present on the slice, and herein lies a difficulty. Placing the metalization on the GaAs creates a zero-bias depletion zone in the GaAs and results in a saturated source-drain current less than that obtained before metalization (Figure 12.9). Therefore, the recess etching process must use a target current greater than the current desired on the completed device. If the doping profile is known, the difference in current can be approximately calculated using

$$\Delta I = qNv_s Ww$$

where q is the electron charge, N the doping level, v_s the saturated electron velocity (2×10^7 cm/s), W the gate width (the long direction!) of the device, and w the depth of the zero-bias depletion zone (obtainable from equation (12.5) with $v_g = 0$). But this is only an approximation. Uncertainties in the value of these parameters, plus geometrical effects in the FET, make this current difference difficult to calculate to sufficient accuracy. In practice, the current difference should be determined experimentally for the specific doping profile used. Because this current difference is crucial to correct device fabrication, it is useful to maintain a permanent record of pre-and post-metalization currents on all slices (Chapter 17).

The gate region of the completed device appears as shown in Figure 12.10. The geometrical configuration of the gate and the recess slot may occupy a distance of less than one micron. But this zone is the most important part of the FET in determining performance. The basic rationale for gate recess was reviewed in Chapter 3. Briefly, gate recess has the following effects. It places the gate below the surface depletion of surrounding material (Figure 12.10), thereby preventing surface depletion from restricting current flow under forward gate bias (channel fully open). It increases breakdown by reducing channel thickness (section 12.1). These two effects make recess necessary for power FETs. Gate recess reduces source-gate resistance by placing thicker material between gate and source. However, gate recess also results in increased gate-to-drain (feedback) capacitance, which tends to make devices unstable and is generally undesirable (oscillator applications would be an exception). As discussed in Chapter 3, the reasons for this effect are still debatable but probably involve the relationship between the recessed slot and the depletion zone. In any event, planar (unrecessed) devices exhibit the greatest isolation. The distance between the gate edge and the slot edge (d in Figure 12.10) may be important for FET performance.

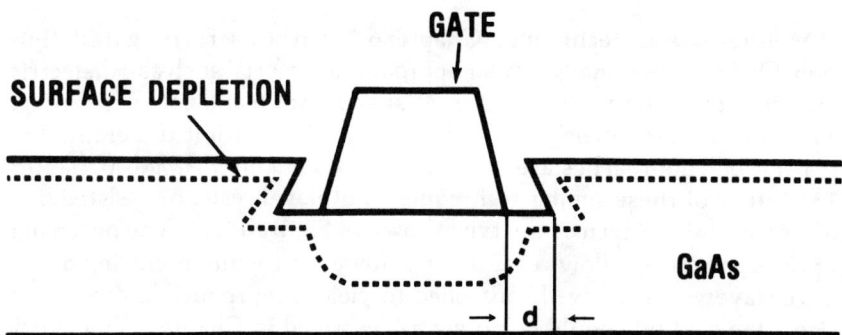

Figure 12.10 Details of the gate and recessed slot, showing the distance between the gate edge and the slot edge.

Again, evidence is not clear and different experts have differing opinions. But it is important to realize that this possibly important spacing is determined by two factors: the undercutting that occurs when wet etching GaAs and the size of the resist opening immediately adjacent to the GaAs, determined by the reverse slope or undercutting of the resist profile. For e-beam resist, the amount of reverse slope is a function of exposure (Chapter 7). As discussed in Chapter 5 (Wet Etching), the choice of etchant and crystal orientation affects the amount of undercutting and the shape of the slot walls. Several possibilities are illustrated in Figure 12.11. The first two examples use the same etchant, but the gates are oriented 90° apart on a (100) slice. The third example uses an isotropic etchant such as one based on HCl. Again, there are differing opinions as to the importance of the slot shape on FET parameters such as breakdown voltage. But the process engineer should be aware of the configuration produced by his chosen process.

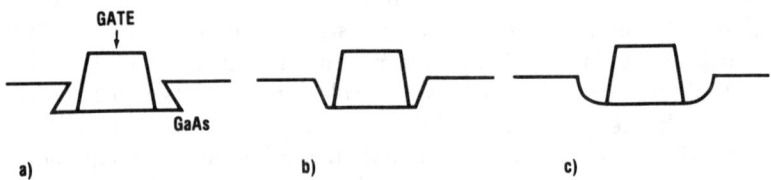

Figure 12.11 Three possible shapes of the etched slot walls as a function of the choice of wet etchant and crystal orientation.

Multilevel resist techniques (Chapter 6,7) can be useful for gate definition. Or, in a functionally similar approach, a material such as a dielectric may be grown on the slice before resist application and gate patterning. Then the exposed dielectric can be etched open until it undercuts the resist. Such approaches are referred to as *assisted liftoff methods* (Chapter 13). Either of these similar techniques (multilevel resist or assisted liftoff) result in a pattern of the type shown in Figure 12.12. The overhang of the upper layer allows easy liftoff. However, the undercutting of the lower layer must be well controlled to yield a reproducible recess slot size. One caution should be noted: as described in Chapter 9, most dry

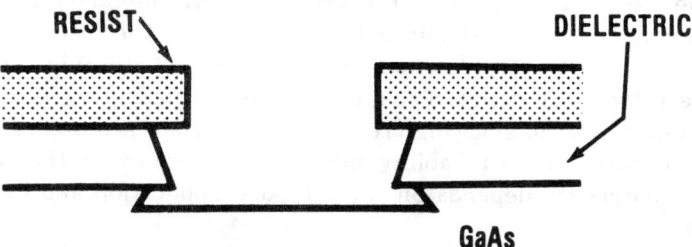

Figure 12.12 Multi-level, or assisted liftoff, technique in gate fabrication. The size of the recessed slot is affected by the extent of undercut of the lower, patterning level.

etching techniques, including plasma etching, can damage the upper layers of the semiconductor. If these techniques are used to open the lower level of a multilevel pattern technique (perhaps by plasma etching silicon nitride), such damage may result. If it is not too deep, it can be removed during the subsequent wet etching used to form the slot. The depth of the damage is highly dependent on the exact processing parameters. To make this issue more complicated, a certain amount of damage may even be useful in special applications. It creates an insulating layer between the gate and semiconductor that results in increased breakdown voltage. However, this damaged region will not conduct current and so will limit rf power under strong drive conditions (forward gate bias). There are clearly many tradeoffs in this issue. It is usually preferable to avoid damage.

Techniques other than wet etching have been used to create the recessed slot. These are mainly attempted for two reasons. First, dry etching processes can generally provide greater uniformity over a slice than wet etching. Second, some device engineers believe the distance between gate edge and slot edge (d in Figure 12.10) should be minimal for some applications such as power FETs. Dry etching avoids the significant undercutting typical of wet etching. Ion milling is one dry etching technique. But such processes tend to damage the material that remains. Consequently, a brief cleanup wet etch is used to remove the damaged material. As above, the depth of the damage and its effect depend on the specific process and application.

Some FET devices also use a dual-recess gate geometry in which the gate is placed in a shallow recess slot, which itself is in a larger recess slot. The trade-offs involved in this geometry were discussed in Chapter 3 (section 3.2.3; see also Figure 3.11). Fabrication of the dual recess geometry requires accurate placement of the second recess within the first one. These two recess slots usually must be defined using two separate lithography steps since high registration accuracy is required. Whether the performance or reliability improvements are worth the extra steps and complexity depends on the intended application and the process yield.

Figure 12.13 shows a scanning electron microscope photograph of a gate between the source and drain of a FET.

12.3.3 Digital FET Gates

GaAs FETs for use in digital logic circuits have gate dimensions different from those used in analog FETs. Gate width is usually 10 to 20 μm. Gate length is usually not less than 1 μm, thereby allowing definition by optical lithography. Optical steppers (Chapter 6) are a popular choice for this lithography step. The feasibility of using e-beam lithography is reduced by the writing time required for the thousands of gates typically on a digital IC. Although some digital logic configurations use FETs that can be fabricated by conventional methods (previous subsection), most digital FETs have special requirements that have led to gate fabrication processes significantly different than those described above. Such digital FETs are usually enhancement mode FETs having highly doped, thin (1000 Å) channels (Chapter 3). The thin channels, combined with surface depletion effects, result in high resistance between the gate and either the source or drain. It is therefore highly desirable to place N^+ material in these regions to enhance conductivity. It obviously is desirable to extend the N^+ material as close to the gate as possible. But if it extends too close to the gate it will cause very low breakdown voltages. It is therefore useful to have a method to self-align this N^+ layer to the gate. Several self-alignment approaches have been developed to achieve that goal. Some require that gate fabrication occur before ohmic contact fabrication. Substantial interest in GaAs digital devices is rather new. There has not been time for substantial evolution of fabrication techniques and the following approaches all have difficulties. On the other hand, all have been used to fabricate digital circuits with hundreds to thousands of gates.

One approach was shown in Chapter 1 as part of a sample process flow. The essential elements are shown in Figure 12.14. After the usual N-layer is present (formed either by epitaxy or ion implantation), a metal is

Schottky Barriers and Gate Formation 279

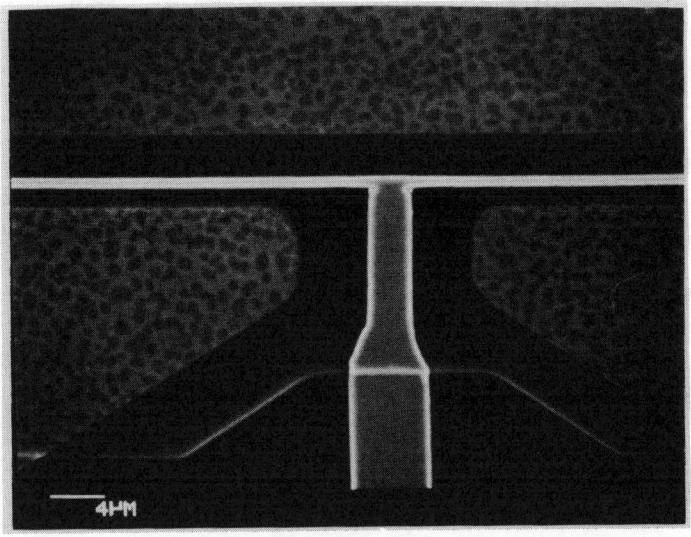

Figure 12.13 A scanning electron photomicrograph of a FET gate placed between source and drain ohmic contacts. Note the typical appearance of alloyed AuGeNi ohmic contacts as they appear in a SEM.

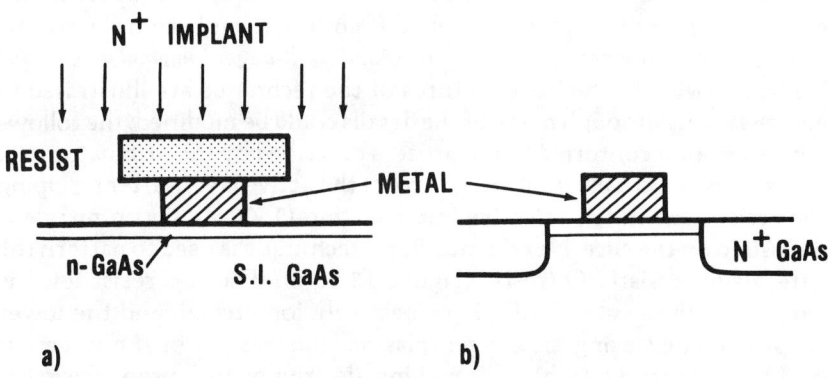

Figure 12.14 Self-alignment method used to form an N^+ region close to the gate. The gate remains on the slice during the subsequent activating anneal. Then ohmic contacts are fabricated.

evaporated or sputtered over the entire slice. Resist is applied and patterned to form a rather long (the short dimension!) gate pattern in resist. Then the exposed metal is etched, undercutting the resist pattern (Figure 12.14(a)). Next, an ion implant step forms the N^+ layer, the resist protecting the gate area. The resist is removed and the slice alloyed to activate all implants (Figure 12.14(b)). The metal must be capable of surviving the ~850°C anneal without damaging the Schottky barrier properties. TiW based alloys have proven suitable for this purpose. TiW alone (10:90 wt. %) can be deposited by sputtering and can be etched with a $CF_4 + O_2$ plasma [4]. Even more stable Schottky characteristics may be obtained using TiW silicides [5]. These compositions are not very conductive, but this is usually not a severe problem for digital FETs in which the gate width (the long direction) may be only 10 to 20 μm. It would be a severe problem for analog FETs having much wider gate fingers. This anneal step also precludes the presence of ohmic contact metal. Therefore, ohmic contacts are formed after gate fabrication.

There are many variations on this basic procedure. For example, resist alone may not suffice to block the ions and so a double layer of dielectric plus resist may be used. A major issue in the technique is reliably controlling the undercutting at the metal etching step, so that gate length and gate-to-N^+ distance are reproducible.

The above process has several restrictions, including the requirement that the gate metal and Schottky barrier be stable at GaAs anneal temperatures. It is also not possible to recess the gate (although this usually is not done for digital FETs anyway). A second approach of growing popularity eliminates these difficulties. It has become known by the (rather cumbersome) acronym SAINT, *self-aligned implantation for N^+ layer technology* [6]. The basic features of the technique are illustrated in Figure 12.15. Although many of the details could be modified, the following description conforms to the process described by the originators [6]. The slice is selectively implanted to form the active layer. After stripping the resist used in the selective implant, thin (1500 Å) silicon nitride is deposited on the slice. Next, a multilevel technique is used to pattern the gate, using resist/SiO_2/resist (Figure 12.15(a). The top resist level is patterned, the exposed SiO_2 is respectively ion etched, and the lower resist is etched using an oxygen plasma; this results in the undercut profile of Figure 12.15(b). This etching also removes the top resist that remained. Then the N^+ layer can be implanted (through the thin silicon nitride), followed by sputtering SiO_2 onto the slice (Figure 12.15(c)). The sputtered SiO_2 film extends to the edge of the side wall to some extent. Etching in buffered HF removes the material on the side wall, and liftoff of the resist results in the geometry shown in Figure 12.15(d). The slice can now be annealed with the SiO$_2$ and Si$_3$N$_4$ remaining in place, activating the N^+ implant. Afterwards, ohmic contacts are fabricated in the conventional manner (although using the SiO$_2$ and Si$_3$N$_4$ for assisted liftoff). Then the gate pattern is aligned to the SiO$_2$ gate opening, the silicon nitride etched open, followed by metalization and liftoff. This results in the completed configuration shown in Figure 12.15(e).

The SAINT process is complicated and careful attention to detail is required to ensure success. Also, it should be clear that numerous variations are possible on the specific steps listed above. But the process does allow any choice of gate metalization, because the gate metal is placed on the slice after anneal. It also allows the gate to be recessed if desired, because ohmic contacts are fabricated before metalization.

In conclusion, it must be remembered that digital FET fabrication techniques are relatively new. It is not yet clear which ones will dominate as GaAs digital fabrication matures.

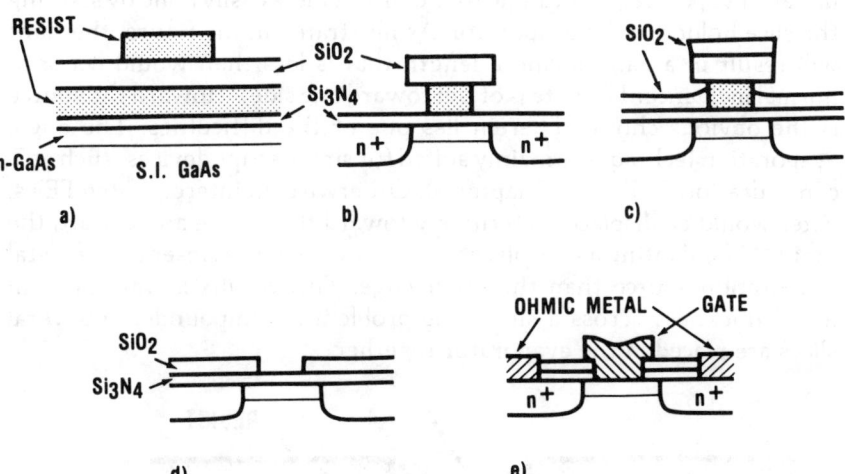

Figure 12.15 Major steps in the SAINT process used to fabricate digital FET gates. This process allows any choice of gate metal and is performed after ohmic contact formation.

12.3.4 Other Gate Processes

The previous two subsections described the basic approach to analog FET gate fabrication, and the present techniques being used for digital FET gate fabrication. But there have been innumerable other approaches to gate fabrication over the years. Many have seemed ingenious, but few have yielded the reliability and reproducibility necessary for production techniques. To repeat a major point of this book: there is a big difference between a process that yields successful results once or twice, and one that can produce the same results in a controllable, reliable, reproducible manner slice after slice. It therefore seems pointless to present a review of all these attempts. But it is instructive to mention a few that have proven less transitory. They should give the reader a flavor of the others.

One of the more reliable of these ancillary techniques is angle evaporation, illustrated in Figure 12.16. A multilevel resist (or assisted liftoff) technique is used and the lower level material (resist, dielectric, etc.) is undercut more than usual. After the recess slot is etched (if desired), the metal is evaporated at an angle to the slice. This is easily done by slanting the slice holder in the evaporator. As illustrated in the figure, the slant will result in a gate having a length that is less than would occur at normal incidence. The gate is offset toward one side — toward the source is the obvious choice. Therein lies one of the difficulties. The angle evaporation technique is mainly suited for single stripe devices (such as T configuration FETs, see Chapter 3). Otherwise, in interdigitated FETs, gates would be displaced alternately toward the source and toward the drain. Also, slanting a slice places one edge of the slice closer to the metal evaporation source than the other edge. This results in variations in metal thickness across a slice. The problem is compounded if several slices are placed in the evaporator together.

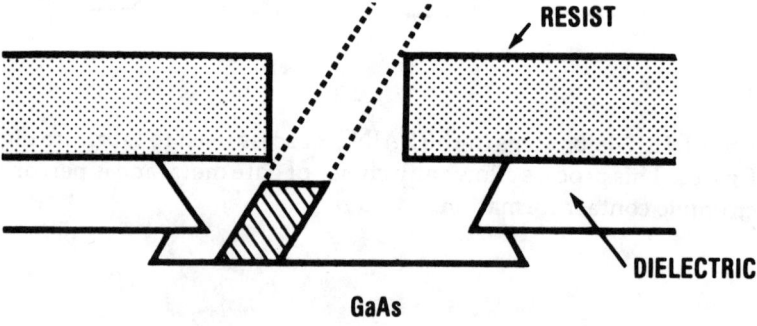

Figure 12.16 Gate produced using angle evaporation.

Another technique that appears from time to time is fabrication of a gate with a large top, as illustrated in Figure 12.17. This approach addresses the problem that smaller and smaller gate fingers (for greater gain at higher frequencies, or for smaller noise figures, see Chapter 3) result in reduced conductivity. These methods attempt to enlarge the gate cross section at the top of the gate, leaving the desired, small length in contact with the slice. The gate in Figure 12.17(a) was produced using two lithographic exposures. One exposed the small slot and was used to etch open the dielectric. A second patterning step was used to define a larger opening aligned to the first. Evaporation and liftoff then result in the pattern shown. There are questions of parasitic capacitance in this technique. The lower patterning level should have a low dielectric constant. It could even be resist and subsequently be removed from the slice. The mushroom shaped gate in Figure 12.17(b) was formed in a method similar to that just described, except it was plated. Still another possibility is to use a bimetal gate and selectively etch the lower metal to undercut the upper metal, as shown in Figure 12.17(c). The difficulty with this process is that any etch slot must be at least as wide as the top portion of the gate (Figure 12.17(c)). The extent of undercutting is also difficult to monitor.

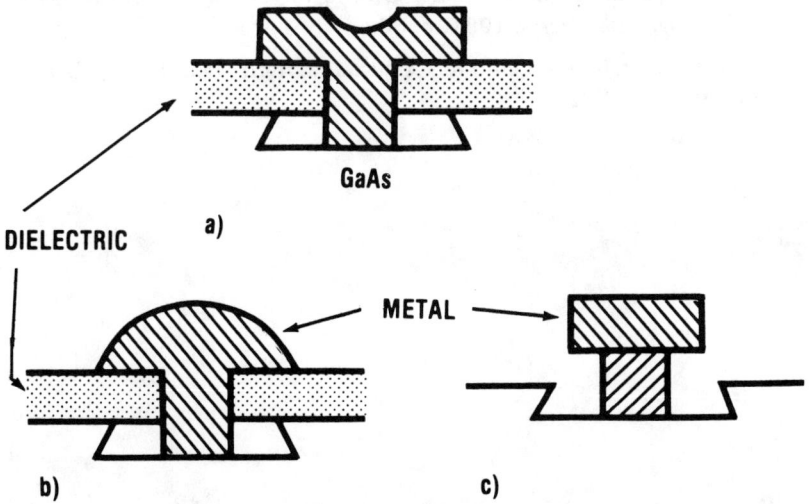

Figure 12.17 Three methods of increasing gate cross section to increase conductivity: (a) Using two separate resist exposures — one to define the opening in a dielectric, the other to define the larger top area; (b) similar process to (a), but using plating; (c) undercut etching the lower of two (or more) metals used to form the gate.

12.3.5 Concluding Comments

The above discussions have described the basic procedures used in gate fabrication. Exact, specific recommendations have been avoided (resist thickness, gate metal thickness, etc.) because these are highly dependent on the specific application and the specific processes. The gate fabrication process utilizes aspects of lithography, etching, and metalization, each one of which has many variations. As in every other aspect of device processing, a certain amount of process development is required which is aimed toward the specific application.

REFERENCES

[1] S.M. Sze, *Physics of Semiconductor Devices*, 2nd edition. New York: Wiley, 1981.

[2] W.R. Frensley, *IEEE Trans. Electron Devices*, 28, 1981, p. 197.

[3] S.H. Wemple, W.C. Niehaus, N.M. Cox, J.V. Dilorenzo, and W.O. Schlosser, *IEEE Trans. Electron Devices*, 27, 1980, p. 1013.

[4] M. Abe, T. Mimura, N. Yokoyama, and H. Ishikawa, *IEEE Trans. Electron. Devices*, 29, 1982, p. 1088.

[5] N. Yokoyama, T. Ohnishi, K. Odani, H. Onodera, and M. Abe, *IEDM Dig. Tech. Papers*, 1981, p. 80.

[6] K. Yamasaki, K. Asai, and K. Kurumada, *IEEE Trans. Electron Devices*, 29, 1982, p. 1772.

CHAPTER 13

FIRST-LEVEL METAL, DIELECTRIC FORMATION, SECOND-LEVEL METAL

13.1 INTRODUCTION

The major components of active devices are in place following the steps described in the three previous chapters: isolation, ohmic contact formation, and gate formation. Nevertheless, a significant amount of processing remains. Overlay metals may be added to increase conductivity or form other features such as inductors and bottom capacitor plates. Dielectric films may be applied for encapsulation and/or capacitor formation. A second major metalization may then be applied to form other connections on the device. These topics are addressed in this chapter. This introductory section discusses general issues common to all metalizations on GaAs, including required characteristics and methods of application (evaporation or sputtering). Subsequent sections discuss first-level metal, dielectric formation, and second-level metal.

13.1.1 Required Metalization Characteristics

Metalization on GaAs serves a number of specific purposes. Ohmic contacts and Schottky barriers are two of these — they were discussed in previous chapters. General metalization must satisfy one or more requirements including high electrical conductivity, good adhesion, effectiveness as a diffusion barrier, or as an etch stop (for reactive ion etching, for example). Usually, no single metal meets all the requirements of a given metalization step. For example, gold is a good electrical conductor but it exhibits poor adhesion to GaAs. Such multiple, conflicting requirements often lead to multilevel metalizations in GaAs processing. This book follows the more common convention of listing metals in the order they are applied (from the substrate upward). That is, TiPtAu metalization has the titanium on the slice and the gold uppermost. There are a few exceptions to this convention in the literature and the reader should beware of these.

In principle, high electrical conductivity could be met by four metals: gold, aluminum, silver, and copper. Copper, however, a rapid diffuser, tends to getter onto crystalline defects, and exhibits high chemical reactivity. These characteristics have caused copper to be regarded as inimical to GaAs material and devices. Although this reputation may not be entirely justified in all circumstances, it has resulted in the fact that copper or copper alloys are almost never used on GaAs devices. Silver tends to be highly reactive and has been used mainly for plated heat sinks on some devices. Gold and aluminum, both of which are used on GaAs devices, have more desirable properties. Aluminum would seem the more attractive because it combines high electrical conductivity with good adhesion (gold does not exhibit good adhesion). However, aluminum has one serious disadvantage — it interacts with gold at high current densities and forms undesirable intermetallic compounds. These can begin as Au_2Al compounds that are whitish and have been called *white plaque*. Further interactions can result in an intermetallic composition that is purple in color and has been called *purple plaque*. Although these deleterious interactions become significant only at very high electrical current densities, GaAs analog devices tend to be very small and carry substantial current. This makes these devices highly susceptible to such aluminum-gold interactions. Because of high current densities and other reasons (see below), gold is the dominant metal used in the GaAs electrical environment; and gold wires (usually 0.001 inch in diameter) or gold ribbon are used to connect chips to the outside circuit. Therefore, if aluminum is used on a GaAs device, there must be a gold-to-aluminum transition at some location on the device. This is usually accomplished by placing a barrier metal (see below) such as chromium between the aluminum and gold.

Other disadvantages of aluminum are lack of chemical inertness (it oxidizes or reacts easily) and an electrical conductivity that, while high, is not as high as gold. Skin depth (see below) and other considerations often place a premium on the very highest conductivity possible. On the positive side, aluminum is the only metal that simultaneously exhibits high electrical conductivity and good adhesion to GaAs. It also forms stable Schottky barriers (gold does not — see Chapter 12). All of these considerations have resulted in aluminum being used on GaAs mainly for Schottky barriers (such as gates on FETs), but for little else.

Gold is the dominant metalization on GaAs circuits because of its high electrical conductivity and relative chemical inertness (more information on the advantageous properties of gold may be found in Chapter 15). Skin depth considerations (discussed below) and high current densities result in the fact that many analog GaAs devices operate at current densities very near the maximum possible. Gold's great electrical conductivity is crucial in this role. Gold does exhibit poor adhesion to GaAs, dielectrics, and some other materials. For this reason, gold is almost never applied alone. It usually follows another metal that is placed on the slice first for adhesion. Of course, the gold must be applied immediately afterward, without breaking the vacuum. Otherwise, oxidation or very minor contaminants on the metal will likely cause poor adhesion of the overlaying gold.

Good adhesion is exhibited by titanium or chromium to almost any surface. These metals are often used as the first component in multimetal layers. Thicknesses generally range from 200Å to 1000Å. Although it is likely that a few monolayers would suffice to assure initial adhesion, reliability considerations suggest greater thicknesses. TiAu and CrAu metalizations are commonly used in GaAs processing. Of course, contaminated surfaces can result in poor adhesion even when these metals are used. Practical and rapid (but destructive) tests for adhesion include scratching the metalization with a probe tip, or attempting to remove the metal pattern using adhesive tape. The latter test is especially severe.

The third major role that a metal may be called upon to fulfill is a barrier to diffusion. Under elevated temperatures and/or high current densities, diffusion can result in intermetallic compounds and/or alloys with gallium or arsenic. Simple diffusion is modeled by the differential equation:

$$\frac{dN}{dt} = D \frac{d^2 N}{dx^2}$$

where the concentration N is a function of position and time and D is the diffusion constant, which has dimensions of cm^2/s. D is highly dependent on temperature. If a diffusing species is placed on another material in sufficient amounts (so that diffusion does not decrease the surface concentration), the above equation leads to the expression

$$N(x, t) = N_o \operatorname{erfc} \frac{x}{2\sqrt{Dt}}$$

where erfc(y) is the complementary error function.

Unfortunately, diffusion involving more than two metals and/or crystalline structures can exhibit far more complex behavior. In the case of GaAs processing, the presence of multi-component metalizations on a binary, imperfect crystal create many opportunities for intermixing. These almost always lead to reliability problems. It is well-known that gold will interdiffuse with GaAs, reducing the electrical conductivity of the gold and destroying the Schottky barrier properties of the gold/GaAs interface. Binary metalizations such as TiAu or CrAu will delay these effects, but do not prevent the diffusion. Therefore, metals that act as good diffusion barriers are placed in appropriate places in multi-component metalizations. Platinum has been found to be a good barrier metal and is used between the titanium and gold for many FET gates (Chapter 12). Palladium has also been used for this purpose. Chromium is not a satisfactory barrier in general, but does seem useful for separating gold and aluminum as noted above. These barrier-preventing metals are typically 500 Å to 1500 Å thick. The TiPtAu metalization has become very popular in GaAs processing and can be used for many purposes including gates, overlay metal, inductors, etc. Titanium provides adhesion, platinum provides a barrier, and gold provides high conductivity.

Other major considerations for GaAs metalizations are skin depth and critical current density. By way of review, dc currents are uniformly distributed throughout the cross section of a conductor. However, ac mutual inductive effects cause ac current to be greatest at the outside surface of the conductor. Current density decreases exponentially from the surface inward. The distance from the surface that the current has fallen to 1/e of its surface value is defined as the skin depth, d. The skin depth of a metal is given by

$$d = \sqrt{\frac{2}{\mu\omega\sigma}}$$

where μ is the magnetic permeability, ω is the angular frequency, and σ is the conductivity. (This expression is often given in Gaussian units in electromagnetic textbooks, in which case it is quoted as $d = c/\sqrt{2\pi\mu\omega\sigma}$, where c is the speed of light.) In fact, it can be shown that the surface resistance of a conductor is simply given by

$$R_{sur} = \frac{1}{\sigma d}$$

At 10 GHz, gold has a skin depth of approximately 0.8 μm. Therefore a gate strip having a gold cross section of 0.5 x 0.5 μm will have the rf current distributed fairly uniformly throughout — the dimensions are less than a skin depth. Conversely, increasing the cross section of metallic conductors more than several skin depths will no longer decrease rf resistance any significant amount. Slightly more complicated situations arise when electric fields exist between the conductor and nearby metalization (as in transmission lines). Nevertheless, the basic considerations still apply and most of the current flows within one skin depth of the surfaces.

The issue of critical current density is important for many GaAs device applications. As pointed out above, GaAs devices can be subjected to substantial current densities. Of course, catastrophic failure will occur in metallic conductors at sufficiently high current densities. But long before this occurs, metal migration can begin. Metal migration is a serious reliability issue for all semiconductor devices. High current flow results in physical movement of portions of the metal (caused by electron scattering), resulting in depletions in one area and accumulations in an adjacent area [1,2]. Once the metal cross section becomes diminished, resistance increases and catastrophic failure can follow. Unfortunately, there is no well-defined "safe" current density. But some idea of safe operating conditions can be inferred from the specifications required of semiconductor devices for military applications. MIL-S-19500F lists the maximum allowable continuous current density (use RMS for pulsed applications) for metal conductors which are in thermal contact with a substrate along their entire length. These are quoted from that specification as follows:

Conductor material	Max current density (A/cm^2)
Aluminum without glassivation	2×10^5
Aluminum glassivated	5×10^5
Gold	6×10^5
All other	2×10^5

Of course, greater current densities can be tolerated if long-term reliability is not required.

13.1.2 Application of Metalization

With the exception of plating (Chapter 15), metal is applied to GaAs slices either by evaporation or sputtering. Evaporation results in metal impinging on the slice from essentially one direction (usually the vertical). Sputtering usually results in metal impinging on the slice from a wide range of angles. These distinctions mean that evaporation techniques are good for liftoff applications but bad for step coverage. The reverse is true of sputtering as shown in Figure 13.1. (Although, there are methods of moving slices within an evaporator during metal deposition to increase the angular range and allow improved step coverage.)

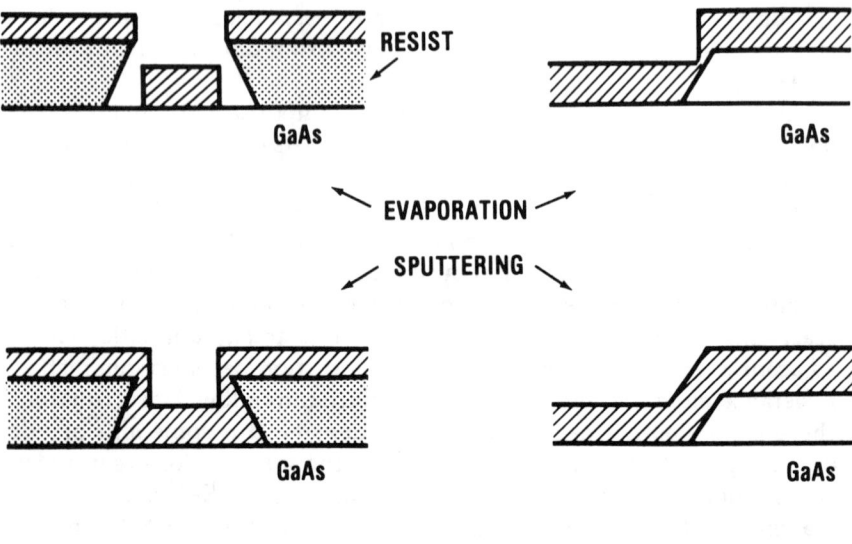

Figure 13.1 Step coverage typically resulting from evaporation and sputtering. Sputtered metal impinges on the slice from a large range of angles, resulting in good step coverage but poor liftoff. Conversely, evaporated metal impinges on the slice from a narrow range of angles, resulting in poor step coverage but good liftoff.

There are other important distinctions. In the evaporation process, the metal is heated to a temperature sufficient to cause substantial vaporization (usually the melting temperature) so it arrives at the slice at high temperatures. Some metals, such as platinum, must be heated to very high temperatures to produce any appreciable atomic flux. Resist and GaAs are not good thermal conductors and the deposition rate must be chosen with care to assure that the resist is not heated to the extent that it flows or becomes unremovable. If the metal application heats the resist to the flowing point, the resulting pattern can be cracked and/or wrinkled (Figure 13.2). Sputtering can result in cooler metalization — neutral species physically blast target atoms off of a target and these atoms can be relatively cool. The temperature of the metal as it is applied to the slice also affects the strain present after cooling. Another distinction between evaporation and sputtering is the vacuum requirement. Evaporators must operate at high vacuum, usually in the 10^{-7} Torr range, while sputtering (being a plasma process) occurs at much lower pressures. Both evaporation and sputtering easily allow multilayer metalizations. Sputtering machines can have different sputtering targets that may be rotated into position without breaking vacuum. Evaporators also may have multiple metal sources (see below). Many of the above considerations would tend to favor sputtering, except that the liftoff process is usually desired and it requires evaporation.

Evaporation is classified in two classes: thermal or electron-beam. Thermal evaporators use high current to heat small coils, boats, or baskets that contain the desired metal (the current is passed directly through the coil, boat, or basket). Heating the coil, boat, or basket melts the metal and vaporization occurs. Electron beam evaporators use an electron beam to bombard a target and locally heat it to melting and vaporization. Electron-beam evaporators may have multiple metal targets that can be rotated into position without breaking vacuum. Thermal evaporators may have multiple sources placed side-by-side and current can be switched from one to another, thereby evaporating a sequence of metals. One possible disadvantage of the latter technique is that metal from different sources impinges upon the slice from slightly different angles. This is usually not a severe problem, but can be troublesome under some circumstances. In all thermal evaporation techniques, a shutter is placed over the source to prevent metal from reaching the slice(s) during the heating process. This prevents "spitting" of metal or other nonuniformities that can occur during heatup.

Figure 13.2 Typical appearance that occurs when metal is applied too rapidly or at too hot a temperature over the resist. Temperature rise causes resist flow that results in wrinkling or cracking.

TABLE 13.1
Bulk Metal Characteristics
(Most data from [3])

Metal	Melting Point (°C)	Density g/cm	Resistivity 10^{-6} Ω-cm	T(°C) at Vapor Pressure			Evaporation	
				10^{-8}	10^{-6}	10^{-4}	E Beam	Thermal
Al	660	2.7	2.7	677	812	1010	E	Coil, Boat, Basket
Au	1062	19.3	2.2	807	947	1132	E	Coil, Boat, Basket
Ag	961	10.5	1.6	574	617	684	E	Coil, Boat, Basket
Be	1278	1.9	3.2	710	878	1000	E	Coil, Boat, Basket
Cr	1890	7.2	12.9	837	977	1157	G	Coil, Boat, Basket
Co	1495	8.9	5.8	850	990	1200	E	Boat, Basket
Ge	937	5.3	—	812	957	1167	E	Boat
In	157	7.3	8.6	487	597	742	E	Boat, Basket
Mo	2610	10.2	5.3	1592	1822	2117	E	—

First-Level Metal, Dielectric Formation, Second-Level Metal

Ni	1453	8.9	7.0	927	987	1262	E	Coil, Boat, Basket
Pd	1550	12.0	10.5	842	992	1192	E	Coil, Boat, Basket
Pt	1769	21.5	10.4	1292	1492	1747	E	Coil, Boat
Si	1410	2.2	—	992	1147	1337	E	Boat
Sn	232	7.3	10.9	682	807	997	E	Coil, Boat, Basket
Ta	2996	16.6	13.2	1960	2240	2590	E	Coil
Ti	1675	4.5	42.7	1067	1235	1453	E	Boat
W	3410	19.3	5.3	2117	2407	2757	G	Coil
Zn	419	7.1	6.0	127	177	250	E	Coil, Boat, Basket

Note 1: Thin film properties will differ from bulk properties.

Note 2: Various sources give different values of some properties.

E = Excellent

G = Good

Metallic alloys can easily be applied by sputtering. Because the process sputters atoms from the surface of the target, the stoichiometry of the alloy tends to be preserved. Alloys are more difficult to apply by evaporation. The differing vapor pressures of metals result in differential atomic fluxes. That is, one component of the alloy (the one with the highest vapor pressure) will evolve faster than other components. Alloys can be thermally evaporated only if preserving average stoichiometry will suffice. In this case, the alloy "charge" in the boat, basket, or coil can be evaporated to dryness. Hence, all atoms are evaporated and average stoichiometry is preserved, but not on a microscopic scale. Evaporation of alloys to form ohmic contacts (Chapter 11) is the only major example of alloy evaporation in GaAs processing.

Table 13.1 lists metals commonly used in GaAs processing, their properties, and evaporation techniques [3]. The table includes vapor pressure as a function of temperature (useful to gauge relative evaporation rates) and bulk properties. Two general cautions apply to using this, or any similar table. First, there are significant differences among existing references as to vapor pressure of metals as a function of temperature. Quotations of temperature at a given pressure can disagree more than 100°C in some cases. Secondly, thin-film properties of metals virtually always differ from the bulk properties. Electrical conductivity is always less. The amount of reduction depends on deposition parameters. Test patterns should be included on slices to monitor metal resistivity.

Virtually all materials can be deposited by sputtering. The sputtering rate will depend on many parameters including pressure, power, and type of material. Relative sputtering rates may be inferred from the representative data in Table 13.2 [4].

Chapter 6 described the liftoff process and it was noted that an undercut resist profile is highly desirable. Similar results may be obtained using "assisted liftoff" techniques during some metalization steps. These are especially appropriate if a dielectric is already present on the slice from a previous step (for example, a dielectric film used for implant anneal). In such cases, the resist pattern may be spun and patterned over the dielectric. Then the exposed dielectric is etched open, undercutting the resist edge (Figure 13.3), resulting in a desirable configuration for achieving liftoff.

13.2 First-Level Metal

First-level metalization is a term derived from silicon processing. Ohmic metalization and gate metalization may or may not be included in the term. In general, *first metal* is generically regarded as metal that rests directly on the slice, under whatever dielectric is subsequently deposited.

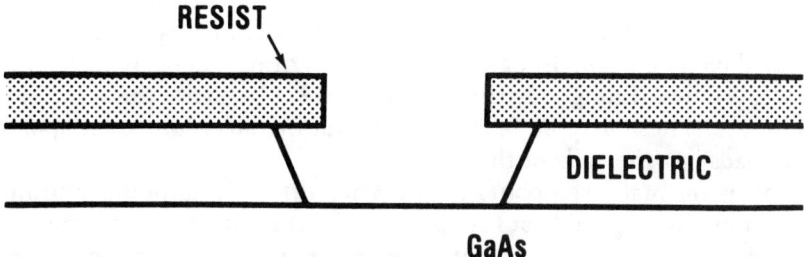

Figure 13.3 Configuration typical of assisted liftoff techniques. Photoresist is used to pattern a second, lower layer (usually a dielectric) that is etched to form an undercut profile suitable for liftoff processes.

In the strictest technical sense, it would also include the alloyed, ohmic metalization and the gate metalization. But common usage sometimes considers these in a class of their own. *First metal* than refers to other, overlay metalization applied afterward for other purposes. These purposes include increased conductivity, formation of lower capacitor plates, or formation of inductors or transmission lines. The exact definition of first metalization is an evolving, semantic issue and will not be pursued here. This section addresses the metalization that occurs after ohmic contacts and gates (if any) have been formed, but before any overlying dielectric is deposited.

One major purpose of first metalization is to overlay ohmic and/or gate metal to increase conductivity. Even if the ohmic metalization included a thick overcoat of gold (AuGeNiAu), the alloy process will result in significantly decreased sheet resistance of this composite film (compared to what would be obtained if the top layer of gold were not subjected to the alloy process). Further, even with the overcoat of gold, the total thickness of the ohmic metalization will often be limited to less than one skin depth to aid subsequent resist coverage at the source-drain spacing (for gate definition). A high slice-to-metal step would impede good resist coverage (Chapter 6). Gate metal thickness will usually be less than one skin depth also (high aspect ratio gates are difficult to fabricate). It is therefore desirable to overcoat the ohmic metalization and any gate pad areas with gold to increase electrical conductivity. This overlay metal extends to within a few microns of the edge of the ohmic or gate metal pattern. Such an indentation allows for reasonable alignment tolerances.

Another major purpose of metalization is to form new patterns on the GaAs surface. These may be lower capacitor plates (which could also be formed over ohmic metalization) or transmission lines. If gates were defined using e-beam lithography, e-beam writing time may have been conserved by defining only the gates themselves, and not the associated pad areas. These may be fabricated at first metalization, overlapping "minipads" on the ends of the gate fingers.

Gold is the major component of first metalization, although another metal must be applied first for adhesion. This means that first metal is often a CrAu or TiAu combination. The gold thickness depends on the application and the lithography technique, but typically is 0.5 to 2.0 μm. Additionally, barrier metals may be used resulting in CrPtAu or TiPtAu metalizations. If the gates are not formed using e-beam lithography, it may be feasible to fabricate both the gates and the first-metal patterns (overlay metal, inductors, etc.) in a common step. The feasibility of this approach obviously depends on the specific case in question.

TABLE 13.2

Sputter Etching Rates of Various Materials
(After reference [4])

Material	Rate (Å/min)	Power Density (W/cm^2)
Aluminum	120-160	1.6
Gold	200-900	1.6
Nichrome	100	1.6
Copper	200-350	1.6
Tungsten	70	1.6
Platinum	900	2.0
Nickel	500	2.0
Titanium	50	2.0
SiO$_2$	120	1.6
Si$_3$N$_4$	60	1.6
Al$_2$O$_3$	20-50	1.6

First metalization is almost always formed using liftoff processes. Patterning the composite metalization by etching techniques (wet or dry) would be difficult. Two or more metals must be etched, including such difficult ones as gold or platinum, all without etching the underlying material (GaAs, ohmics, etc.) or the masking material. Processes such as ion milling are usually eliminated by the rapid ion milling rate of GaAs (Chapter 9) — it is not possible to ion mill the metal and stop before removing GaAs material.

13.3 DIELECTRIC FORMATION

Dielectric films are used in GaAs processing for environmental encapsulation, for capacitor dielectrics, and for cross-over insulators. Dielectrics are formed on the slice by plasma assisted deposition (Chapter 8), sputtering, or anodization of metals. Chemical vapor deposition of dielectric films is not possible because the high temperatures ($> 500°C$) are incompatible with GaAs material (decomposition) and many metalizations used on GaAs slices (ohmic contacts). Dielectric formation usually follows the formation of ohmics, gates, and first metalization.

The first purpose of dielectric formation is simply protective encapsulation. This protects the surface of the slice from environmental contamination and mechanical damage. The small size of most GaAs device features (such FET gates) make them prone to mechanical damage during handling or probing. The dielectric also seals the surface, keeping chemicals, gases, and particles from sensitive areas of the device. GaAs FETs are especially sensitive to surface effects in the gate area. The IV characteristics of an unprotected FET can be affected merely by blowing on it. Long term degradation can occur through other processes such as oxidation or shorting by contaminating particles. Dielectrics are also used as an insulating spacer between two levels of metalization, allowing one to cross over another (the bridge process also accomplishes this goal — Chapter 15). The encapsulating dielectric is usually also used for MIM capacitors if these are present, although other dielectrics may be employed to fabricate capacitors (Chapter 14).

With the exception of polyimide (discussed below), the encapsulating dielectric is usually silicon dioxide or silicon nitride. These two materials are easily applied to the slice either by sputtering or by plasma enhanced chemical vapor deposition (PECVD). Chapter 8 describes the PECVD films in some detail; their properties are reviewed in that chapter and in Chapter 14 (in the discussion of capacitor dielectrics). That information will not be repeated here. But it is worth emphasizing the fact that the dielectric constant of these films is a function of deposition method and parameters, and can vary over a wide range. Silicon dioxide films have a lower dielectric constant than silicon nitride films, making them preferable for a cross-over dielectric (lower cross-over capacitance). Conversely, the higher dielectric constant of the silicon nitride makes it preferable for use in capacitor formation. Further, silicon nitride is less permeable to ions and therefore makes the superior encapsulant.

The thickness of these dielectric films represents a trade-off between conflicting requirements. If the dielectric is too thin, the number of pinholes will be large and mechanical strength less. If the dielectric is too thick, parasitic capacitances will increase, as will the capacitance per unit area of MIM capacitors (Chapter 14). The parasitic capacitance problem is especially evident in FETs. The feedback capacitance between gate and drain is undesirable (see Chapter 3). Placing a dielectric film on the slice increases that capacitance; the thicker the film, the greater the capacitance. These considerations usually result in dielectric film thicknesses between 1000 Å and 4000 Å. The thickness and index of refraction of these films may be measured using an ellipsometer. It is generally desirable to measure film thickness accurately after every process run (using a test slice included for this purpose) to aid in process control and monitoring. It is possible to judge approximate dielectric thickness by the color of the dielectric film when viewed perpendicular to the slice; interference effects give rise to differing color as a function of thickness. Table 13.3 gives these colors as a function of thickness for silicon nitride and silicon dioxide [5]. Of course, the slice must be viewed using white light rather than the yellow illumination typical of most clean rooms. These tables should be mostly of historical interest — good process control requires accurate measurements such as are possible using ellipsometry.

An important issue is the effect of dielectric formation on device characteristics. Application of dielectrics on GaAs FETs can affect saturation current and breakdown voltage, among other parameters. Although these issues have been rarely mentioned in the literature, they are well known to anyone fabricating devices. The minimal representation in the literature reflects the almost complete lack of any quantitative understanding of these issues. Some highly qualitative observations may be made. The use of sputtering or plasma CVD to apply dielectrics results in ion bombardment of the exposed surface of the slice. This ion bombardment can render the top portions of the material insulating and reduce the FET current. The magnitude of the effect seems to depend on many parameters [6]. Under conditions where the effect is large, the applicable energies do not seem to explain the extent and depth of the damage. As one extreme example, the author has exposed GaAs slices to a low frequency nitrogen plasma in a small plasma reactor and used low energy electron scattering (in conjunction with etching) to profile the resulting crystal damage. The non-crystallinity extended almost a thousand Angstroms into the slice. Such damage may also involve high energy electrons. Also, damage depths in GaAs are known to be up to a factor of

TABLE 13.3

Color of SiO_2 and Si_3N_4 Films as a Function of Thickness*
(in Angstroms, after [5])

Order	Color	Silicon Dioxide Thickness Range	Silicon Nitride Thickness Range
	Silicon	0-270Å	0-200Å
	Brown	270-530	200-400
	Golden brown	530-730	400-550
	Red	730-970	550-730
	Deep blue	970-1000	730-770
1st	Blue	1000-1200	770-930
	Pale blue	1200-1300	930-1000
	Very pale blue	1300-1500	1000-1100
	Silicon	1500-1600	1100-1200
	Light yellow	1600-1700	1200-1300
	Yellow	1700-2000	1300-1500
	Orange red	2000-2400	1500-1800
1st	Red	2400-2500	1800-1900
	Dark red	2500-2800	1900-2100
2nd	Blue	2800-3100	2100-2300
	Blue-green	3100-3300	2300-2500
	Light green	3300-3700	2500-2800
	Orange yellow	3700-4000	2800-3000
2nd	Red	4000-4400	3000-3300

*Assuming Refractive Index of: SiO_2 = 1.48
Si_3N_4 = 1.97

two greater than the corresponding ion implant range. This damage has a greater affect on devices such as thin-layer enchancement mode digital FETs than on recessed gate depletion mode FETs. Techniques to deposit dielectrics at low temperatures without plasmas, such as photo-CVD [7], may find use in processing sensitive devices.

The effect on breakdown voltage is even less understood. Breakdown tends to be dominated by the sharp curvature of electric field lines that occur at the edge of the gate (on the drain side for gate-drain breakdowns; see Chapter 12). It is very plausible that any surface condition (presence of oxides, free arsenic, etc.) could affect these fields and also affect breakdown. Although there is little definitive information to impart about this subject, the process engineer should be aware that changes in device characteristics are possible following dielectric formation.

Polyimides are also useful dielectrics for encapsulation or metal cross-overs. These organic materials are applied to the slice by spinning and baking. If suitably cured, they become a highly stable, permanent feature that is almost impossible to remove. But they also preserve some pliability and hence resist cracking if subjected to stress. When cured to a lesser extent, they may be patterned or removed by ashing or use of appropriate solvents. These materials have a low dielectric constant (~3.5) that makes them highly suitable for low-capacitance metal cross-overs. Dielectric strength is on the order of 10^6 V/cm. Polyimides are similar to resists in that there are many types, each having different properties, and users tend to keep exact processes proprietary. However, the manufacturer's data sheets are excellent guides to their use. Curing procedures usually consist of several baking steps at different temperatures. Preliminary bakes are on the order of 150°C. Permanent bakes may be near 300°C. After a preliminary bake, either positive or negative resists may be applied and patterned on top of the polyimide. The exposed material may be etched by an oxygen plasma (ashing) or solvents such as tetramethyl ammonium hydroxide in either aqueous or alcohol solution. The etching characteristics depend critically on the baking history of the material and the manufacturer's data sheet should be consulted.

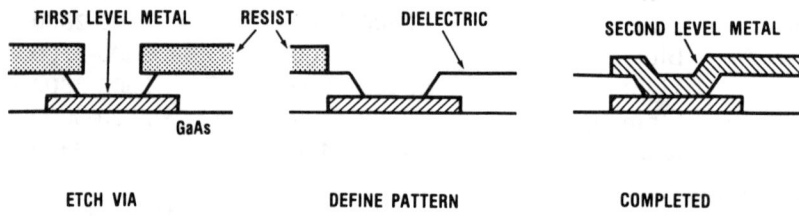

Figure 13.4 Major steps to form a via connection between first- and second-level metalization.

13.4 SECOND-LEVEL METAL

After the dielectric is formed on the slice, another layer of metal may be needed to form interconnections. One method of accomplishing this is the use of plating techniques; these are described in Chapter 15. Other approaches are considered in this section.

A basic approach typically used in digital device fabrication is to first etch via holes in the dielectric (using etching procedures that yield obtuse slope) and then define the second-level metalization. This requires two lithography steps. In the first step, resist is spun and exposed to define the via hole pattern. The vias are etched (either by wet or dry etching processes) and the resist removed. In the second step, photolithographic techniques are used to define the second metalization pattern. These steps are shown in Figure 13.4. Second-level metal connects with first level metal at the via holes. It is crucial that the via hole formation process yield appropriately slanted edges; otherwise, good electrical connections may not be obtained.

As usual, liftoff techniques can be used to define the metal pattern. But, unlike first-level metal, dry etching techniques such as ion milling may be used to define second-level metal. Etching slightly into the dielectric will generally cause no harm. Second-level metal is usually thick gold (0.8 to 2 μm) with the usual adhesion layer (Ti or Cr) applied first.

REFERENCES

[1] R.E. Hummel and H.B. Huntington, editors, *Electro-and Thermo-Transport in Metals and Alloys*. New York: TMS-AIME, 1977.

[2] F.M. D'Heurle, "Electromigration and Failure in Electronics: an Introduction," *Proc. IEEE*, 59, 1971, p. 1409.

[3] *Sloan Technology Corporation Handbook of Thin Film Materials*. Sloan Materials Division, El Segundo, CA.

[4] C.M. Melliar-Smith and C.J. Mogab, in *Thin Film Processes*, J.L. Vossen and W. Kern, editors. New York: Academic Press, 1978, p. 497.

[5] S.M. Sze, *Physics of Semiconductor Devices*, Second Edition. New York: Wiley, 1981.

[6] A.K. Gupta, D.P. Siu, K.T. Ip, and W.C. Petersen, *IEEE Trans. Electron Devices*, 30, 1983, p. 1850.

[7] J.W. Peters, F.L. Gebart, and T.C. Hall, *Solid State Technol.*, Sept. 1980, p. 121.

CHAPTER 14

CAPACITORS, INDUCTORS,
AND
RESISTORS

14.1 INTRODUCTION

Monolithic microwave integrated circuits (MMICs) are complete electrical circuits fabricated on a single chip. They can be multistage amplifiers as indicated by the circuit diagram and MMIC contained in Figure 14.1. In addition to the active devices (FETs and diodes), these circuits will generally include capacitors, inductors, and/or resistors. The ability to fabricate MMICs (and realize the advantages of size, ruggedness, reproducibility, and cost) demands that methods exist to fabricate the required passive components. Economic considerations require that the MMIC chip size be minimized as much as possible to maximize the number of chips on a slice. The passive components, especially inductive elements, tend to utilize most of the chip area. The area occupied by active devices is relatively small. Careful attention to layout can result in compact realizations of the circuit, as indicated in Figure 14.2. This MMIC is designed to have the same performance as the chip in Figure 14.1, but is approximately 60% as large.

These passive elements may be either distributed or lumped. Distributed elements are physically large enough that transmission line characteristics play a significant role in their function. An example of a distributed element is a transmission line that acts as an inductor, or a transmission line stub which functions as a capacitor. Distributed elements have inductive, capacitive, and resistive aspects, all of which are taken into account by the transmission line treatment. Lumped elements are physically small enough that transmission line effects do not play a significant role in their function. This generally requires that they be less than 0.1 wavelength in size. Nevertheless, even lumped elements are not purely inductive or resistive or capacitive, but have aspects of all three. Failure to take account of this complexity leads to significant error in circuit design. These issues will become clearer as specific cases are examined in the following sections.

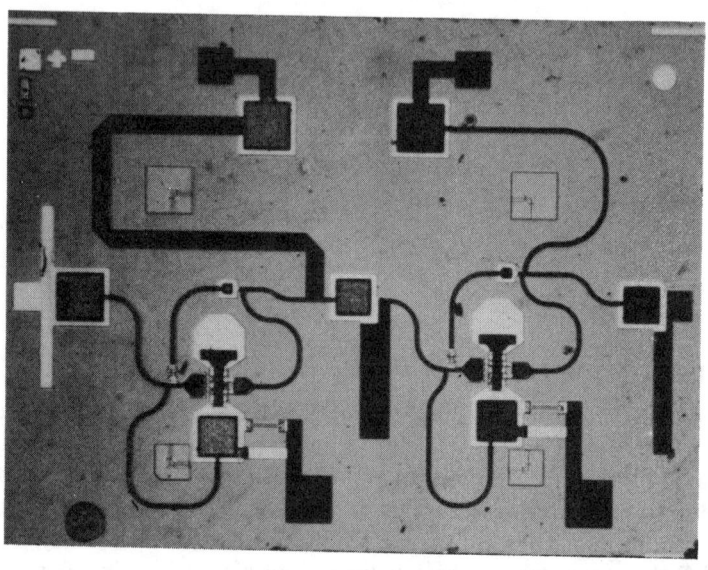

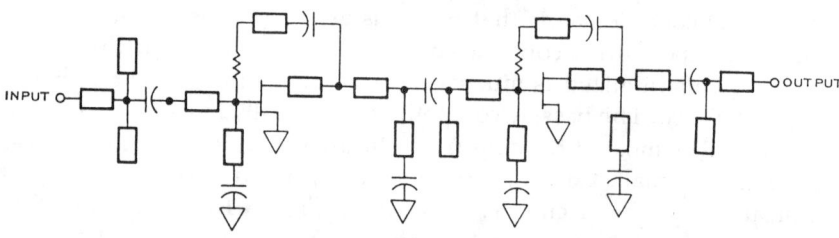

Figure 14.1 Photograph of an MMIC circuit and the equivalent circuit.

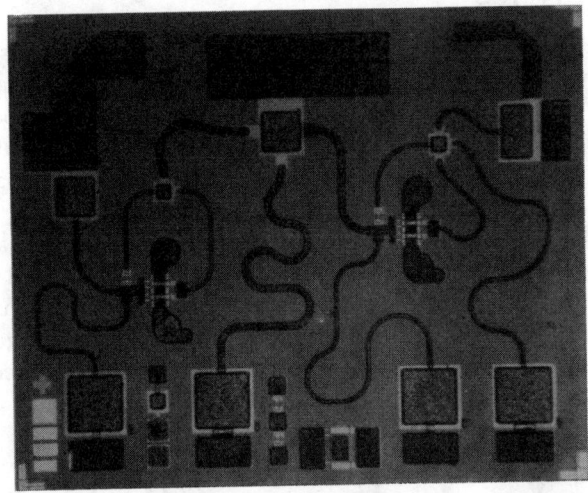

Figure 14.2 MMIC design which performs the same function as the chip in Figure 14.1, but in less area.

The complex, multifaceted nature of passive elements on monolithic circuits would make circuit design exceedingly difficult, were it not for the existence of *computer-aided design* (CAD) programs. Modern CAD programs, such as SUPERCOMPACT (a product of Comsat General Integrated Systems, Inc.), can accurately model the various passive elements and are used by electrical engineers to create and optimize MMIC designs. Such modeling indicates that a tolerance of ± 10% on passive component values (capacitance, inductance, resistance) is adequate for most present MMICs. This applies both across a slice and from slice to slice. With care, this is a reasonable goal. Stricter tolerances may be desired in the future and seem achievable.

All passive components must be fabricated on insulating material. This usually is the semi-insulating GaAs substrate itself (although Schottky diodes and GaAs resistors will utilize conductive GaAs). The definition of these insulating areas occurs during the isolation process step (Chapter 10).

Detailed examination of the electrical engineering aspects of passive elements and their relationship to circuit design is a formidable task and is not attempted here. But a basic understanding of the fundamentals is useful in defining the parameters important in fabrication. The function and fabrication of each passive element (capacitor, inductor, resistor) is described in the following sections.

14.2 CAPACITORS

Capacitors fabricated as part of an MMIC are referred to as *monolithic capacitors*. They are used in MMICs for blocking, bypass, or tuning purposes. These functions are illustrated in Figure 14.3. Both blocking and bypass capacitors are used to pass rf (microwave) signals, but block dc voltages. The term *blocking* is used when the emphasis is on separating dc biases. For example, blocking capacitors are used between amplifier stages to isolate the drain bias of the first stage FET from the gate bias of the second stage FET (Figure 14.3(a)). The term bypass is used when the emphasis is on "bypassing" the rf current to ground. This function is illustrated in Figure 14.3(b) in which the second gate of a dual gate FET is dc biased but rf grounded using a bypass capacitor. High accuracy in

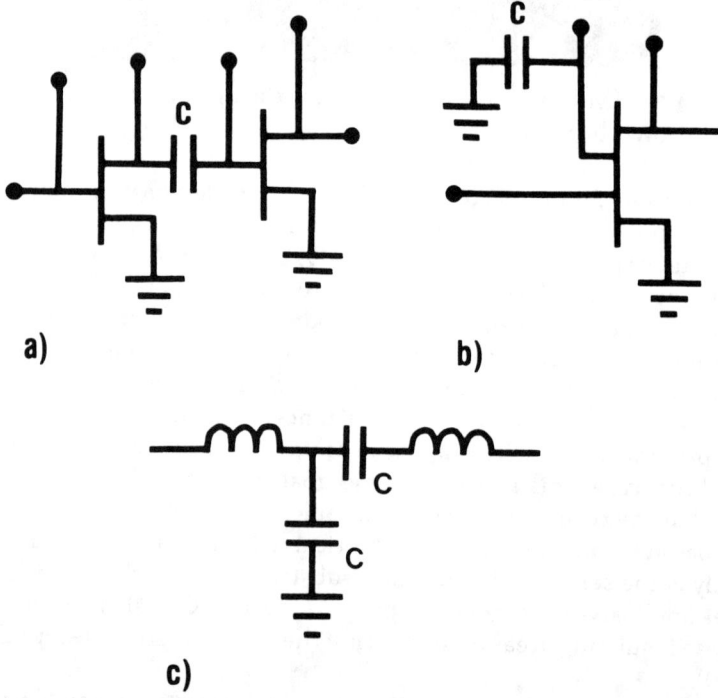

Figure 14.3 Examples of the uses of capacitors in MMIC circuits: (a) blocking: to pass the rf signal between amplifier stages but block dc biases; (b) bypass: to allow the second gate of a dual gate FET to be dc biased but rf grounded; (c) tuning: part of an interstage matching network (tuned curcuit).

achieving a specified capacitance value is not required for these purposes. The capacitance must simply be sufficiently large to provide low impedance at the frequency in question,

$$Z = \frac{-1}{j\omega C}$$

where Z is the impedance, C the capacitance, and ω the angular frequency. Note that at 10 GHz an 80 pF capacitor has a capacitive impedance of approximately 0.2 ohms.

The third application is tuning. In this case the capacitance value plays a role in a tuned circuit; the design value must be accurately obtained. Examples include interstage impedance matching networks (Figure 14.3(c)) or feedback networks. Monolithic capacitors typically have values between 0.2 pF and 100 pF.

14.2.1 Overview of Capacitor Types

Capacitance may be included in the MMIC with four basic configurations. Three are illustrated in Figure 14.4. These are a stub in a transmission line (Figure 14.4(a)); coupled lines, usually formed as an interdigitated capacitor (Figure 14.4(b)); and a metal-insulator-metal (MIM) configuration (Figure 14.4(c)). The fourth possibility is a Schottky diode. Each of these are described below. The MIM structure is the most common configuration on modern MMICs.

Schottky diodes on GaAs have a capacitance that is a function of diode area, GaAs doping profile, and applied voltage (Chapter 12). The voltage dependence makes these structures unsuitable for use in most analog MMIC applications. *Voltage controlled oscillators* (VCOs) are an exception. If the rf voltage across the diode is small compared to the dc voltage, the capacitance is determined by the dc voltage (Chapter 12). Placed in appropriate circuit locations, such a voltage controlled capacitance (a *VARACTOR*) can determine the frequency of the oscillator. Schottky diodes are also used in digital logic circuits, but they function principally as voltage level shifters in this application.

A stub in a microstrip transmission line (section 14.3) can act as a capacitor or an inductor. An open (no termination) stub less than a quarter wavelength in length is capacitive. A shorted stub less than a quarter wavelength is inductive. Such stubs are distributed elements and are designed and accurately modeled using CAD programs. Such techniques can require substantial chip area at lower frequencies (less than 8 GHz). They are fabricated as part of the transmission line and will not be considered further here.

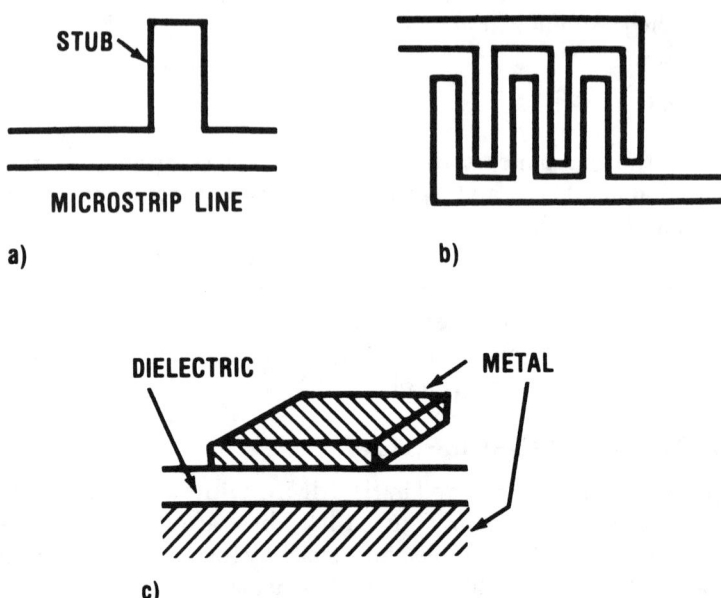

Figure 14.4 Three implementations of capacitors on MMICs: (a) a stub in a microstrip transmission line; (b) an interdigitated capacitor; (c) a metal-insulator-metal (MIM) capacitor.

14.2.2 Interdigitated Capacitors

The capacitance of interdigitated capacitors (Figure 14.4(b)) arises from capacitive coupling between adjacent conductors, aided by the substrate dielectric. They are useful principally for capacitances of 1 pF or less. Larger values would require significant chip area and involve distributed effects. These structures have been described in several major papers [1-3]. They are not a pure, series capacitance, but have associated inductance, resistance, and shunt capacitance. Figure 14.5 shows an equivalent circuit used for modeling interdigitated structures [3]. The capacitance, C, is the interelectrode capacitance. But the other elements must be considered for accurate modeling [3]. The capacitance C can be calculated using

$$C = (N_f - 1)C_g L \tag{14.1}$$

where N_f is the number of fingers, L is the finger length, and C_g is the static gap capacitance per unit length. C_g has been calculated and is given in Figure 14.6 for substrate thickness of 101 μm and 203 μm [3]. Note that the data presented is based on a dielectric constant of 12.5 for GaAs rather than the presently accepted 12.9. The unloaded Q of these capacitors is in the range of 35 to 50 at 12-14 GHz [3,4]. The losses associated with the current crowding near the conductor edges prevent attainment of significantly greater Q values. The Qs quoted above are for a conductor thickness of approximately two skin-depths. Reducing metal thickness to a skin-depth or less significantly reduces Q. The major fabrication consideration for interdigitated capacitors is accurate achievement of the design size and spacing of the conductors.

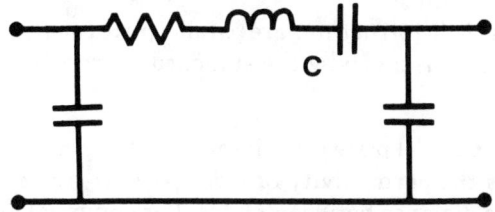

Figure 14.5 Equivalent circuit of an interdigitated capacitor. The shunt capacitance is to the backside ground plane.

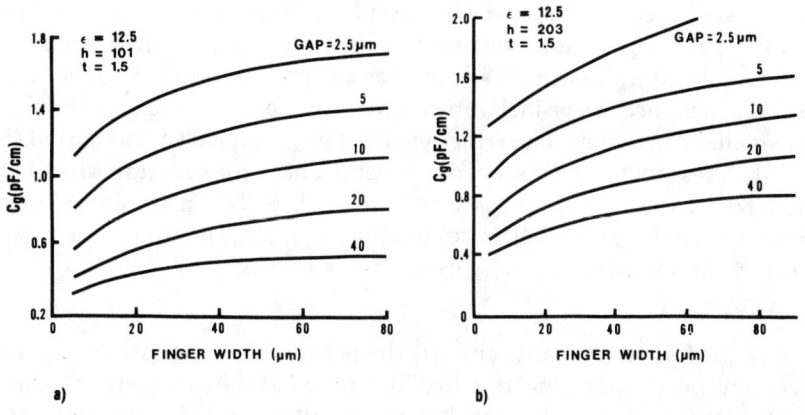

Figure 14.6 Single finger gate capacitance (per unit length) of an interdigitated capacitor: (a) 101 μm substrate thickness; (b) 203 μm substrate thickness. (Note the dielectric constant of 12.5 instead of the presently accepted 12.9. After reference [3].)

14.2.3 Metal-Insulator-Metal Capacitors

The most popular type of capacitor for MMICs has become the MIM configuration shown in Figure 14.4(c). The thin film capacitor is composed of two metal plates separated by a dielectric material. The dielectric is typically 0.1 to 0.4 μm thick. The metal plates should be at least two skin-depths thick to provide adequate Q. These plates are mostly gold (skin-depth at 10 GHz $\approx$ 0.8 μm) and so a few microns thickness is desired. The top plate may be plated to increase thickness. The lower plate must have a smooth surface and so cannot be plated; it may be only one to two skin-depths thick. Roughness in the lower plate would cause greater defects (pinholes) in the dielectric grown upon it, would decrease Q, and would increase the lower plate area (and hence the capacitance) by an unknown amount.

Because these MIM structures are parallel plate capacitors, their capacitance to first order is given by the standard expression:

$$C_o = \epsilon A/d$$

where C_o is the parallel plate capacitance, ϵ is the permittivity (dielectric constant times the permittivity of free space, $k\epsilon_o$), A is the area of one plate, and d the distance between plates. For commonly used dielectrics and thicknesses, this capacitance is typically in the range 50 pF/mm^2 to 300 pF/mm^2. However, these series capacitors also have other components (resistive, inductive, and shunt capacitive) as illustrated in the equivalent circuit in Figure 14.7. In fact, all but the very smallest ones have distributed effects, and it is useful (often necessary) to model the lower capacitor plate as a transmission line. There are also fringing fields at the plate edges that increase the basic capacitance given above. This extra capacitance can be insignificant for large capacitors, but substantial for smaller capacitors. The fringing effect has been well documented for metal plates (such as contact pads) on semiconductor substrates having a metalized back side. The capacitance of such pads can be significantly greater than the parallel plate component, C_o [5,6]. The total capacitance has been found to be given by expressions such as

$$C = K_1 C_o + K_2 P$$

where the Ks are constants and P is the perimeter of the capacitor plate. K_2 was found proportional to $1/\log(T)$, where T is the substrate thickness [5,6]. MIM capacitors have substantially thinner dielectrics than the substrate thickness that applies in the above case, and so the effect is not as great. Nevertheless, physically small MIM capacitors exhibit signficant fringing effects which are a function of the perimeter. As always, the process engineer is well advised to conduct experiments to determine the magnitude of the effect for specific process parameters (dielectric type, thickness, etc.). Otherwise, the capacitor fabricated on the chip may not meet the design value. In this regard, test capacitors that can be easily probed should be included on the slice (perhaps in plug bars, see Chapter 17 for in-process verifications.

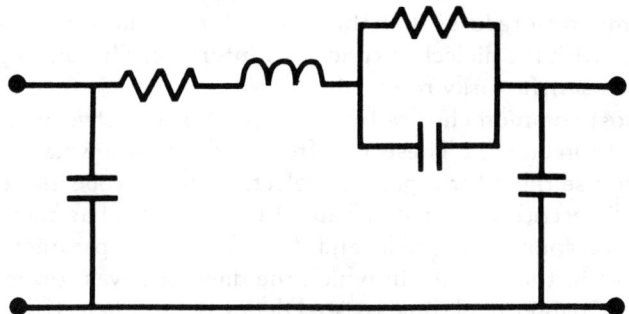

Figure 14.7 Equivalent circuit of a metal-insulator-metal (MIM) capacitor.

The Q of MIM capacitors is important for microwave applications. The Q depends on the size of the capacitor. At 10 GHz, the Q of a large MIM capacitor ($>$ 10 pF) can be 10 or even less, while the Q of a small MIM capacitor ($<$ 1 pF) can exceed 100. The losses are either dielectric losses or conductive losses. The Q arising from dielectric losses is given by

$$Q_d = 1/\tan \delta$$

where $\tan \delta$ is the dielectric loss tangent. The loss tangent is generally in the range of 10^{-1} to 10^{-4}. The Q arising from conductor losses is given by [4]

$$Q_c = \frac{3}{2\omega R_s (C/A) l^2} \tag{14.2}$$

where ω is the angular frequency, R_s is the surface skin resistivity, C is the capacitance, A is the plate area, and l is the length of the plate in the direction from which microwave current enters. If one electrode is thin (with respect to skin-depth), R_s will increase and Q_c will decrease significantly. Assuming adequate metal thicknesses, Q_c is a function of frequency and capacitor shape. If the capacitor is square in shape, and the numerical factor is taken as one, equation (14.2) becomes the often quoted expression:

$$Q_c = \frac{1}{\omega R_s C} \tag{14.3}$$

The total Q is given by

$$\frac{1}{Q} = \frac{1}{Q_d} + \frac{1}{Q_c}$$

In practice, Qs actually obtained are nearly ten times less than would be predicted if there was no dielectric loss, implying that thin-film dielectrics are much more lossy than the bulk material, and/or there are losses associated with the dielectric-conductor interface. This discrepancy has never been satisfactorily resolved.

The most common choices for the capacitor dielectric are silicon nitride or silicon dioxide. These are often used for encapsulation and it is natural to use them for capacitor dielectrics also. Properties of several possible dielectrics are listed in Table 14.1. The data has been obtained from several sources (see table) and the ranges in the parameters reflect differences in the manner in which the dielectric was formed. As described in Chapter 8, plasma assisted deposition of dielectrics results in nonstoichiometric films whose properties will vary as a function of deposition parameters. Dielectric constant can also vary as a function of film thickness (caused by nonuniform composition and/or strain). Hence, the process engineer must characterize his own dielectric to obtain accurate values for use in circuit design. The breakdown electric field of the dielectric is not a limiting factor in practice. The breakdown field of most dielectrics is above 10^6 V/cm. In theory, higher breakdown fields would allow thinner dielectrics, yielding greater capacitances in the same area. But applied voltages are only on the order of ten volts, and defects limit film thickness well before the breakdown voltage imposes a limit.

TABLE 14.1
Properties of Dielectrics for Capacitors

Dielectric	Dielectric Constant	Temperature Coefficient (ppm)	Formation
SiO_2	4 - 5 (5)	50 - 100	PECVD, Sputter, Evaporate
Si_3N_4	5.5 - 7.5 (7.5)	25 - 35	PECVD, Sputter
Ta_2O_5	20 - 25 (21)	200 - 400	Anodize, Sputter
Al_2O_3	6 - 10 (9)	100 - 500	PECVD, Sputter, Anodize
Polyimide	3 - 4.5 (3.5)	-500	Spin-on and Cure

NOTE 1: The ranges reflect different formation conditions (see text). They have been gathered from several sources ([4,9,10,13]) and from the author's experience. Nominal values are indicated in parentheses.

NOTE 2: PECVD = plasma enhanced chemical vapor deposition (Chapter 8).

Most of the dielectrics are formed either by plasma assisted deposition or sputtering (Chapter 13). However, dielectrics such as Ta_2O_5 can be formed by reactive sputtering [7,8] or anodization of Ta films [9,10]. Anodized Ta capacitors have been used in other thin-film hybrid circuit activities, and have also been used for GaAs MMICs. It has been found that applying the Ta by sputtering in an argon gas that includes a few percent of N_2 is important; incorporation of nitrogen into the film decreases the temperature coefficient of capacitance and the loss tangent [9,11,12]. The anodization process also tends to undercut the mask, resulting in dielectric formation even under dust particles that would cause defects in other processes. Another advantage of forming this dielectric by anodization is that the thickness of the anodized film is determined by the applied voltage and so can be accurately controlled. Still another advantage of the anodization approach is that the sputtered Ta can be used to form resistors (section 14.4). The high dielectric constant of this material (~21) would seem to make it very attractive. Yet the process engineer must always consider trade-offs. Is the decreased capacitor area worth the time, expense, and complexity of extra process steps? Or is it preferable to use a dielectric already required to be on the slice (such as Si_3N_4). There is no general answer, except to point out that adding extra process steps is not a trivial matter with regard to yield or cost (Chapter 17). Generally, decreasing the capacitor area of some MMICs will not significantly reduce the total chip area. Yet, future MMICs may have very large capacitors, and therefore, the Ta_2O_5 dielectric will be very attractive. Another consideration to include in such decisions is the reduction of discontinuities (transmission line to capacitor) made possible by smaller capacitors.

The yield of good capacitors on a slice is an important issue. Some MMICs make liberal use of capacitors; one pinhole which ruins one capacitor also ruins the entire chip. A complex MMIC could utilize as many as ten reasonably large MIM capacitors. The problem is illustrated by noting that a fabrication yield of 0.95% per capacitor would produce a chip yield of only 0.95^{10} = 60% yield on capacitor defects alone. Hence, capacitor yield must be very high. The major yield limiting factor for MIM capacitors is shorts caused by pinholes in the dielectric or sharp points on the metal plates. Pinholes are almost impossible to eliminate completely, but can be minimized by good cleaning and deposition processes (dust on the slice, or particles formed by contaminant gases in a plasma deposition reactor, can cause pinholes in the dielectric film). The problem will be worse for thinner dielectrics, yet thinner dielectrics are desirable to decrease capacitor size for a given capacitance value. Again, the tradeoffs require an engineering judgement based on the experimentally determined yields of the process in question. It might seem reasonable that capacitor yield would be only a function of plate area. This would certainly be true if only pinholes were the problem. However, several investigators have noted that experimental yields from test slices tend to exhibit a yield pattern that is also dependent on capacitor perimeter. At least, plate area alone is not sufficient to explain the yield pattern. The reasons are not clear, but could be related to the metal edges (or corners) of the plates, roughness, or stresses that increase dielectric pinholes, perhaps. In any event, because of these effects, experimental capacitor studies are very useful to the process engineer to ascertain the yield for a particular capacitor size, shape, and fabrication process.

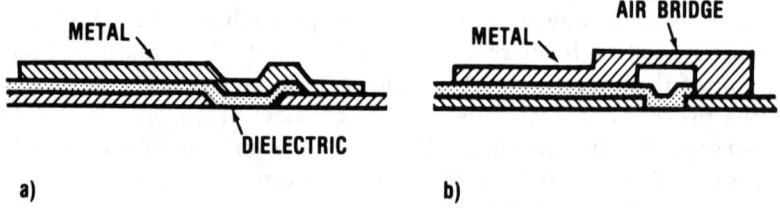

Figure 14.8 Two basic methods of top plate connection: (a) directly on the insulator; (b) a bridge connection that avoids edge or step-over problems.

It is necessary that connection be made between the top capacitor plate and adjacent metalization. The two basic means of achieving this are illustrated in Figure 14.8. The most straightforward would seem to be the case illustrated in Figure 14.8(a), the metal connection to the top plate rests on the dielectric as both step over the edge of the lower plate. Unfortunately, this process tends to result in many defects. Shorts are prone to occur at the stepover, caused either by rough edges on the metal plate, or stresses in the dielectric at the crossover. Also, smooth metal coverage at a stepover is critically dependent on the slope of the edge profile. The defects resulting from that type of connection drove development of other approaches. The almost universal choice for analog MMICs has become the bridge connection of the type illustrated in Figure 14.8(b). (The fabrication of such bridges is described in Chapter 15.) This type of connection completely avoids problems of edges or slopes. Also, because such bridges are usually present for other purposes (Chapter 15), it does not require extra process steps to implement this feature.

14.3 INDUCTORS

Inductors are necessary elements in MMICs where they function as part of tuned circuits. They are either lumped of distributed. The distributed form are treated as transmission lines. Inductors on GaAs MMICs are usually between 0.5 and 20 nH.

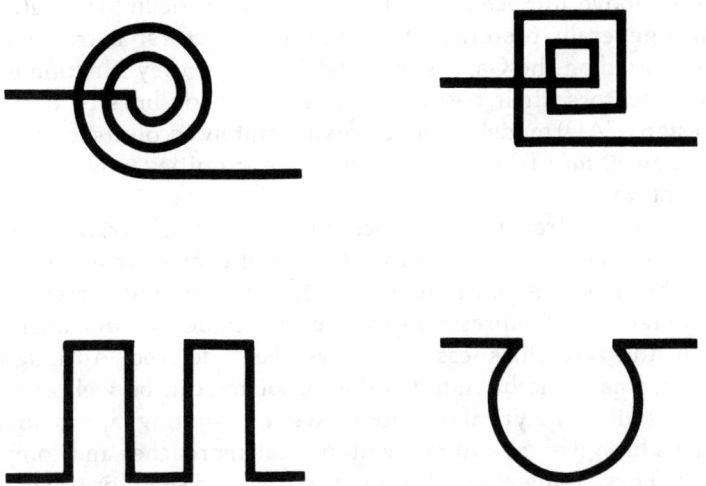

Figure 14.9 Lumped inductor configurations.

14.3.1 Lumped Inductors

Figure 14.9 illustrates several types of lumped inductors. These are lengths of metal line that have mutual inductance between segments because of electromagnetic interactions. But they also have capacitive components, especially to the backside ground plane. In fact, because inductive and capacitive impedance are given by

$$Z_L = \omega L \text{ and } Z_C = -\frac{1}{j\omega C}$$

(where ω is the angular frequency), such a lumped inductor will be more capacitive than inductive at higher frequencies. This is tolerable as long as the effect is included in the circuit modeling. These monolithic, lumped inductors have been described several places [4,13,14,15]. Their small size makes modeling difficult, although reasonably accurate equations have been developed [14]. Rectangular spiral inductors are treated in reference [14], circular spiral inductors in reference [15]. Many of these can also be modeled as transmission lines, using sophisticated CAD programs. Although these inductors are usually placed directly on the substrate, an air bridge spiral inductor has been used [16] in which most of the inductor length is elevated above the slice surface by air bridge technology (these were modeled as transmission lines). The Qs of lumped inductors tend to be near 50 at 10 GHz, larger inductors having smaller Qs.

14.3.2 Distributed Inductors (Transmission Lines)

As noted above, lumped element inductors are difficult to model accurately. It is generally preferable to form the "inductors" as transmission lines fabricated on the GaAs substrate. Although they function essentially as inductors, their treatment as transmission lines by computer aided design (CAD) modeling programs accurately accounts for associated capacitance, mutual couplings, and discontinuities where they join other metalization.

The physics of transmission lines on dielectric substrates is rather complex, especially considering the realities of composite metalizations of finite thickness, discontinuities, radiation effects, and current distribution. In fact, exact expressions for the current density of a microstrip line with non-zero thickness have never been derived. Although the basic electromagnetic boundary value problems can be well specified, they do not yield analytical solutions except in limiting, approximating cases. This has given rise to many numerical approaches and computer solutions. These numerical solutions are often too slow for use in CAD circuit modeling programs. Analytical approximations have been developed that closely match the exact, numerical solutions under appropriate circumstances; these are used in the CAD programs. The general topic of microwave transmission lines on dielectric substrates could fill several books. (Reference [17] is an excellent introductory text on such transmission lines.) It is not the purpose of this section to delve into these complexities — fortunately, this is not necessary. For the process engineer, it is usually sufficient to understand a few basic concepts and let the CAD programs and the designers worry about the complexities. Even the design engineer would be hard-pressed to include all effects accurately. Although he (or she) must have a sound understanding of the fundamentals to effectively design MMICs, the CAD programs are enormously useful in accurately accounting for the subtleties.

These basic concepts may be summarized briefly. The major characteristics of the transmission line are its impedance and its length. These must be implemented accurately on the chip. The impedance is mainly determined by simple, fundamental geometrical relationships (discussed below). Other variations in impedance can be affected by parameters such as conductor thickness or interactions with adjacent metalization.

The Q of the conductor is usually determined by resistive losses in the metal, dielectric losses being minimal.

For MMIC purposes, these transmission lines are almost always microstrip although coplanar transmission lines have been used. Both are shown in Figure 14.10. The transmission line impedance of the coplanar configuration depends on the dimensions of the center strip, the spacing to the lateral ground planes, and the dielectric constant of the substrate material. This is indicated in Figure 14.11. Coplanar lines have two disadvantages. First, they occupy more space than microstrip lines because of the lateral ground planes. Second, a pure coplanar mode is difficult to obtain. MMIC chips are usually rather thin for thermal purposes (100 to 150 μm) and the back is usually metalized or in contact with metal. This results in a mixed-mode transmission line, with some (often significant) energy coupled to the backside metal in a microstrip mode. The author recalls one experience in monitoring coplanar line impedance while slowly narrowing the ground planes by scratching metal from the slice. Complete removal of the lateral ground planes resulted in only a small impedance change, indicating the dominance of the associated microstrip mode on the 100 μm slice. A potential advantage of the coplanar configuration is that the ground plane is available from the front of the slice. This can be useful for some applications.

Microstrip is the most popular choice for transmission lines. The impedance of the line is determined by the ratio of the conductor width to the substrate thickness, the dielectric constant of the substrate, and (to a minor extent) by the thickness of the conductor. These features are indicated in Figure 14.12. An impedance of 50 ohms is generally preferred for input and output transmission line; this matches the usual impedance of microwave circuity and test instrumentation. But other impedances may be useful for interstage matching networks and other applications.

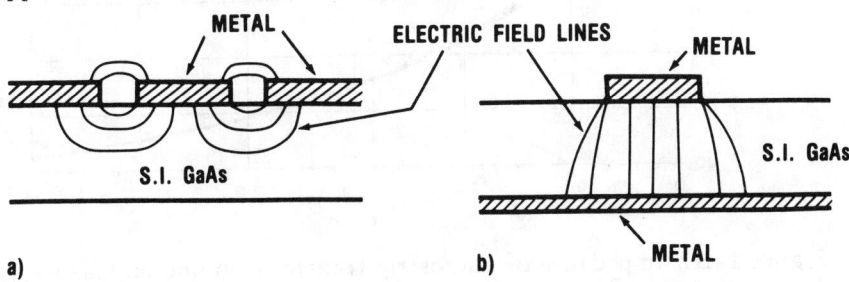

Figure 14.10 Two basic types of transmission lines used on MMICs: (a) coplanar; (b) microstrip.

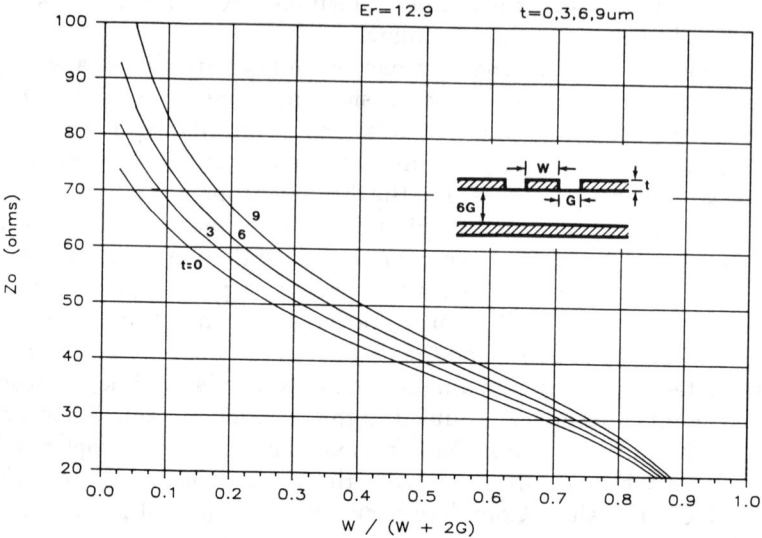

Figure 14.11 Impedance of coplanar transmission line on GaAs as a function of the geometrical parameters.

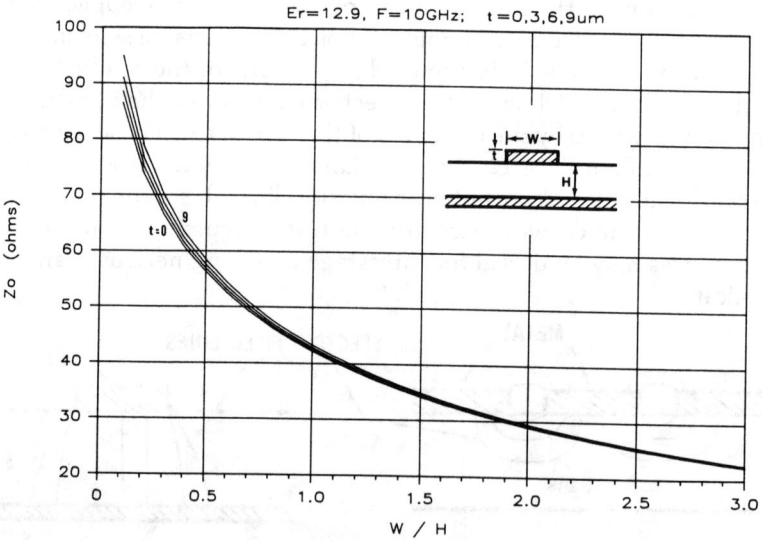

Figure 14.12 Impedance of microstrip transmission line on GaAs as a function of the geometrical parameters.

Electromagnetic coupling between transmission lines can be significant. Such coupling is usually modeled rather accurately by the modern CAD programs. Reference [18] gives experimental data on the coupling of many configurations, including coupling between lines separated by a third, "shielding" metal line. The results are compared to predictions of several CAD modeling programs.

Losses in microwave transmission lines are dielectric, conductive, magnetic, and radiative. Losses in complanar transmission lines have been treated [19], as have losses in microstrip lines [17,20,21]. Reference [20] is especially definitive. The major loss element under control of the process engineer is the conductive loss, which is greater in the front side conductor than on the back side ground plane. The transmission line must be sufficiently thick to carry the required current without significant ohmic losses. This means that it should be at least two skin-depths thick. This, of course, applies for rf currents. If transmission lines are used to carry dc bias, dc currents will also be present. In this case, the cross section must be sufficient to support the current density. The critical current density for gold is on the order of 10^5 to 10^6 A/cm^2 (Chapter 13). Sometimes transmission lines are gold plated to increase thickness (Chapter 15). In this case, the lateral extent of the plating must be controlled to minimize conductor "swelling" (even if the lateral extension of the plated gold is not in contact with the substrate, the effective inductance of the line will be changed). The process engineer must accurately maintain the width of the transmission line and the thickness of the substrate it is placed on. The desirable tolerance on substrate thickness depends on the substrate thickness; it can be assessed using the CAD modeling programs. For 100 μm (0.004 inch) substrates, this tolerance can be as stringent as $\pm$ 2.5 to 5 μm.

There is also analysis to show that conductor loss can be moderately increased if the conductor is too thick [21]. There exists an optimum thickness of approximately three skin-depths (near 3 μm for gold at 10 GHz). These studies also show that even two skin-depths provide reasonably adequate thickness.

14.4 RESISTORS

Resistors are used in MMICs for several purposes including feedback, isolation, self-biasing, terminations (power combiners), and voltage dividers in bias networks. Resistors on MMICs are divided here into two classes: those fabricated using the GaAs material, and all other types. Each class will be discussed below.

14.4.1 GaAs Resistors

The simplest method to fabricate a resistor on the slice is to use the resistivity of the GaAs itself. This approach requires no extra process steps because the two major requirements, isolation and ohmic contacts, must be performed anyway. Such a resistor is indicated in Figure 14.13. The resistance will consist of both the resistance of the GaAs material and the contact resistance of the two ohmic contacts. It is important that the latter not be overlooked; it can be a nontrivial fraction of the total resistance. The resistance is given by

$$R = R_s \frac{L}{W} + \frac{2R_c}{W} \tag{14.4}$$

where R_s is the sheet resistance of the GaAs (ohms/square), L is the distance between the ohmic contacts, W is the width of the resistor (both the GaAs material and the ohmic contacts), and R_c is the contact resistance for a unit length (ohm-mm). (These concepts were discussed in Chapter 11.) The first term in equation (14.2) represents the resistance of the GaAs, the second term the two contact resistances. Sheet resistances typically found in GaAs FET fabrication are 300 to 450 ohms/square. Sheet resistance of uniformly doped layers of GaAs are indicated in Figure 14.14. This figure is useful as a general guide. However, it does not include surface depletion effects, nor allow for nonuniform doping profiles. Therefore, the actual sheet resistivity should be determined from actual measurements on the specific type of material to be used. The same caveat applys to R_c. Both of these can be determined using the techniques described in Chapter 11. Typical values will serve for an example: let R_s = 400 ohms/square and R_c be 0.25 ohm-mm. Then a GaAs resistor with length 30 µm and width 20 µm, would give

$$R = 400 \left(\frac{30}{20}\right) + 2 \left(\frac{0.25}{0.020}\right) = 600 + 25 = 625 \text{ ohms}$$

Note that neglecting the contact resistance term would create an error of about 4% in this particular case.

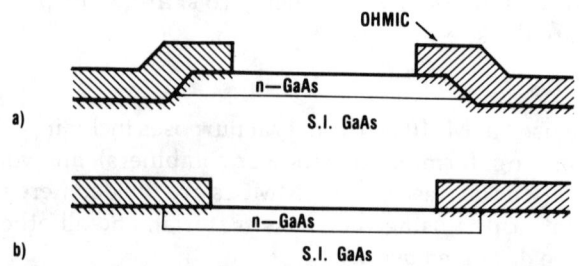

Figure 14.13 MMIC resistor composed of active GaAs material: (a) defined by mesa etching; (b) defined by implant isolation or selective implantation.

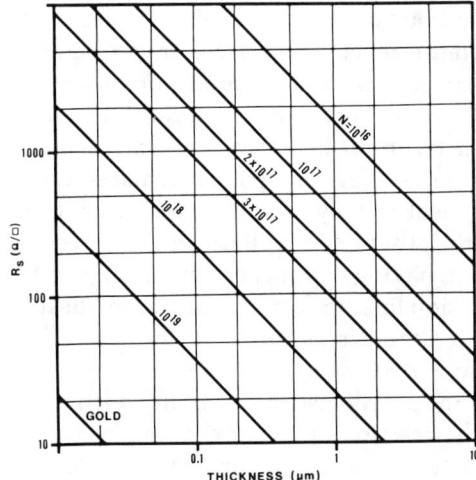

Figure 14.14 Sheet resistance of GaAs as a function of (uniform) doping concentration and active layer thickness. Surface depletion effects are not included.

GaAs resistors have several potential problems: current saturation, Gunn domain formation, and temperature coefficient. Above a critical electric field, the current in GaAs will saturate (Chapters 2 and 3). This must be avoided if the resistor is to act like a resistor. In practice, this is not a severe limitation; the length of the resistor is usually sufficient to prevent the electric field from reaching its critical value. For example, taking the critical field as approximately 3.3 kV/cm, a four-micron spacing typical of an FET would saturate at a voltage difference between source and drain of

$$V = 3300 \text{ V/cm} \cdot 0.0004 \text{ cm} = 1.3 \text{ volts}$$

However, a 50 μm space (0.002 inch) between ohmics would require a voltage differential of 15.5 volts before saturation is approached. GaAs devices and MMICs operate in the 10 volt range. The Gunn domains (Chapter 3) could lead to severe oscillation problems if they were formed. But these occur only above the critical field also.

The last potential disadvantage is more serious. The temperature coefficient of the GaAs resistivity is approximately +3000 ppm/°C. That is, the temperature coefficient is both large and positive. This can result in significant resistance changes over temperature. Many MMICs intended for military systems must operate over large temperature ranges. Computer modeling can determine the amount of resistance change that is tolerable for a given circuit in a given application. Of course, it is possible that temperature compensation techniques could benefit from the temperature coefficient of such resistors.

There is also the issue of current handling capacity, thermal dissipation, and distributed effects. The resistor must be able to handle the current passing through it. Some applications have low current, but resistors in the dc source or drain circuity of FETs may have to carry significant current (without becoming saturated). Such currents also lead to temperature rise. GaAs is a poor thermal conductor and will not be able to remove heat rapidly enough if too much is dissipated in too small an area. On the other hand, a physically large resistor will become a distributed element, having shunt capacitance (Figure 14.15) and acting as a lossy transmission line. In fact, all but the smallest resistors should be modeled as distributed structures (using the CAD programs) or effective microwave resistance (impedance) may differ significantly from the dc resistance. Any discontinuity in width between the transmission line and the resistor should also be modeled.

Although GaAs resistors for analog MMICs should be designed to prevent current saturation, such saturated loads are used in digital logic circuits. These "active" loads are used in the drain circuits of common-source FETs instead of true resistors, as shown in Figure 14.16. The rf signal resistance of the active load is substantially greater than the dc bias resistance. These structures are not passive resistors and will not be discussed further.

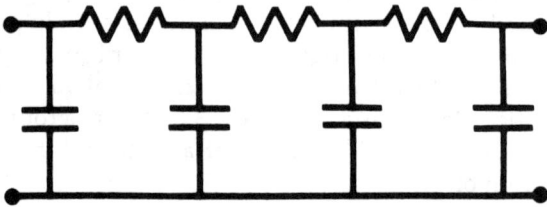

Figure 14.15 Distributed resistor.

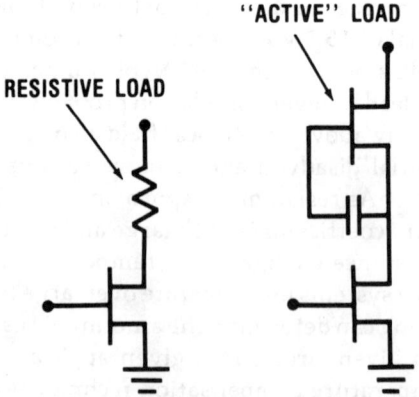

Figure 14.16 Active load as used in digital logic circuits.

14.4.2 Thin-film Resistors

The use of GaAs material as resistors has several disadvantages as listed above, including rather high values of sheet resistance for some application. It is therefore useful to consider other thin-film materials to provide resistance. These must, of course, be compatible with GaAs processing. For example, they must exhibit good adhesion to the substrate. Materials that have been used for this purpose are listed in Table 14.2. The data in this table only should be taken as representative information. Exact resistivities and temperature coefficients depend on the deposition conditions. In addition to the materials listed in the table, cermet resistors have been considered for MMIC applications. These consist of a mixture of a metal and a dielectric. These can be prepared by cosputtering. However, these do not seem completely satisfactory as yet. Cermet films are expected to exhibit an RC frequency dependence similar to carbon resistors, which is likely to be a problem at microwave frequencies [4].

TABLE 14.2

Materials for Use as Thin Film Resistors

Material	Resistivity (μohm-cm)	TCR (ppm)
Cr	13 (bulk)	3000
Ti	55-135 (43 bulk)	2500
Ta	180-220 (13 bulk)	-100 to +500
TaN	280	-180 to -300
Ta_2N	300	-50 to -110
Ni	7 (bulk)	-
NiCr	60-600	200
GaAs	300-450 ohms/sq	3000-3200
AuGeNi (alloyed)	2 ohms/sq	-

NOTE: The exact value that is obtained is dependent on deposition parameters. In all cases, thin-film resistivity will be greater than bulk resistivity. Therefore this data should be used only as a guideline. The data has been gathered from several sources [4,13] and the author's experience.

From a process perspective, the parameters that must be controlled are resistivity and thickness — this may be difficult to achieve to a high degree of accuracy. One approach to this problem is to use a four-point probe technique (Chapter 18) on an insulating surface placed in the evaporator to monitor resistivity of the metal film as it is applied. Deposition can be stopped when the desired resistivity is reached. Use of such techniques indicates the difficulty of achieving tight tolerances without such monitoring.

TaN resistors are usually produced by sputtering Ta in an argon gas with some amount of nitrogen present. This nitrogen becomes incorporated into the film. The resistivity and temperature coefficient of the resulting film is very sensitive to the nitrogen pressure, as indicated in Figure 14.17 [13].

Resistors other than GaAs resistors generally require at least one additional masking level to define the pattern (GaAs resistors can be fabricated using the existing isolation and ohmic contact masks). One exception is the use of alloyed ohmic metalization as resistors. For example, alloyed AuGeNi can produce resistivities of approximately 2 ohms/square. If gold overlayers are used on ohmics (AuGeNiAu), the AuGeNi resistors may be formed using a separate masking step. This also allows a wider range of resistivities by allowing thinner ohmic metal for the resistor.

As is always the case, the process engineer must conduct experiments to determine the values that apply to his particular process.

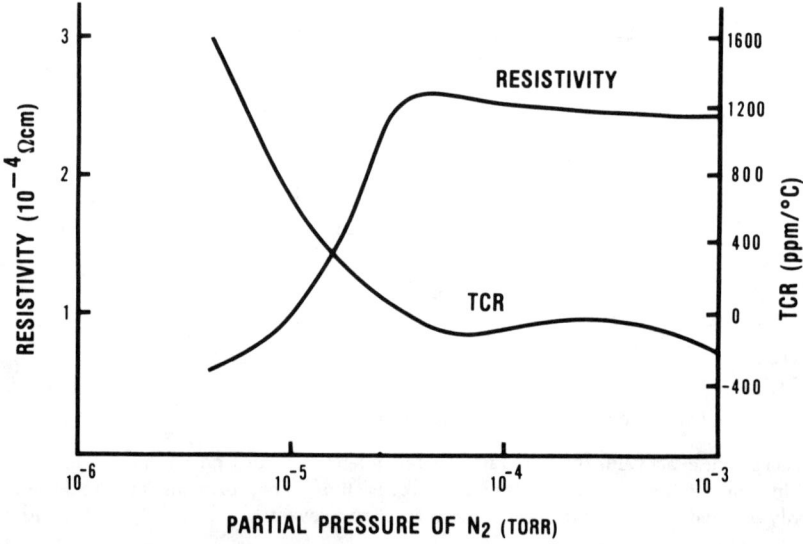

Figure 14.17 Resistivity and temperature coefficient of TaN as a function of nitrogen partial pressure (from reference [13]).

REFERENCES

[1] G.D. Alley, *IEEE Trans. Microwave Theory Tech.*, 18, 1970, p. 1028.

[2] J.L. Hobdell, *IEEE Trans. Microwave Theory Tech.*, 27, 1079, p. 788.

[3] R. Esfandiari, D.W. Maki, and M. Siracusa, *IEEE Trans. Microwave Theory Tech.*, 31, 1983, p. 57.

[4] R.A. Pucel, *IEEE Trans. Microwave Theory Tech.*, 29, 1981, p. 513.

[5] A. Higashisaka, and F. Hasegawa, *Electron. Lett.*, 16, 1980, p. 411.

[6] W.C. Chew and J.A. Kong, *IEEE Trans. Microwave Theory Tech.*, 28, 1980, p. 98.

[7] F. Vratny, *J. Am. Ceram. Soc.*, 50, 1967, p. 283.

[8] M.E. Elta, A. Chu, L.J. Mahoney, R.T. Cerretani, and W.E. Courtney, *IEEE Electron. Device Lett.*, 3, 1982, p. 127.

[9] M. Durschlag and J.L. Vorhaus, *1982 Technical Digest Papers, GaAs IC Symp.*, IEEE, Piscataway, NJ, p. 146.

[10] A. Chu, L.J. Mahoney, M.E. Elta, W.E. Courtney, M.C. Finn, W.J. Piacentini, and J.P. Donnelly, *IEEE Trans. Microwave Theory Tech.*, 31, 1983, p. 21.

[11] M.H. Rottersman, M.J. Bill, and D. Gerstenberg, *IEEE Trans. Components, Hybrids, Man. Tech.*, 1, 1978, p. 137.

[12] P.K. Reddy and S.R. Jawalekar, *Phys. Status Solidi*, 54, 1979, p. K63.

[13] R.S. Pengelly, *Microwave Field-Effect Transistors — Theory, Design, and Applications*. New York: Wiley, 1982.

[14] H.M. Greenhouse, *IEEE Trans. Parts, Hybrids, and Packaging*, 10, 1974, p. 101.

[15] R.L. Remke and G.A. Burdick, *Proc. Electronic Components Conference*, May 1974, p. 152.

[16] S. Moghe, T. Andrade, G. Policky, and C. Huang, *1983 Technical Digest Papers, GaAs IC Symp.*, IEEE, Piscataway, NJ, p. 7.

[17] K.C. Gupta, R. Garg, and I.J. Bahl, *Microstrip Lines and Slotlines*. Dedham, MA: Artech, 1979.

[18] H. Finlay, J. Jenkins, R. Pengelly, and J. Cockrill, *1983 Technical Digest Papers, GaAs IC Symp.*, IEEE, Piscataway, NJ, p. 16.

[19] A. Gopinath, *IEEE Trans. Microwave Theory Tech.*, 30, 1982 p. 1101.

[20] E.J. Denlinger, *IEEE Trans. Microwave Theory Tech.*, 28, 1980, p. 513.

[21] R. Horton, B. Easter, and A. Gopinath, *Electron. Lett.*, 17, 1971, p. 490.

CHAPTER 15

PLATING
AND
BRIDGE INTERCONNECTS

15.1 INTRODUCTION

Plating operations are usually the last major steps of front side fabrication. Plating may be used to create a second level of metalization, to increase metal thickness and improve conductivity, to fabricate bridge interconnects, to provide a desirable surface for wire bonding, or in back side processing (Chapter 16). The plating is typically 3 μm to 8 μm in thickness for front side applications, and up to 50 μm thick for back side applications. The substantial thickness of plated patterns make subsequent lithographic operations difficult. This is one reason that plating is usually the final front side process step.

Increased electrical conductivity is obtained by plating features such as top capacitor plates and transmission lines. This plating should be several skin-depths thick (Chapter 13,14). Plated air bridge interconnects are commonly used in GaAs devices and MMICs; these are discussed in section 15.3. Improved bonding is obtained by plating bonding pads. The thick, relatively soft plated gold is especially suitable as a surface on which to bond gold wires, either by ball bonding, ultrasonic bonding, or thermocompression bonding. Back side plating is used to form plated heat sinks and through-substrate via connections; these are described in Chapter 16. These applications use gold thicknesses much greater than typically applied by evaporation or sputtering. Plating is the only economical means to provide such thick coatings. Formation of such thick layers by evaporation or sputtering would use enormous amounts of gold, most of which would be deposited on places other than the slice and which would have to be recovered. Even ignoring economic and reclamation difficulties, such thick films are very difficult to pattern using liftoff or etching processes.

Plating on GaAs devices is almost always gold (silver is sometimes used for plated heat sinks on microwave diodes). Gold has many advantageous properties for electronic components — it plates well, has high electrical conductivity, is easily soldered or welded, is resistant to oxidation, and is ductile. Gold is an extremely noble metal (standard electrode potential = +1.68 V). It does not react with oxygen, nitrogen, sulfur, selenium, or carbon at any temperature. Hydrogen is essentially insoluble in gold. Tellurium forms a number of intermetallic compounds with it (the only compounds of gold found in nature are the tellurides). At room temperature, completely dry halogens exhibit little or no reaction with gold — except for bromine. In the presence of moisture, bromine reacts vigorously with gold; chlorine and iodine react with gold to a lesser extent. Fluorine reacts strongly with gold at red heat, but reacts very little below 110°C.

Gold is also resistant to attack by most acids; for example, sulfuric acid below 250°C, hydrochloric acid below its boiling point, hydrofluoric acid (if free of oxidizing agents), and nitric acid if it has specific gravity less than 1.46 (80% HNO_3) and is free of halogens. However, if gold is made anodic, or if oxidizing agents are present, hydrochloric acid will attack gold rapidly. Hydrocyanic acid will react slowly with gold in the absence of oxygen, and rapidly if oxygen is present. Phosphoric acid, even when hot, will not etch gold. The only pure, single acid that will etch gold is selenic acid at about 225°C. The most active etchant for gold is aqua regia, a mixture of three parts hydrochloric acid and one part nitric acid. Gold resists attack by alkali hydroxides and carbonates at all temperatures. Mercury will amalgamate readily with gold (such compounds are used in dentistry).

Gold used in microelectronics is highly pure, usually 99.99% or better. One reason microelectronics requires such purity is the rapid increase in electrical resistance as a function of impurity concentration. Table 15.1 lists the effect on resistivity resulting from a 1% alloy with different metals [1]. Thermal conductivity and many other parameters also exhibit degradation as other metals are alloyed with gold.

The basics of gold plating are discussed in section 15.2. The use of plating to construct bridge interconnections is described in section 15.3.

TABLE 15.1
Effect of Impurities (alloy) on the Resistivity of Gold
(From reference [1])

Alloy	Resistivity (micro-ohm/cm)	% Increase In Resistivity
Pure	2.2	-
1% Silver	2.82	28.2
1% Palladium	2.9	31.8
1% Cadmium	3.07	39.5
1% Platinum	3.3	50.0
1% Copper	3.59	63.3
1% Indium	4.5	104.
1% Zinc	4.93	124.
1% Nickel	5.1	132.
1% Tin	7.6	245.
1% Colbalt	17.8	710.
1% Iron	26.9	1220.

15.2 GOLD PLATING

Plating in general is a formidable topic and the subject of much theoretical research and enormous engineering development [2]. Entire books have been written about gold plating alone [1]. This section will review the basic concepts of gold plating, especially those related to microelectronics. Silicon processing rarely uses gold plating directly on slices, but has used this technology in packages. GaAs processing often makes extensive use of gold plating for the purposes listed in the previous section. Electroplating of gold is discussed in subsection 15.2.1. Electroless methods are briefly addressed in subsection 15.2.2. A light assisted technique useful for plating via holes in GaAs substrates is described in subsection 15.2.3.

15.2.1 Electroplating of Gold

Two basic terms used in electroplating are cathode efficiency and throwing power. Cathode efficiency (or current efficiency) is the ratio of the weight of metal actually deposited to that which would be deposited if all the current had been utilized in depositing the metal. In general, a portion of plating current may be involved in electrochemical reactions other than metal deposition, such as depositing gases or reducing ionic species. The type of baths used to plate pure gold for microelectronic purposes usually have cathode efficiencies near 100% (cathode efficiencies can be much lower in other plating processes). Throwing power is the ability of the plating bath to produce deposits of uniform thickness on cathodes having irregular surfaces.

Most gold plating baths (for all purposes — not just microelectronics) utilize a gold cyanide complex. Other systems exist [1,2], but are rarely used in commercial operations. Gold plating compositions may also include other ingredients such as buffers, brighteners, current carrying species, and chelating agents. Buffers stabilize the chemistry and increase conductivity. Brighteners promote grain refinement, resulting in a brighter appearance (all brighteners tend to reduce the density of the deposits). Chelating agents increase conductivity, improve throwing power, and/or serve as inhibitors for the deposition of any base metals present. A great many commercial, proprietary gold plating solutions exist. These have been optimized for various applications (one vendor offers 27 different solutions [3]). Cyanide gold plating baths are of three types: unbuffered alkaline, buffered neutral, and buffered acidic. Historically, the alkaline type was used. Many metals are plated using cyanide solutions, but gold is unique in that all other metal cyanides become unstable in acidic solutions, decomposing and liberating poisonous hydrocyanic acid. For this reason, gold cyanide plating paths were historically held alkaline. However, the gold cyanide complex is stable in acidic solution, down to a pH of about 3, and acidic solutions have become very popular. Gold plating baths for microelectronic purposes are usually the acidic type. Potassium gold cyanide is often the source of the gold cyanide complex, and an appropriate phosphate may be used as the buffer. $KAu(CN)_2$ is, of course, poisonous.

The non-alkaline (acidic or neutral) baths exhibit a wider range of properties than is obtainable from the alkaline systems. This flexibility in pH allows freer use of photoresists than is possible using alkaline baths, which attack many such materials. Acidic baths can also produce the purest gold deposits.

The chemistry of gold is mainly the chemistry of complexes; few simple compounds exist. This is true of both oxidation states of gold, +1 and +3. Many +1 gold complexes are stable in aqueous solution, especially the cyanide used in gold plating, $Au(CN)_2^-$. This gold cyanide complex is very stable. The equilibrium constant has been listed as 4×10^{-28} [3], but other sources have placed it as small as 5×10^{-39} [2]. In either case, it is clear that the Au is tightly bound in the cyanide complex and few free Au^+ ions exist in solution. This means that gold plating does not proceed in the simplified manner typified by plating free ions. The electrochemical reaction involves the entire $Au(CN)_2^-$ complex. Fundamental, well-documented studies of the kinetics and electrochemistry are surprisingly sparse, as has been noted by others [2]. The use of commercial plating compositions intended for semiconductor use is highly recommended. The manufacturer's recommendations should be followed.

Such plating baths are intended to produce pure gold (>99.99%); they operate at essentially 100% cathode efficiency. They are usually operated near 60°C and agitation is recommended. Acidic baths readily plate alloys so contaminants must be rigorously excluded — codeposition results in decreased resistivity (Table 15.1). Cathode efficiency will fall if such codeposition is occurring. Inclusion of impurities can result in color changes of the plated gold. Observation of any such color should cause immediate concern and investigation (although some darkening naturally occurs as roughness increases and plasma etching can darken gold). Other properties such as pH and specific gravity must be maintained also. Commercial suppliers generally provide other chemicals that can be used to replenish gold content, adjust pH, or adjust specific gravity. The properties of acidic baths are sensitive to these parameters and care should be taken that they remain within the recommended range. For small scale operations, it may be more convenient to simply discard and replace the plating solution at regular intervals, rather than use the above chemicals.

Platinum is the main choice for anode material. Other anodes, such as platinized titanium or even carbon anodes, can be used. Usually this is largely an economic (or even a security) decision in shops that plate large parts. For the size and quantities found in GaAs processing, the platinum anode is economically suitable. Current density is recommended by the commercial supplier (often in rather inappropriate units such as A/ft^2) and typically is on the order of 5 mA/cm^2.

Photoresist is a suitable masking substance in plating operations. It must be baked to assure adequate adhesion in the warm (~60°C), acidic solution. The back of the slice may be painted with resist to prevent plating, which may occur to some extent even on semi-insulating substrates. Achieving correct current density requires accurate knowledge of the total exposed area on the slice(s). This "plating factor" can be determined from knowledge of the mask pattern. Large current densities can rapidly destroy the ability of the photoresist to effectively mask the slice. Special circumstances may require plating of sparsely spaced, very small areas — areas so small that photolithographic variations result in large fluctuations in the plating factor. Further, in these cases the plating current can be significantly smaller than usually required and may be difficult to control accurately using equipment and meters in place for other, less critical applications. An old but useful approach in these situations is to simultaneously plate a much larger area (such as an evaporated gold surface on a glass slide). The larger area then determines the current density, almost independent of variations in the pattern on the slice.

Plating current may be dc, ac (swinging positive and negative), or ac superimposed on a dc bias (always remaining positive). The ac methods generally yield decreased grain size. However, the pure gold films and acid baths used in microelectronics tend to result in fine grains anyway.

Electroplating requires that plating current be conducted to all portions of the slice. This task is a major concern when using the commonly employed semi-insulating substrates. A grid of active GaAs material extending across the slice (used, for example, to prevent charging during e-beam exposure — Chapter 7) is not sufficient for this purpose except on very small slices. If this approach is attempted, plating thickness will vary across the slice as a function of distance from the electrical contact. Usually, metal patterns must be used to distribute the plating current. This can be a grid of metal formed at some step (such as first metal) for this purpose. Alternatively, the approach described in the next section to plate bridge interconnects may be adopted; a very thin layer of metal is applied to the slice to carry the current. It is then removed after the plating operation is completed.

Another issue in plating is use of a "strike" to initiate plating. The strike is a very thin, plated layer used to prepare the surface for subsequent, normal plating. It is usually only thick enough to cause a clear change in coloration. In general, such strikes are used to seal a surface (to prevent contamination of the plating solution) and to enhance adhesion. For arbitrary metals, these strikes can be essential. However, the benefit in plating gold on gold is less clear. In general usage, gold strikes are formed in baths having lower gold concentration and using higher currents than employed in plating. Again, for gold on gold, it is often sufficient to forgo the separate strike bath and initiate plating with a higher current than normal.

Good electrical contact to the slice is necessary. It may be useful to fabricate (as part of another process step) a small strip of metal near one edge of the slice (perhaps along the flat) to provide a well-defined area for electrical contact. The surface to be plated must be very clean. Any oxide present would be disastrous. Most cases of gold plating in GaAs processing consist of plating gold on (evaporated or sputtered) gold and oxides are therefore not a problem.

15.2.2 Electroless Gold Plating

In some applications it is useful to plate gold simply by immersing the part into an appropriate solution, without use of electrical currents. (A special case of electroless plating, involving photogenerated electron-hole pairs in GaAs, does not belong in this discussion and is treated in the next subsection.) Electroless processes are especially appropriate for plating nonconductive materials, such as semi-insulating GaAs. Such a process has been used to plate via holes and the back side of GaAs MMICs [4,5]. These techniques can be complex and troublesome, requiring surface activation by compounds such as $PdCl_2$ [5]. Only a very brief discussion of these is intended here.

The term *electroless plating* has been over-generalized to refer to any plating process performed without imposing an external voltage. Technically, such plating techniques fall into two categories: immersion plating, and autocatalytic or electroless plating [1]. Immersion plating proceeds by the chemical replacement of the base metal by the more noble (higher electrode potential) metal coating. The reaction will essentially cease once the base metal is effectively covered by a thin coating. Autocatalytic or electroless plating proceeds by the continuous chemical reduction of the noble metal to form a coating on the base metal. The coating metal may or may not act as a catalyst for the reduction.

In immersion plating, the more noble metal can replace any less noble metal (lower electrode potential). Because gold is one of the most noble metals, it should immersion plate onto many substances. In practice, however, the situation is very complex and it may be difficult to obtain dense, adherent gold coatings. The process engineer who elects to explore this process should expect to spend some significant time in process development.

Autocatalytic/electroless plating has a larger body of literature, much of which is listed in Chapter 11 of reference [1], along with several suggested plating baths. Another process is described in reference [4]. These processes are also rather tricky and exhibit a low plating rate. They are not extensively used in GaAs processing.

15.2.3 Light Assisted Plating on GaAs

A special method useful for plating via holes in GaAs devices has recently been described [6]. The process will be referred to in Chapter 16 (Back Side Processing), but is described here because of possible use in other applications. For convenience, the process will be described in the context of the original article — to plate via holes in GaAs substrates. The basic principle is the use of photogenerated electrons to reduce the solvated $Au(CN)_2^-$ complex, resulting in gold plating. In the arrangement shown in Figure 15.1, the front side of the thinned slice is mounted on a transparent plate. The GaAs substrate is semi-insulating. The back side has via holes etched in the GaAs exposing front side metal (see Chapter 16 for a discussion of via holes and their formation). The wafer is immersed in a gold plating bath (a buffered $KAu(CN)_2$ solution) and light from a tungsten-halogen lamp irradiates the front of the slice through the transparent plate, generating electron-hole pairs. The electrons and holes are separated by the space-charge field in the semi-insulating GaAs. The space-charge region (depletion region) extends completely through the low-doped substrate. Hence the holes drift to the back side of the GaAs substrate, resulting in etching of the GaAs material (holes are needed for etching — Chapter 5). The electrons drift to the front side of the slice and onto the (exposed) metal pads where they reduce the solvated $KAu(CN)_2^-$ complex, resulting in plating on these pads. The related etching of the GaAs substrate is quoted as being minimal (≤ 1 μm) [6]. This general approach could be used in other circumstances, especially those in which it is difficult or impossible to supply electrical current directly.

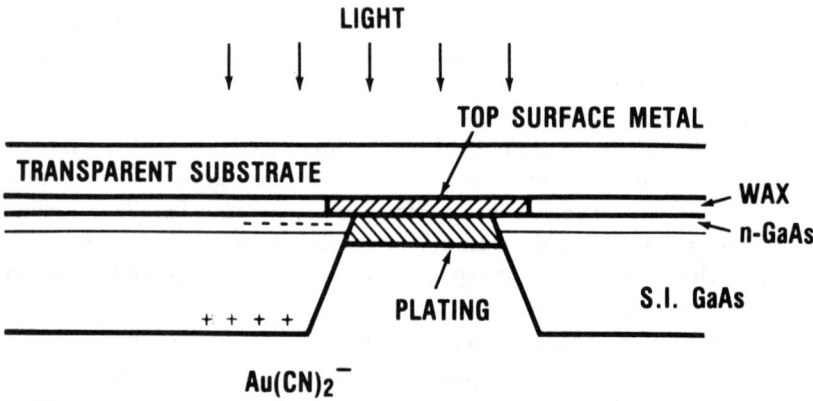

Figure 15.1 Photoelectrochemical plating of via holes in a semi-insulating GaAs substrate.

15.3 BRIDGE FORMATION

Bridges of the type illustrated in Figure 15.2 are used extensively in GaAs analog devices and MMICs for interconnections. They may be used to interconnect sources of FETs, to cross over a lower level of metalization, and/or to connect the top plate of a MIM capacitor (Chapter 14) to adjacent metalization. Often, there is no material (other than air) between the bridge and the slice beneath. Such structures are sometimes called air bridges. These bridges have several advantages over the dielectric crossover typical of digital devices (Figure 15.3). These include low parasitic capacitance, immunity to edge profile problems (see below), and the ability to carry substantial current.

Analog GaAs devices may operate at high current density and therefore benefit from the metal thickness of plated bridges. Low parasitic capacitance (between the bridge and any metalization beneath) follows from the large spacing and low dielectric constant of the intervening medium. This capacitance is a function of the thickness, and the dielectric constant, of the intervening material. Air has a much lower dielectric constant than any dielectric, and the space under the air bridge tends to be greater than the thickness of typical dielectrics. These considerations mean that air bridge crossovers are less capacitive than the dielectric type by a factor (typically) of five to twenty. The problems related to edges in crossovers were discussed briefly in Chapter 14. A step-over of the type indicated in Figure 15.3 requires a smooth, obtuse slope of the dielectric edge. Appropriate dry etching processes can form such edges with good reliability. However, the dielectric also crosses over the lower metalization. That metal is often formed by liftoff processes in GaAs processing (rather than etching) and typically has an ill-controlled edge profile. It

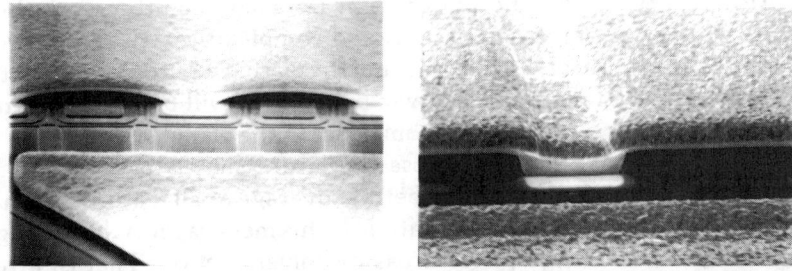

Figure 15.2 Scanning electron microphotograph of a plated air bridge structure.

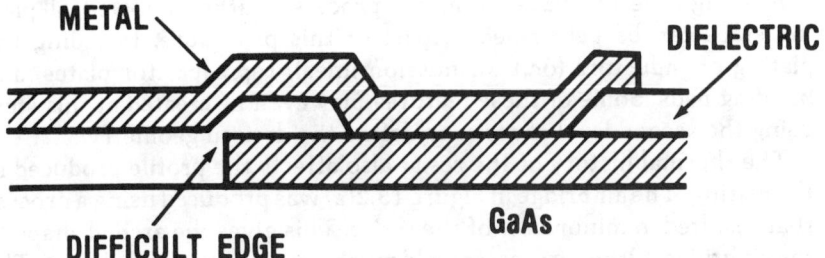

Figure 15.3 Dielectric crossover, typically used in digital circuits instead of a plated bridge.

may also be rough. This causes difficulties in GaAs processing, leading to increased probability of shorts between the two metalizations. It was these difficulties that led most of the industry to adopt the air bridge technique to connect top capacitor plates of MIM capacitors to adjacent metalization. The same considerations recommended this technique for other crossovers. Of course once a single air bridge is used, it is no more difficult (other than yield considerations) to include many others. Hence, many MMICs make liberal use of air bridges.

Although these bridge structures may be fabricated using evaporation rather than plating, plating is the usual procedure and is described here. There are many variations possible, but the major steps typical of most processes are illustrated in Figure 15.4. A layer of resist is spun and patterned to open areas over metal pads. Then a thin coating (100 — 500 Å) of metal is applied to the entire slice. Next, a second coating of resist is applied and patterned, as indicated in Figure 15.4(b). Then the slice is plated, the thin metal layer conducting the plating current to all parts of the slice. After the plating operation, the top resist, thin metal, and lower resist are removed, leaving the plated bridge.

Although the basic concept of this process is straightforward, there are a number of process variations and complexities. All resist layers must be sufficiently baked to withstand the plating environment. Yet the lower resist level must not be over-baked or it will be difficult if not impossible to remove the resist from beneath the plated bridge. The thin level of metal that extends across the entire slice is best applied by sputtering. Sputtering produces better coverage of all features (such as resist sidewalls) than does evaporation. If this metal layer is thin enough, it can be torn from the plated areas by solvent action. That is, liftoff procedures can remove both levels of resist and the intermediate layer in a single operation. Alternatively, each of the three levels can be removed individually.

Although the air bridge formation process is rather involved, all plating steps can be performed as part of this procedure, including the plating of inductors (or transmission lines), top capacitor plates, and bonding pads. Some of these features may even be created at this step, using the second level of resist to define the desired geometry.

The thermal history of the resist also affects the profile produced in the plating. The air bridge in Figure 15.2(a) was produced using a process that resulted in minor flow of the resist. This gives the arched shape to the air bridge. Other processes yield much more rectangular shapes. The thickness of the first layer of resist also determines the spacing between the bridge and the material beneath (usually a dielectric). Hence, this layer of resist is usually rather thick — on the order of 2 to 4 μm. This first layer of resist can also be used to open any dielectric film on the slice. That is, after the resist is patterned, the exposed dielectric can be plasma etched to expose the metal below. The resist then remains on the slice for the subsequent application of the thin metal layer. This layer is usually gold or a two level metalization such as CrAu or TiAu. The top level should be gold for the subsequent plating (titanium alone has been reported in the literature for the current spreading layer — but evaporated gold was subsequently applied in locations to be plated). Although gold alone does not usually adhere well, sputtering processes tend to clean the slice somewhat and sputtered gold tends to stick to other gold surfaces without the use of an intermediate metal to provide adhesion. However, sputtered gold exhibits poor adhesion to most other surfaces.

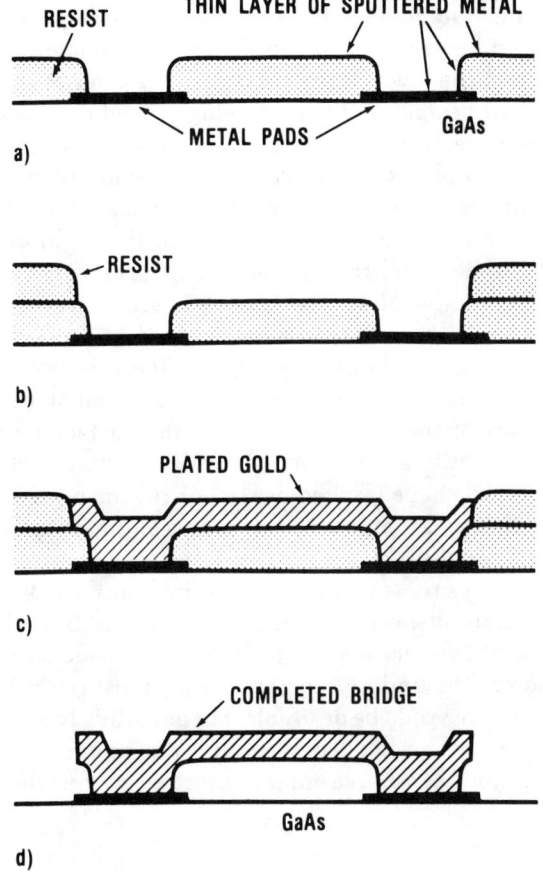

Figure 15.4 Major steps in plated bridge formation.

This metal layer is usually chosen to be very thin — only thick enough to give good stability and to carry the necessary plating current. A few hundred angstroms is usually sufficient. Such a thin layer is also semi-transparent, easing alignment of the subsequent resist level. The top layer of resist is spun and patterned in the conventional manner. It should be baked to improve resistance to the plating bath. However, baking should not result in resist flow, which at this stage would compromise the integrity of the current-spreading metal layer between the two resist levels.

After plating, the resists and thin metal must be removed. As indicated above, this can be performed in some circumstances by conventional liftoff procedures, the solvent attacking the resist and simply tearing the thin metalization away. Alternately, the top layer of resist can be removed by soaking in solvents or by dry etching (ashering). Then appropriate etchants can be used to remove the thin metal layer, followed by solvent removal of the lower layer of resist. This layer of resist extends underneath the air bridges and the stripping procedure should be sufficient to remove all resist beneath the bridge. Material that remains can lead to reliability problems. In this regard, very wide air bridges can be troublesome and are best replaced by two parallel air bridges.

Another aspect to consider is the strength of the air bridge — this is not a reference to the strength required to support its own weight. The strength-to-weight ratio of such microscopic structures is enormous. Rather, the strength to withstand external forces is referred to. Back side operations (Chapter 16) usually require that the slice be mounted face down on another surface. Reasonable force must be applied to assure the surface of the slice is parallel to the surface on which it is mounted. Such operations can deform air bridges. This effect is sensitive to the thickness and shape (arched or flat) of the air bridge, and to the mounting materials and methods. The process engineer should be alert to this possible difficulty.

Air bridges, or any crossover, have one significant disadvantage. It is impossible to visibly inspect what is beneath them. Such bridges are often used to interconnect source pads of FETs which then prevent visual inspection of the underlying gates — the most critical part of the FET. Such inspection would be desirable, but generally is not a sufficient reason to forgo the air bridge geometry. It is a consideration though, if other viable options of source-interconnect exist for the particular design.

REFERENCES

[1] F.H. Reid and W. Goldie, eds., *Gold Plating Technology*. Ayr, Scotland: Electrochemical Publications Limited, 1974.

[2] F.A. Lowenheim, ed., *Modern Electroplating*, 3rd Edition. New York: Wiley, 1974.

[3] F.A. Lowenheim, *Electroplating*. New York: McGraw-Hill 1978.

[4] L.A. D'Asaro, J.V. DiLorenzo, and H. Fukui, *IEEE Trans. Electron Devices*, 25, 1978, p. 1218.

[5] Y. Okinaka, *Plating*, 57, 1970, p. 914.

[6] P.A. Kohl, L.A. D'Asaro, C. Wolowodiuk, *IEEE Electron Device Lett.*, 5, 1984, p. 7.

CHAPTER 16

BACK SIDE PROCESSING

16.1 INTRODUCTION

After the front side processing of a GaAs slice is completed there may be significant back side processing remaining. Digital ICs usually require a minimum of such processing; but analog devices and MMICs (monolithic microwave integrated circuits) will require more extensive and difficult operations. Back side processes include wafer thinning, via hole formation, back side metalization, plated heat sink formation, and dicing into individual chips.

Back side operations are important for several reasons. In a production environment, significant economic investment has been made in the slice by the time the front side processing is completed. Therefore, any difficulties or mistakes that occur in the back side operation (resulting in scrapped slices) are much more costly than those which occur earlier in the process flow. Also, several of the back side operations critically affect device performance. MMICs require accurate control of the thinned substrate thickness to control transmission line impedance. Via holes must be accurately located and formed to result in low impedance grounds.

Unfortunately, back side operations are the least explored and least mature of the GaAs fabrication processes. Yet, they are critical to yield and economics. The following sections explore present methods used in wafer thinning, via formation, and die separation. (Plated heat sinks usually are intimately associated with via holes and are discussed in that section.) However, fewer specific details are given in this chapter than typically found in other chapters for two reasons. First, as with the topics of other chapters, precisely detailed step-by-step instructions have been avoided for practical reasons. As has been emphasized repeatedly, variations in equipment and applications make it almost useless to specify such details. Processes must be individually optimized, based on the fundamentals and guidelines given in this book. Secondly, back side operations are still in development with little information available in the technical literature. The importance of these developing processes to commercial applications make the individuals involved reluctant even to discuss these techniques in private among colleagues, much less to have them published. For these reasons, some material in this chapter describes difficulties and applicable considerations without giving specific solutions.

Back side operations may require that a back side pattern be aligned to a front side metalization. Examples are via holes or a back side plating pattern. Of course, after one back side pattern is in place, alignment marks on it can be used to align other back side patterns. The back side patterns are usually defined using contact photolithography and special exposure towers — either an infrared exposure tower, or an "over-under" exposure tower. An infrared aligner uses infrared light that passes through the GaAs slice and is detected by an appropriate video detector and displayed on a TV screen. Hence, the metal pattern on the front of the slice (facing downward in the exposure tower) can be seen on the screen. The pattern on the photomask (in contact with the back of the slice, which is facing upward) also appears on the screen (it is illuminated by reflected light from above). Hence, the back side mask can be aligned to the front side pattern. The "over-under" aligners use optical techniques to image both surfaces of the slice simultaneously. Optical microscopes of either type (infrared, "over-under") are also available to examine slices.

Another approach is to etch holes from the front of the slice to the rear, using these holes for subsequent back side alignment. In practice, the approach can be awkward, requiring extra chip area, causing mounting difficulties, and/or resulting in ill-defined alignment markers. It is worth noting still another method for achieving back-to-front alignment that can be used for small scale laboratory work, although it is not suitable for production quantities. In this approach, the GaAs slice is mounted face down on another substrate material. Most of the slice is painted off with resist, leaving only portions of the edge exposed. Then the exposed GaAs material is etched away (using wet etching — Chapter 5), exposing the front side metalization. This visible, front side metal at the edges is then used to align the back side pattern. This method is clearly cumbersome and also is not compatible with precise alignment (because the mask and exposed metal are separated by the thickness of the slice).

Metal is often placed on the back of slices for numerous reasons, one being to allow the chip to be soldered to a mounting block. Several different metalizations have been used for this purpose, including CrAu, TiAu, TiPtAu, and AuGeNiAu. Various aspects of back side metalization are addressed in each of the following sections. But one caution applies to all cases: soldering tends to leach gold from the back of the slice. If the gold is not sufficiently thick, or a barrier metal is not present, this can result in voids between the chip and the mounting block. Even if these voids do not result in poor mechanical attachment, they will increase the thermal impedance and result in decreased reliability.

16.2 WAFER THINNING

GaAs devices such as FETs or microwave diodes can generate significant heat. In operation, GaAs devices are usually mounted with the front side upward and the bottom soldered or epoxied to another material which serves as a heat sink. (Another approach, rarely used, is to mount the chip front-side-down on the heat sink. This "flip-chip" technique causes difficulty in mounting, bonding, and inspection. It does allow heat to be removed from the front of the slice, but is unlikely to gain any popularity.)

Hence, most of the heat generated at the front of the slice must be conducted through the GaAs substrate and into the heat sink. Unfortunately, GaAs is a poor thermal conductor (0.55 W/cm-K). This requires that the GaAs substrate be thinned from its original thickness, 0.017 to 0.030 inches, to a significantly lower value, 0.002 to 0.006 inch (50 μm to 150 μm). Such thinning is absolutely necessary for power FETs and medium to high power MMICs; these devices require a low thermal impedance.

Another reason to thin GaAs substrates is to allow reasonable size transmission lines on MMICs. The physical size of these transmission lines is a function of substrate thickness (Chapter 14) — thick substrates would require wide transmission lines, resulting in significantly larger (and more expensive) chips than presently exist. Also, via holes of any practical size can only be fabricated in thin substrates.

Low power GaAs devices that do not employ transmission lines or via holes can be completed on reasonably thick substrates. Examples include digital circuits. Nevertheless, even these devices can generate non-trivial heat. It is also more difficult to cut or scribe thick substrates. Hence, they are generally thinned to some extent.

The above considerations mean that virtually all GaAs slices are subjected to a thinning operation after front side processing. The required accuracy depends on the device and the application. MMICs that include transmission lines have the most stringent requirements. Transmission line impedance will depend on substrate thickness. Generally, an accuracy of ± 3 μm or even less may be required. This is a difficult specification to achieve over a 2 or 3 inch slice. The specification for other devices can be less restrictive, being as liberal as 10 to 20 μm. But even this accuracy is non-trivial to obtain across a large slice and requires care.

The reduction in substrate thickness is usually accomplished by lapping techniques, although precision grinding operations (typical of modern optical fabrication) have been considered. Lapping is usually used to remove most of the thickness. The basic mechanism is similar to that traditionally used in optical fabrication for decades. A slurry of water and grit is used between the slice and a flat plate. The slice is moved with respect to the plate and the grit mechanically removes GaAs material (Figure 16.1). The slurry must be liquid enough to prevent the slice and lapping plate from sticking to each other. The slurry may also include lubricants (essentially soaps) or other components. The grit is made of a very hard compound such as silicon carbide (carborundum) or alumina. Such grits are commercially available in a wide range of sizes. The relevant size range for GaAs slice lapping is 5 μm to about 0.3 μm. Figure 16.2 shows a photomicrograph of 5 μm grit. Even though the grit is very hard, it rapidly loses its sharp edges during lapping and must be replaced regularly or continuously.

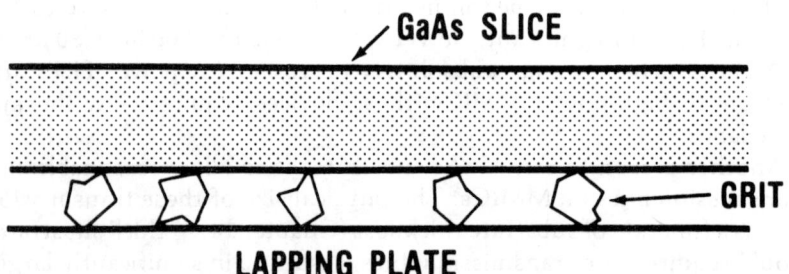

Figure 16.1 Use of grit to thin a GaAs substrate by lapping.

Figure 16.2 Photomicrograph of 5 μm silicon carbide grit, used in lapping operations. Note the irregular shape and sharp edges.

The lapping plate must be flat. The plate may be glass or another substance. Such surfaces are much more resistant to the grit than is GaAs material. Nevertheless, the plate will slowly be lapped and will eventually depart from flatness. It must then be replaced or resurfaced. Although these lapping plates must be close to optical flatness, the techniques developed for optical figuring make this obtainable with relative ease and expense. The slice is mounted on a flat holder, usually by wax or some similar substance. The mounting operation must result in slice flatness and uniform thickness of the adhesive. There must be provision to keep the slice holder parallel to the lapping plate. Otherwise, the slice tends to develop a wedge as it is lapped. Note that there are a number of requirements for flatness and parallelism. As a typical example, the lapping process must be capable of lapping a 3 inch slice to 0.004 inch thickness, ± 0.0002 inches.

In the research environment, such processes were usually performed on small, individual slices by a skilled technician who might take several hours to mount and lap one slice, continually checking thickness and parallelism. Such approaches obviously are not suitable for production volumes. The exact methods being developed by institutions to address these volumes are generally proprietary.

The lapping operation introduces damage into the GaAs crystal substrate. The depth of this damage is not well documented, but extends at least as deep as the dimension of the grit being used and probably much more. The effect of such damage is also largely undocumented. Damage in crystalline materials is known to getter dopants (such as Cr in GaAs substrates) and the damage also physically weakens the material. Definitive studies on these questions are lacking. Nevertheless, the damage that occurs from routine lapping operations does not appear to cause either electrical or mechanical problems in present devices. But it is prudent to minimize the damage. The damage can be reduced by several methods. One is simply to use a finer grit to complete the last part of the lapping operation. This is relatively easy to do. A second approach is to follow the lapping operation with a chemical etch to remove some (possibly most) of the damaged material. Wet etches cannot be used to remove a significant thickness because of the difficulties of etching perfectly uniformly. Such a wet etch also has the advantage of removing any fine pieces of grit that are lodged in the GaAs surface.

Another issue is the smoothness that should be obtained on the back of the slice. The Q of microwave transmission lines will be improved by smoothness in the back side ground plane (which is determined by the smoothness of the GaAs back side surface). However, this effect does not seem to be great, and the back side of GaAs slices are rarely brought to a polish. The surface left by the fine grit is often sufficient. Most GaAs slices will require a subsequent metalization to be applied on the back side. Hence, the back side must be very well cleaned after the lapping operation.

Lapping is obviously a "dirty" process and should not be performed in the same room as other operations. The grit and debris can play havoc with clean room environments. Clothing (aprons, etc.) used in the lapping area should be prohibited from the clean rooms.

After a slice is lapped, the subsequent treatment depends on the type of device being fabricated. In all cases, however, difficulties are caused by the thin, brittle GaAs slice. It is not easy to handle a 2 or 3 inch GaAs slice that is 0.004 or 0.006 inchs thick. Not only is the material brittle, but the thinner substrates can exhibit curvature resulting from internal stress. Again, techniques to dismount and/or handle such material in production volume are still in the development stages and are generally held proprietary. Nevertheless, a few generalizations about such handling will be made at appropriate points in this chapter. With care, a lapped slice can be removed and handled to some extent even without the support of an underlaying substrate.

16.3 VIA FORMATION

Many discrete GaAs devices and MMICs employ via holes through the (thinned) GaAs substrate to provide a connection from front side metalization to the back side ground plane. Low inductance grounds are critical

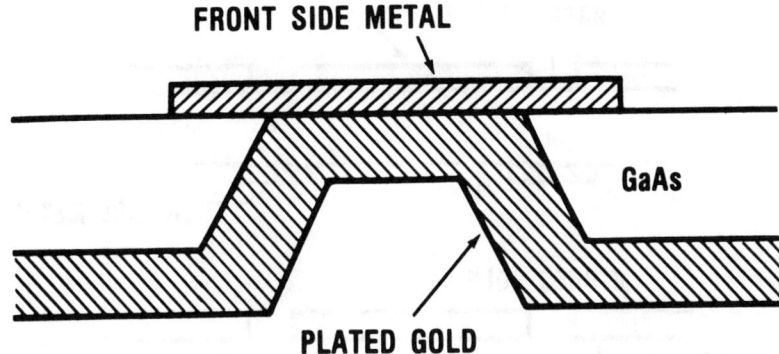

Figure 16.3 Cross section of a plated via hole through the GaAs substrate, connecting front side and back side metal.

for state-of-the-art FETs and/or terminations on MMICs transmission lines. Hence, the ability to reach "ground" at any place on the chip is important. (Note that the term "via" is used in two contexts in GaAs processing. In digital work especially, vias can be openings in the front side dielectric coating, allowing connections between first and second level metalization. Or, as in the case here, vias are holes through the GaAs substrate, connecting front side and back side metal. The meaning is usually clear from the context.)

Figure 16.3 shows a cross section of a typical via in a GaAs device. The basic process is illustrated in Figure 16.4. After the slice is thinned, the back side is patterned using photolithographic techniques to open holes corresponding to the desired via locations (Figure 16.4(a)). Then the exposed GaAs is etched away until a hole is etched completely through the substrate, exposing the lower level of the front side metalization (Figure 16.4(b)). The masking pattern is removed and the slice is metalized to form a continuous metal coating that extends into the via holes (Figure 16.4(c)). This type of geometry usually is plated afterward as shown in Figure 16.4(d). Such a plated, rear metalization is called a plated heat sink.

The most critical step of the process is the etching. This is usually performed using dry etching procedures, especially reactive ion etching (RIE — Chapter 9). Wet chemical etching can be used, but the large undercutting rate typical of wet etchants (Chapter 5) makes this approach suitable only for large vias. As shown in Figure 16.5, smaller via holes may be formed using RIE because of reduced undercutting. High aspect ratio holes can be formed. Such narrow holes are often required, for example, to form vias beneath every source pad of an FET (Figure 16.6). RIE is the major dry etching technique used for via etching. Other processes, such as ion milling might be attractive, except for the low etch

348 *GaAs Processing Techniques*

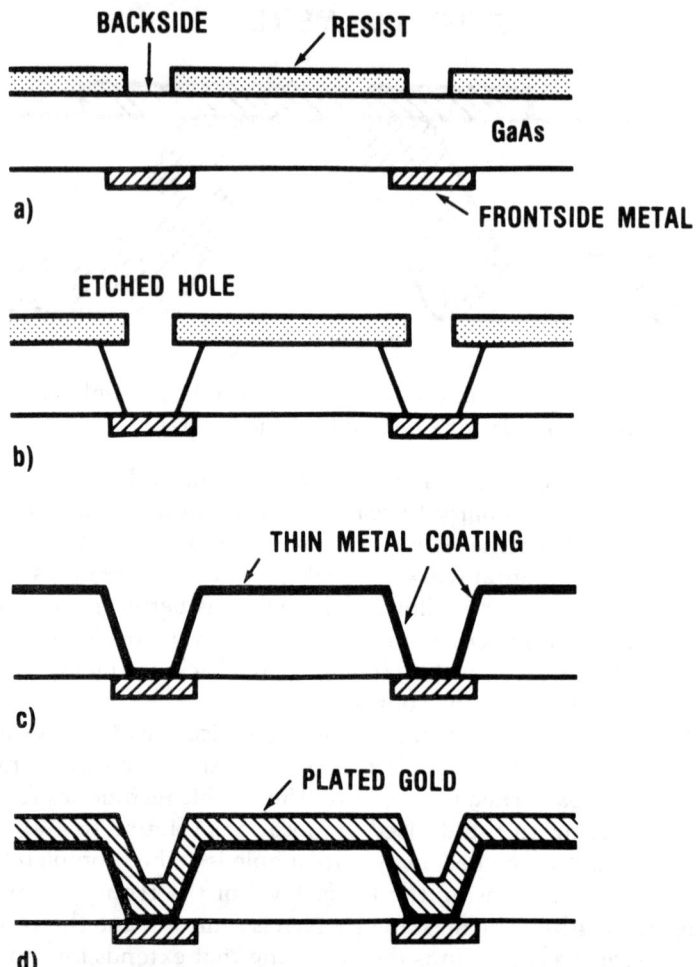

Figure 16.4 Basic steps in the via fabrication process (see text).

rate. RIE vias can be etched in tens of minutes. Ion milling techniques would take many hours and generate significant heat. The exact size and shape of the RIE holes will depend on the etching parameters, as noted in Chapter 9. Hole shape will also depend on crystal orientation; these holes usually do not etch symmetrically. The shape and size of the masking pattern (circular, rectangular) also affects the resulting size and shape. There may be difficulties if different size or shape vias are attempted on the same slice. As is always the case, some amount of experimentation and process development is needed to optimize individual applications.

The masking material must satisfy several requirements. It must stand up to the etching environment without significant degradation or damage, at least for the time required to etch the via holes. It must be capable of being patterned to sufficient resolution. It must be capable of being removed after subjection to the etching process (which may generate heat if dry etching is used). Photoresists can satisfy these requirements if properly selected and processed. Riston (Chapter 6) is a very tough resist that can withstand extreme environments. However, its thickness makes it difficult to pattern to small dimensions. The patterning and etching process must be reliable — after etching is started, it can be very difficult to re-work the slice if trouble develops.

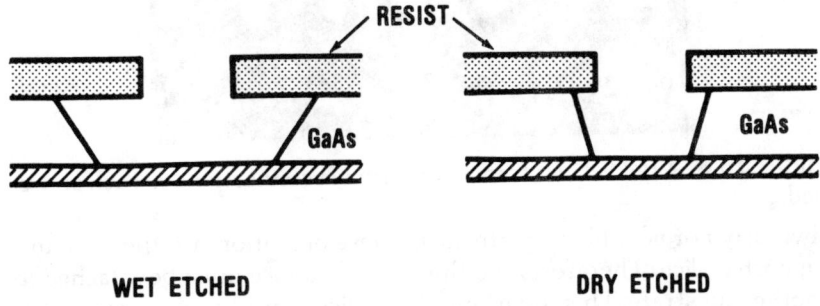

Figure 16.5 General cross sections resulting from wet etching as opposed to dry etching of via holes.

The etching must stop when the front side of the slice is reached. This is relatively easy if wet etching is used — GaAs etchants attack metals much slower than the GaAs material itself. However, in some cases an increased lateral etch rate occurs at the GaAs-to-metal interface, resulting in poor control of the process. Dry etching techniques are more difficult to stop — most will also etch the front side metal at reasonable rates. Some metals, such as platinum, etch slower under RIE conditions than other metals such as gold, for example. The completion of GaAs etching (and the initiation of metal etching) can be detected using appropriate optical detection techniques (monitoring optical emission from the plasma). In any case, it is usually necessary to cease etching before significant damage has been done to the front side metal.

The via holes are etched after the slice has been thinned, in some cases to as little as 0.002 inch (50 μm) or less, although 0.004 or 0.006 inch are more common. The via process requires significant amounts of handling — cleaning, patterning (resist spinning, clamping in an exposure tower etc.), etching, pattern removal, and more cleaning. It is

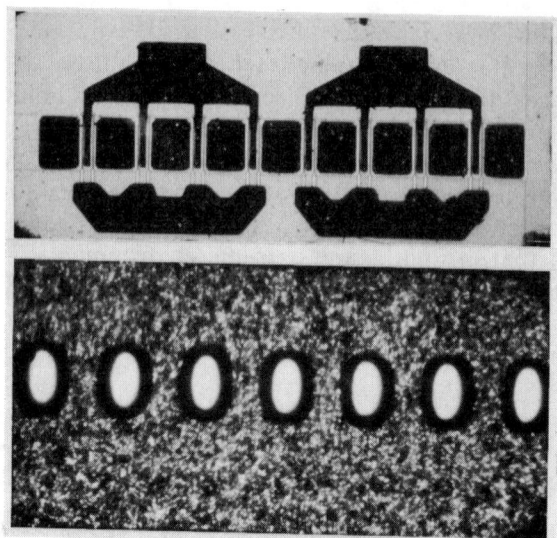

Figure 16.6 A field effect transistor with via connections to each source pad.

obviously not possible to perform all these operations on the thin, unsupported slice. Therefore, the thinned GaAs slice must be attached to another substrate. This could be a silicon slice, glass, or other material. It could be the same supporting substrate the slice was attached to during the thinning operation (if infrared or over-under aligners are used, the supporting substrate must be transparent to the appropriate light). In any case, the substrate and the adhesive agent used to attach the GaAs must be capable of surviving all processing associated with via fabrication.

After the via holes are etched and the masking substance removed, the back side is metalized. Sputtering is the best technique to assure that this metal enters the holes and covers the slopes. In this regard, a minor slope on the hole edge profile is desirable. Vertical holes would be more difficult to cover and plate. (However, such holes can be plated using the photoelectrochemical technique described in Chapter 15, section 15.2.3.) Then the back side is plated to increase conductivity in the via holes. Such plating may also lend strength to the slice if it has been thinned to 0.002 inch or less. If the entire back side is plated, the slice can be diced only by sawing through both the GaAs and the plated gold. This is not desirable (section 16.4). Hence, a back side plating pattern may be defined before the plating step, to prevent plating in the saw streets (Figure 16.7). The sputtered metal provides a method of spreading the plating current. The resulting via structure was shown in Figure 16.3.

The process should not be operated marginally — it is important that vias be fabricated with high yield. This is because it is difficult to test for electrical continuity of the vias (front side to back side). If only one of several vias is "bad," it will not be easily detected from dc measurements, but will cause degradation when the device is operated at microwave frequency where the change in inductance is serious.

Finally, it should be noted that other, completely different approaches to via formation are under investigation in the laboratory. These include the use of lasers to form the vias. But it should be emphasized still again that most of these techniques will not evolve to the point that they are technically and economically suitable for reliable production operations.

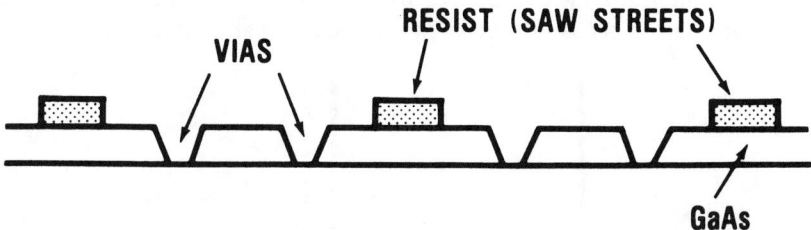

Figure 16.7 Masking pattern to plate the back side in all locations except the saw streets.

16.4 DIE SEPARATION

After all processing operations are completed on the entire slice, it remains to cut the slice into individual bars, chips, or dies (these terms are used synonymously). This can be done in several ways, including scribing, sawing, or etching. Laser scribing is still another approach. In many cases, it is also desirable to maintain the identity of individual bars, to know what location on the slice they are from. This requirement can complicate the issue.

Scribing consists of dragging a diamond stylus along the surface of the slice, parallel to one of the cleavage planes. This creates a scratch on the surface and, due to the crystalline nature of the slice, initiates fractures along these lines. The fracturing is completed during a "breaking" operation, the entire process being referred to as scribing and breaking. The breaking can be performed by flexing the material on which the slice is mounted. Scribing only works along cleavage directions. However, most GaAs devices are fabricated on (100) slices and have the devices oriented parallel to a cleavage direction. The two cleavage directions are 90° apart (Chapter 2) and are therefore suitable for breaking the slice into rectangular die. Scribing has some disadvantages in GaAs work. If any metal is on the back of the slice, if may be difficult to separate the chips. The GaAs may fracture, but leave the metal to act as a hinge between chips. The scribing operation can also damage the brittle GaAs surface by chipping. Scribing was developed mainly for the silicon industry. However, commercial scribes that claim to be optimized for GaAs have recently been introduced.

Another approach is sawing, also popular in the silicon industry. High speed circular saws (about 3 inch blade diameter) are used to cut the slice as the table on which it is mounted moves beneath the blade. These blades typically rotate on air bearings at speeds near 30,000 rpm. Cutting speed is controlled by the table movement. Because of their development for the silicon industry, these saws are very accurate and sophisticated. The blades have a cross section (near the edge) as shown in Figure 16.8.

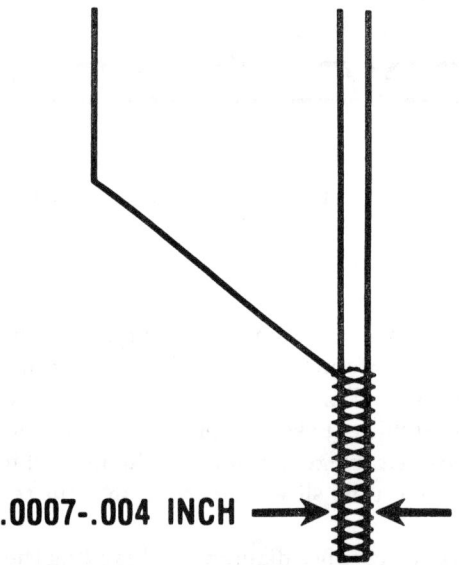

Figure 16.8 Cross section of the outer radius of a saw blade used to saw semiconductor slices.

The width of the blade can be 0.0007 to 0.004 inches. The ends are coated with diamonds, or the entire end is composed of a composite material that includes diamonds in its structure. These blades can cut into material quite deeply (~10 thousandths of an inch). Results depend critically on saw speed, coolent, and the material being cut. Some chipping of the GaAs is difficult to avoid under any conditions. Blade life depends on the type of material being cut. Metals are harsher on blades than is GaAs, especially gold which is soft enough to fill in the blade's rough edge. Thin film metals are not prohibitively difficult to saw, but thick metals (such as plating) can be very difficult. It is generally helpful to design chips so that all extraneous material (metals, dielectrics, etc.), are out of the saw streets.

Another approach is to etch the GaAs to separate the slice into chips. If the back side is plated (except in the saw streets), that plating can be used as an etch mask. Spray etching of wet etchants has been used for this purpose. The spray etching technique results in less undercutting than typically occurs in wet etching. In this case, the resulting chip will likely be handled by the plated heat sink (if handled with conventional tweezers) and so the plating should be sufficiently thick if this is so. But vacuum tweezers can also be used to handle chips. In general, if conventional tweezers are to be used to handle chips, the edge profile that results from the separation process must be considered.

In many cases, devices on the slice will be probed by automatic equipment after front side processing is completed (Chapter 17). It is therefore desirable to be able to associate chips (after separation) with this data. One method is to place numbers on each device by having numbers on a photomask (if using contact lithography), or by defining them during electron beam lithography. This approach is most appropriate to MMICs where there are only a few hundred devices on a slice. If there are thousands of smaller chips, it is less feasible to place these numbers on the chips and to find the chips again after they have been separated. Therefore, another useful approach pioneered in the silicon industry is to place the slice on a sticky tape before it is sawed (or scribed). After sawing, the tape can be stretched and clamped in a frame, thereby separating the chips far enough apart to allow removal of one without disturbing others. These tapes have been especially developed for this purpose and do not leave any residue. This approach allows all chips to stay in the same relative location and to be picked off according to the data taken earlier.

CHAPTER 17

PROCESS INTEGRATION
AND
CONTROL

17.1 INTRODUCTION

All other chapters in the book address specific, well-defined portions of the fabrication process. But there are general issues that relate to the entire fabrication sequence. These global issues fall into two classes: interrelationship of process steps, and in-process controls. There is not a large amount of information to impart, but the importance of these issues and their global relevance make it appropriate to treat them in a separate chapter.

Establishing a viable, reliable, and reproducible fabrication sequence for GaAs devices requires more than developing and optimizing individual process steps. These steps must be integrated into a whole. Each process step invariably influences aspects of other process steps. These mutual influences and interactions must be considered, and processes developed in the context of the entire fabrication sequence. Once an entire process is established, changes in one step can require complicated adjustments in other steps, or even changes in the photomasks. (These issues are addressed in section 17.2.)

Control of a process flow requires continual monitoring of results, which allows the process engineer to detect and correct deviations rapidly. The monitored parameters also generate a data base which allows realistic determination of normal process variations. (These issues are addressed in section 17.3.)

In order to avoid continual reference to previous chapters, it is assumed that the reader is familiar with that material.

17.2 INTERRELATIONSHIP OF PROCESS STEPS

Several of the major interrelationships among process steps have already been discussed in previous chapters. For example, one of the self-aligned gate processes for digital FETs requires that an n^+ implant be annealed after the metal gate is in place on the slice. This dictates that the gate be composed of a refractory metal (such as TiW) to yield a stable Schottky barrier under high temperature anneal. The procedure also dictates that the ohmic fabrication step follow the gate fabrication step (because the ohmic metalization cannot tolerate the high anneal temperature). Thus, the choice of using the self-aligned n^+ implant and anneal technique dictates aspects of other process steps.

Other good examples arise from considering the process flow illustrated in Figure 17.1, used by a Japanese manufacturer to produce MMICs [1]. The process requires only five photomasks, the first of which is for isolation. The second photomask is for gate definition. In this case, aluminum is applied over the entire slice and the gate pattern is formed (in resist) over the metal. Then the exposed aluminum is etched, undercutting the resist mask (Figure 17.1(a)). That resist remains in place during the subsequent evaporation of AuGeNi for ohmic contact formation, causing the ohmic metal to be self-aligned to the gate stripe (Figure 17.1(b)). Then all resist is stripped and the ohmic alloyed (Figure 17.1 (c)). A third mask is then used to pattern and etch away other portions of the aluminum metalization, forming features such as bottom capacitor plates. The dielectric is applied and a forth mask used to open vias in the dielectric. A thin layer of Ti is applied over the entire slice (for subsequent use in spreading plating current) and a fifth mask is used to define the TiPtAu metalization which forms upper capacitor plates, inductors, and interconnects. This same mask is used to plate these metals to greater thickness. The Ti layer is removed. Then the backside is lapped and metalized. This process has very attractive aspects. It requires only five photomasks, while other MMIC process flows may require several more for frontside processing. It also has a gate that is self-aligned to the ohmic contacts, and thus no high registration lithography is required.

However, nothing comes free. The price that is paid for the above process is substantial inflexibility. Each step is tightly related to others and few changes can be made. For example, the gate metal must tolerate the ohmic alloy temperatures (without degrading the Schottky), must be reasonably electrically conductive and must be readily etchable (without the etchant attacking the GaAs or the photoresist). These requirements essentially specify aluminum. If the process engineer should want

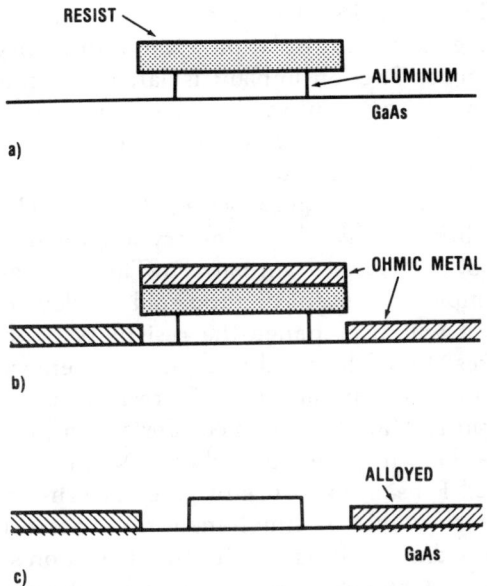

Figure 17.1 Major features of a simple process flow used to manufacture MMICs.

TiPtAu or any other choice, many other changes would be required. Further, the gate cannot be recessed and it cannot be purposefully offset in the source-drain channel. Airbridges are not used, and so metal and dielectric edge quality becomes critical in achieving good metal stepovers when connecting to top capacitor plates. There are many other such restrictions, but of course this is not meant to imply that the process is undesirable. Although the process is somewhat inflexible, the simplicity makes it economical and, in fact, it obviously works very well for the specific application. That is the point. The simplicity or complexity and the flexibility or inflexibility of the process should be a conscious decision — not the random result of assimilating separate process steps.

Other examples of possible interrelationships among process steps are listed below on a step-by-step basis. The process steps are ordered to correspond to that followed in Part III of the book, representing the process order used for many discrete devices and MMICs. These examples are certainly not meant to be exhaustive, but they illustrate the type of considerations needed.

1) *Isolation*. If isolation is by mesa etching, any metal step-overs must occur across the mesa edge having the desirable, obtuse angle, rather than across the undercut slope (although, plated metalization can cross

such steps). Thus, a device or circuit designed for mesa isolation can tolerate a change to implant isolation without difficulty. However, the reverse is not true. The ion implant isolated structure is planar and metalization can cross over onto active GaAs from any direction. Changing to a mesa technology may cause step-coverage problems. Problems may also occur from resist nonuniformity near such edges.

2) *Ohmic contacts.* If a subsequent gate pattern is to be exposed using electron beam lithography, the necessary alignment marks must be formed as part of the source drain-pattern. The thickness of the metalization (for example, the extent of a top gold overlay, if any) will affect resist spin-on; it will also change the resistivity of the metalization. Thus, MMIC resistors fabricated using ohmic metalization would require mask changes to maintain the same resistance.

3) *Gate fabrication.* Many self-aligned fabrication processes are completely incompatible with gate recess (the SAINT process is an exception, see Chapter 12). E-beam exposure of gate patterns may require that large areas (which would be simultaneously exposed in optical lithography) be exposed in a subsequent first-metalization step.

4) *First-level metal.* Gates and first-level metal can be formed in the same step if the same type of lithography can be used for both patterns.

5) *Dielectric.* Changes in dielectric type or thickness can change capacitive parasitics in devices. Movement of the dielectric application step to another point in the process can result in numerous modifications.

Any change in a plasma cleaning procedure when resist patterns are present can result in changes in the edge profile that affect liftoff operations.

Another consideration in any process change is reliability. A change of cleanup or metalization, for example, can result in unforeseen reliability problems. This possibility must always be in the back of the process engineer's mind.

17.3 IN-PROCESS CONTROLS

It is essential to monitor the fabrication process continuously in order to detect deviations and to determine normal process variations. Another major reason to employ in-process monitoring is to identify out-of-specification slices as early in the process as possible (note that a specification may require both an absolute range and a given uniformity across a slice). It is obviously a waste of time, effort, and therefore money, to continue processing a slice that cannot meet requirements. Thus, it is desirable to remove such defective slices from processing as soon as possible after the deviation occurs. Further, some processing errors can be corrected (and the slice reworked) if identified immediately:

for example, incorrect dielectric thickness. (It is another issue as to whether testing of individual slices, to determine rework or scrap, is economically feasible. For large quantities of slices, these tests and decisions can be based on sampled data from groups of slices being processed together.)

Such process monitoring usually requires that special test patterns be present on the mask set. These test patterns can be used to monitor parameters such as the source-drain current of FETs, Schottky diode characteristics, contact resistance, metalization resistivity, sheet resistance of the GaAs, capacitance of MIM capacitors, or many other items. Many of the test procedures and patterns are described in Chapter 18. Others have been described in the appropriate chapter (e.g., those describing ohmic contacts and gates).

These test patterns can be located in essentially two places: on the device bar or on special bars referred to as *plug bars*. In this context, *bar* refers to the entity that occurs repetitively across the slice. Hence, the device bar may consist of one MMIC, or a large number of discrete FETs. The plug bar is a different set of patterns that replaces the device bar at a few locations. A typical configuration is illustrated in Figure 17.2, in which one plug bar is located in the center of the slice (i.e., the center of the photomask) and four others are located in each of the four quadrants. It is relatively easy to implement plug bars if contact photolithography is used. They are simply placed at appropriate places on the photomasks. Many e-beam machines are also capable of exposing plug bar patterns at the appropriate locations. If optical steppers are used, recticles must be exchanged in the step-and-expose operation when the plug bar is to be exposed, which can be cumbersome. Optical stepers, presently used in GaAs processing primarily for digital ICs, usually consist of rather large bars. It is often feasible to place test structures within the device bar, avoiding the issue of plug bars.

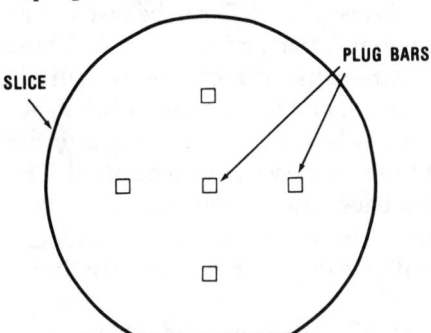

Figure 17.2 Typical location of plug bars on a slice. These bars contain test patterns.

The following list enumerates items that may be useful to monitor on either each slice, each lot of slices, or at regular intervals. These examples are limited to nondestructive measurements that do not preclude further processing. The list is not intended to be exhaustive.

1) *Active layer formation*. After active layer formation (implant and anneal, epitaxy, etc.), contactless conductivity measurements can ascertain sheet resistance.

2) *Isolation*. Resistivity of local areas can be approximately determined by placing two probes on the surface of the slice near to each other, and applying an ac voltage (a curve tracer is appropriate for this). The two probes act as back-to-back Schottky barriers, and a Schottky breakdown should occur at a voltage corresponding to the resistivity of the material below the probes. For example, an active layer doped near 1×10^{17} cm^{-3} will exhibit a breakdown voltage near 12 volts. However, properly isolated material (either implant isolated or etched into the buffer or substrate) will exhibit a breakdown voltage of hundreds of volts. Surface effects, and difficulties in making good contact between the probes and the surface, can introduce anomalies that cloud but do not obscure these results. However, isolation processes are usually highly reliable, and it is questionable whether this test is worth doing routinely. The isolation can be unequivocally determined after the ohmic contacts are in place.

3) *Ohmic contacts*. The isolation produced in a previous step can be determined by applying a fixed voltage between two closely spaced ohmic contacts, which are separated by isolated GaAs, and noting the resulting current. Appropriate test patterns can be used to ascertain contact resistance, sheet resistance of the GaAs active layer, and the sheet resistance of the ohmic metalization. Field effect transistors can have the (ungated) source-drain current measured. Critical dimensions, such as source-drain spacing on FETs, can be measured.

4) *Gates (all Schottky contacts)*. FETs can be tested for I_{dss}, I_{max}, transconductance, pinch-off voltage, various parasitic resistances, and breakdown voltages. The dc gate resistance can be measured, gate length can be checked (using a SEM), and the location of the gate within the source-drain spacing also can be checked. If necessary, the barrier height and ideality factor of the Schottky can be measured. The gate resistance can be determined and backgating examined.

5) *First-metal*. Any critical features, such as inductors, can be measured to assure correct dimensions. The sheet resistivity of the metal can be determined.

6) *Dielectric coating*. The thickness and index of refraction can be ascertained (using an ellipsometer), and the effect of plasma processes on

active devices can be assessed by reprobing the appropriate test structures.

7) *Plating*. The plating thickness and resistivity can be determined.

8) *Lapping*. The thickness (and its uniformity across the slice) can be measured.

In addition to the above in-process tests, photoresist patterns can always be inspected to assure proper dimensions before metalization, etching, etc. Photoresist steps are the easiest to rework. The resist can be removed, respun, and repatterned. As a generalization, however, any reworked slice tends to lead to lower yield. Visual inspection is useful at many stages in the process to note anomalies such as poor liftoff, misaligned patterns, or inadequate metal adhesion.

The quantities presently being fabricated in GaAs production lines make it feasible to use automatic probing techniques to dc probe every active device on every slice. The resulting data is also valuable in characterizing the result of the entire process (at least up to the step at which the slice is probed). Such probing is simplified if done before the slice is thinned. On-slice rf probing is also becoming a reality.

Good record keeping is essential to diagnosing problems. Each slice (or in quantity, each lot of slices) should be accompanied by a document (a *traveler*) that records completion of each step (with the date), any anomalies noted, and the results of in-process tests. This document could be written, or stored on computer and modified from terminals.

In conclusion, the complexity of a complete process flow requires constant monitoring to assure success. Otherwise, deviations in individual process steps can rapidly change an orderly operation into chaos.

REFERENCE

[1] S. Hori, K. Kamei, K. Shibata, M. Tatematsu, and K. Mishima, *IEEE Trans. Electron Devices*, 30, 1983, p. 1867.

PART IV.
CHARACTERIZATION
AND
MEASUREMENTS

CHAPTER 18

ELECTRICAL CHARACTERIZATION

18.1 INTRODUCTION

Electrical characterization is used for material and device evaluation. In-process electrical characterization is essential for maintaining good process control. Virtually every process step can alter electrical device properties; process development or modification must be closely coordinated with electrical parameters. This chapter will review the major measurement techniques used for these applications. A few of these measurements are described in the appropriate chapter in Part III (contact resistance in Chapter 11, Schottky barrier height and ideality factor in Chapter 12).

Section 18.2 reviews the common electrical techniques used for material characterization. These include measurement of conductivity (four point probe, contactless methods), doping profile (CV measurements), and mobility (Hall effect, "FATFET" technique).

Section 18.3 reviews electrical techniques used to characterize devices. This section is heavily weighted toward the field effect transistor (FET) — the dominant GaAs device. It reviews precautions necessary to obtain accurate results, including taking proper account of resistance on the source side of the measuring system. Techniques are described to measure parasitic resistance, characteristic voltages, and other FET parameters.

18.2 CHARACTERIZATION OF MATERIALS

There are a large number of analytical tools to evaluate GaAs material properties. Many of these are discussed in Chapter 19. This section is limited to non-destructive, electrical tests commonly used to evaluate the basic material properties of conductivity, doping profile, and mobility.

18.2.1 Measurement of Sheet Conductivity

A common method that has been employed to measure sheet resistance is the four point probe [1]. In principle, the conductivity of material could be determined using only two probes. In practice, this is virtually impossible to do accurately because of the uncertainty of the contact resistance between the probe and the slice. Therefore, four probes are used: two outer probes to pass the current I and two inner probes to measure potential difference, V (using a high impedance voltmeter to avoid affecting the current). Although many configurations are possible [2], the most common approach uses a linear array of equally spaced probes, as shown in Figure 18.1. If the probes rest on a homogeneous, semi-infinite medium, the sheet resistance is given by

$$R_s = 2\pi S \frac{V}{I}$$

where S is the spacing of the probes. For application to GaAs slices, two corrections must be made. First, the material is not infinitely thick; in fact, the conductive layer is usually very thin (with respect to the probe spacing). Secondly, the lateral extent of the slice is not infinite either. If the thickness of the active layer is much less than the probe spacing, these two corrections are independent of each other. But both corrections are generally combined into one correction factor, C_f, and the equation is

$$R_s = C_f \frac{V}{I} \text{ ohms/square}$$

In this case, the correction factor for the first effect (thin layer) is $\pi/\ln 2 = 4.54$. Correction for the finite lateral extent is more complicated, and a number of different cases have been treated [1,2]. However, if the distance from any edge of the slice to the nearest probe is over ten times the probe spacing, the correction factor will be nearly 1.

Contactless methods have also been developed to measure sheet resistance and commercial instruments exist for this purpose. These are very attractive because they do not require fabrication of test patterns or modification of the surface. Hence, they may be used for rapid, routine evaluation of all slices. A typical method uses two magnetic coils. A conductive material placed between the coils changes the Q of the circuit. Such methods require calibration using a slice of known sheet conductivity and are generally not useful for low doping concentrations. They also average conductivity over an area of approximately 1 cm².

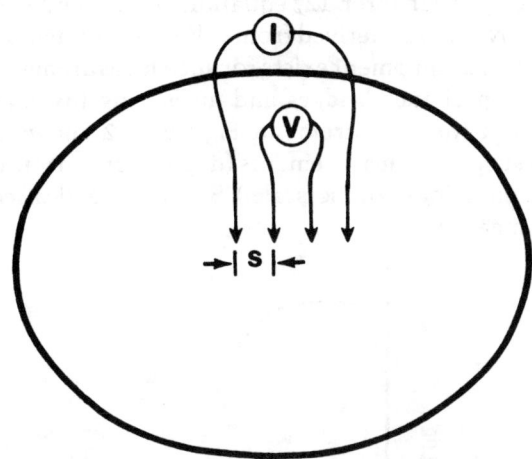

Figure 18.1 Typical configuration of a four-point probe used to measure contact resistance.

18.2.2 Measurement of Doping Profile (CV Method)

The doping profile of semiconductors is typically measured using a Schottky diode and using the relationship between applied voltage and diode capacitance that was derived in Chapter 12. Equation (12.10) is repeated here:

$$N(x) = \frac{2}{q\epsilon A^2}\left(\frac{-1}{\frac{d(1/C^2)}{dV}}\right) \tag{18.1}$$

where N is the doping concentration, q is the electron charge, ϵ is the permittivity, A is the diode area, C is the capacitance, and V is the applied voltage. For profiling purposes, this equation is usually written with the differentiation completed:

$$N(x) = \frac{C^3}{q\epsilon A^2}\left(\frac{dC}{dV}\right)^{-1} \tag{18.2}$$

And by equation (12.8), the depth x is given

$$x = \epsilon A / C \qquad (18.3)$$

(Different authors may or may not place a negative sign on the right-hand side of equation (18.2). This depends on how q and V are treated. C is decreased by making V more negative; q is the electron charge. Regardless of the convention being used, $N(x)$ should be a positive number!). As described briefly in Chapter 12, equations (18.2) and (18.3) allow the doping profile, $N(x)$, to be derived from C-V measurements on a Schottky diode. Commercial equipment exists for such measurements, or computer controlled capacitance bridges and measuring instruments may be used to perform the measurement. Figure 18.2 shows a $N(x)$ profile derived from such C-V measurements (doping concentrations are usually presented on a logarithmic scale.) Several practical considerations apply to this measurement.

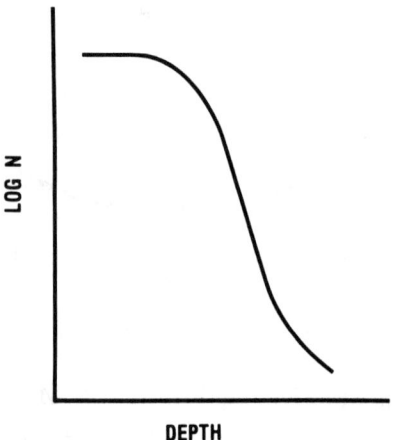

Figure 18.2 Doping profile obtained from capacitance-voltage measurements of a Schottky barrier.

Because of the built-in voltage, $V_{bi} \approx 0.8\ V$ (Chapter 12), the depletion region under the Schottky contact extends some finite distance into the material. This means it is not possible to profile material very near the surface. This inability is represented by the $N(x)$ plot in Figure 18.2 which does not begin at the surface, but at some depth into the slice. The Schottky diode can be forward biased to a minor extent to shrink the depletion region and profile a portion of this zone. But forward leakage current in the diode prohibits much improvement from the zero bias condition. In practice, shining light on the slice (as from a microscope light) generates enough electron-hole pairs to essentially yield the same effect.

The above considerations indicated a minimum depth that could be profiled using the C-V procedure. There may also be a maximum depth set by reverse breakdown of the Schottky diode. As the breakdown condition is approached, current flows and the measurement technique becomes invalid. This is a serious restriction. GaAs slices suitable for some applications (such as recessed-gate depletion mode FETs) will have active layers and doping concentrations that result in breakdown before the entire layer can be profiled. A destructive method to avoid this difficulty is to etch closely spaced portions of the slice to different depths, fabricate diodes at each level, and make C-V measurements at each depth. The various data can be combined into a composite profile as indicated in Figure 18.3. The procedure assumes that no significant lateral variations exist within the extent of the array of diodes. On slices used to fabricate recessed-gate FETs, the most relevant doping concentration is under the gate. Often, C-V Schottky diodes will be fabricated at the same step as the recessed gate, and so will be recessed into the material. These usually allow the entire remaining thickness to be profiled without breakdown problems (otherwise, the FET would not pinch off).

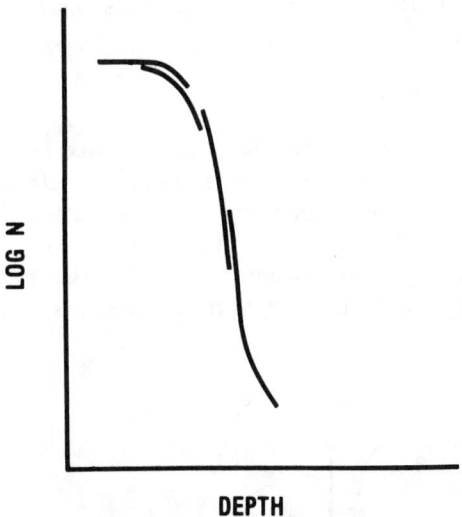

Figure 18.3 Piecewise doping profile obtained by step etching.

Even if breakdown does not limit profile depth, series resistance effects can limit accuracy [3]. As indicated in Figure 18.4, planar Schottky diodes will exhibit as a series resistance that is a function of the depletion

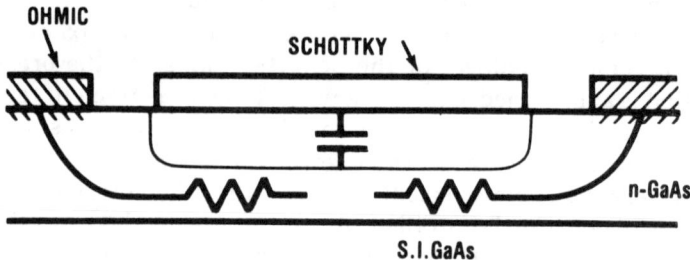

Figure 18.4 Series parasitic resistance that limits C-V profiling accuracy.

depth into the material. It is clear that this resistance will rise as the depletion region extends into low doped regions. Almost all capacitance meters and profiling instruments measure capacitance by applying a constant *rf* voltage (much smaller than the *dc* voltage that determines the diode capacitance) and monitoring the imaginary component of the resulting *rf* current. This technique is completely appropriate if there is no series resistance. However, if there is series resistance, the measured capacitance will be given by [3]

$$C' = \frac{C}{1 + \omega^2 R^2 C^2} \tag{18.4}$$

where C is the actual depletion layer capacitance, ω is the measurement frequency, and R is the series resistance. This effect leads to serious errors in the tail of doping profiles, over-estimating the doping concentration (Figure 18.5). In principle, these errors can be corrected if the phase angle of the *rf* measurement signal is also monitored [3]. Such corrections are crucial if accurate measurements of the doping profile

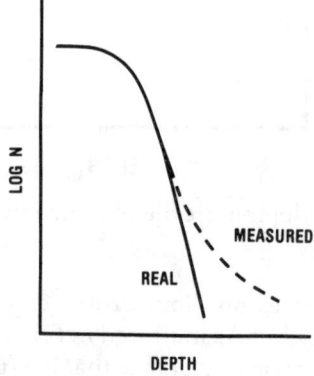

Figure 18.5 Actual doping profile and measured profile, the inaccuracy being caused by series resistance.

tails are desired. However, it is useful and practical to ignore these errors for routine in-process measurements made on test diodes. For slices having peak doping near 10^{17} cm^{-3}, such measurements generally will accurately yield the doping profile above approximately 10^{16} cm^{-3}. Serious error will occur below 10^{15} cm^{-3}.

The test diode should have a Schottky barrier large enough to allow accurate measurement of the area. The square of the area is used in the formula, so accuracy is important. The Schottky must also have a capacitance large enough for accurate measurement using capacitance bridges. High aspect ratio geometries should also be avoided to reduce edge effects (parasitic capacitances). These considerations eliminate modern FETs as suitable test devices. A reasonable range for diode area is approximately 0.01 to 0.04 mm^2. The ohmic contact should completely surround the Schottky contact to minimize series resistance.

Resolution of the C-V method is limited to a distance on the order of the Debye length [4]. The Debye length arises from electron screening effects and represents the departure from the sharp boundary usually assumed at the bottom of the depletion zone. The Debye length is given by

$$L_D^2 = \frac{kT\epsilon}{q^2 N}$$

and is on the order of 0.02 μm for doping levels near 10^{17}.

It is also possible to use the C-V method on unpatterned slices using a mercury probe. This instrument uses two or more closely spaced tubes which contain mercury. These are placed against the slice and pressurized gas used to press the mercury in intimate contact with the slice, forming Schottky diodes. The absence of an ohmic contact is not severe — two Schottky diodes may be used with appropriate modification of the formula (one Schottky is forward biased). Alternatively, if one Schottky is significantly larger than the other, it may be treated as an ohmic. However, there are a number of practical difficulties in using mercury probes (such as keeping the mercury clean and confined) and they have not become as popular as originally anticipated.

18.2.3 Measurement of Hall Mobility

There are two types of mobility and two corresponding methods of measurement. The Hall mobility represents the response of the carrier to a driving magnetic force (the Hall effect). The drift mobility represents the response of the carrier to a driving electric field (section 18.2.4). The two types of mobility are generally not equal, but are related. For most semiconductor applications, the two agree to within 10 to 20%.

Both Hall and drift mobility are commonly used to assess material quality — mobility is reduced by increased impurity concentration. Both are also a function of temperature (Chapter 2).

The basic Hall effect is illustrated in Figure 18.6. An electric field causes carrier drift in the x direction. A magnetic field is applied in the z direction. The Lorentz force causes the carriers to develop a velocity component in the y direction. No dc current is allowed to flow in the y direction so the deflected carriers set up an electric field in the y direction that exactly balances the Lorentz force. This is designated the Hall field; it is measured externally. If E_y is the Hall field, J_x the current density in the x direction, and B_z the applied magnetic field, then the Hall coefficient, R_H is defined by the equation

$$E_y = R_H J_x B_z$$

The Hall mobility μ_H is defined to be the product of the Hall coefficient and the conductivity:

$$\mu_H = |R_H \sigma|$$

Figure 18.6 shows a test pattern appropriate for measuring Hall mobility of conductive layers formed on semi-insulating GaAs substrates. The side arms allow good contact to the Hall structure without significantly distorting the current flow. Accurate Hall effect measurements entail a number of detailed, practical considerations [2]. Specifications and precautions for making such measurements are specified by the American Society for Testing and Materials in the document ASTM F 76.

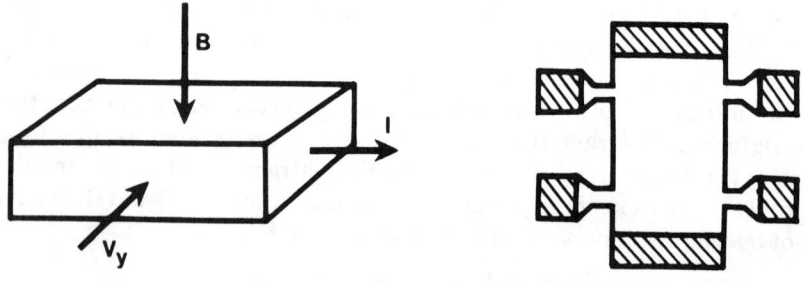

Figure 18.6 Hall mobility, and typical pattern used to measure it.

18.2.4 Measurement of Drift Mobility (FATFET)

The above subsection defined the difference between Hall and drift mobility. Drift mobility is the type relevant to FET device physics and performance. An approximate determination of the average drift mobility of active layers formed on semi-insulating substrates may be obtained by measuring the resistance between two ohmic contacts spaced a distance L apart. If the voltage is well below that causing current saturation (electric field = 3.3 kV/cm), then

$$I = qvNWa = q\bar{\mu}_d ENWa \tag{18.5}$$

where $\bar{\mu}_d$ is the average drift mobility, v is the electron velocity (= $\mu_d E$), E is the electric field ($E = V/L$), N is the average doping concentration, W is the width of the ohmic contacts, and a is the thickness of the active layer. Then

$$\mu_d = \frac{L}{qNWa}\left(\frac{1}{V}\right) \tag{18.6}$$

However, the two contact resistances must be taken into account. If V_a is the applied (external) voltage, then the voltage V that determines the electric field in the semiconductor is

$$V = V_a - 2IR_c$$

where R_c is the contact resistance of one contact. Hence,

$$V = V_a(1 - 2IR_c/V_a) \tag{18.7}$$

Equations 18.6 and 18.7 yield

$$\mu_d = \left(\frac{L}{qNWa}\right)\left(\frac{1}{V_a}\right)\left(\frac{1}{1 - \frac{2IR_c}{V_a}}\right) \tag{18.8}$$

If the channel resistance is much greater than the contact resistance, then $2IR_c \ll V_a$ and the correction terms can be ignored. In any event, the average mobility calculated in the above manner gives a very approximate answer: the thickness of the active channel must be known accurately and surface depletion effects are ignored. Also, mobility is a function of doping concentration (Chapter 2) so it is more useful to obtain mobility as a function of depth.

The most popular method used to measure drift mobility profiles uses a long gate-length FET [5] (other methods exist [6], but are more cumbersome and will not be described here). The technique uses a FET geometry that has become known as a *FATFET*, so-called because it is a planar FET with a very fat gate, as shown in Figure 18.7. Gate length is typically 20 μm to 250 μm (compared to ≤ 1 μm in most FETs). By measuring the C-V and I-V (or transconductance — see below) profile of this FET, the mobility profile can be calculated. The doping profile, of course, is simultaneously obtained using the C-V information (section 18.2.2). This method may be summarized by noting that a differential increase in gate voltage (more negative) will increase the depletion zone depth an amount dx and reduce the source-drain current, I, a differential amount, dI. The C-V data yield the depth, x, and the doping concentration at that depth, $N(x)$. The mobility can then be obtained using

$$dI = q\mu_d(x) EN(x) W dx \qquad (18.9)$$

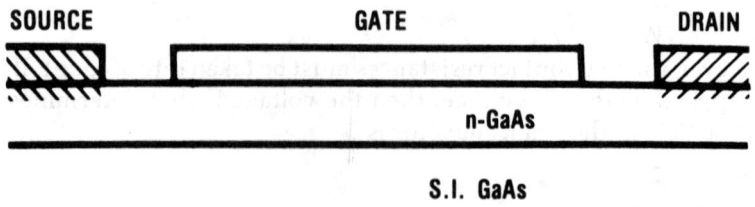

Figure 18.7 FATFET geometry used to measure drift mobility.

This process assumes that the electric field is well below that causing current saturation. It is also small enough that there is little skewing of the depletion region toward the drain contact. That is, the depletion depth at the source end of the gate is equal to the depletion depth at the drain end. The relevant expressions will now be derived in more detail. Equation (18.8) yields

$$\frac{dI}{dV_g} = q\mu_d ENW \frac{dx}{dV_g} \qquad (18.10)$$

where V_g is the gate voltage. Equation (18.3) gives

$$\frac{dx}{dV_g} = \left(\frac{-\epsilon A}{C^2}\right)\left(\frac{dC}{dV_g}\right) = \frac{C}{qAN} \qquad (18.11)$$

where the last step is obtained using equation (18.2) (assuming for this purpose a negative sign in equation (18.2) — see comment immediately following equation (18.3). Equation (18.10) and (18.11) give

$$\frac{dI}{dV_g} = \mu_d \frac{EC}{L} = \mu_d \left(\frac{VC}{L^2}\right)$$

or

$$\mu_d = \left(\frac{dI}{dV_g}\right)\left(\frac{1}{C}\right)\left(\frac{L^2}{V}\right) = \left(\frac{g_m}{C}\right)\left(\frac{L^2}{V}\right) \tag{18.12}$$

where the definition $g_m = dI/dV_g$ has been used. (An alternate derivation of this expression may be found in reference [5].) Hence I-V_g data and C-V_g data yield both the mobility profile and the doping profile. Analysis could proceed using the data to compute $N(x)$, x, and then $\mu_d(x)$ using equation (18.9).

Alternatively, the information may be used in equation (18.12), which is the format usually quoted. The I-V data supplies dI/dV; the C-V data supplies $N(x)$ and x (equations (18.2) and (18.3). As in the case above, the contact resistance may have to be included. As shown in Figure 18.7, there are the contact resistance and the channel resistance between the Schottky contact and each ohmic contact. If the FET gate is "fat" enough, and the gate edge to ohmic spacing small enough, these end resistances will amount to errors of less than 2% and can be ignored. Otherwise, they must be taken into account in the same as described above.

These measurements could be obtained using separate C-V and I-V structures placed side-by-side (data would be correlated using the same gate voltage). However, it is preferable to perform both measurements on the same test structure. This approach requires that the FATFET gate be sufficiently large to allow accurate capacitance measurements. Accuracy is reduced as the FATFET approaches pinch-off, caused by inaccuracies in the C-V data resulting from series resistance as described in the previous subsection.

18.3 DEVICE CHARACTERIZATION

There are a number of standard and useful electrical measurements made on GaAs devices, especially Schottky diodes or FETs, that can and should be used for in-process monitoring. The type and complexity of the measurements depend on the application, as can be exemplified by considering the pinch-off voltage of a FET. For the device user (and hence, the device processor), V_{po} is usefully defined as the gate voltage that results in a source-drain current equal to 2% of I_{dss}. This is easily measured and specified. Other definitions are more appropriate in detailed studies of device physics; one definition is the intercept of the $\sqrt{I_{ds}}$ vs. V_g curve (using the linear part of the data — Chapter 3). Still more complex definitions exist [7]. Results will also depend on whether parasitic resistances are included (the "extrinsic" or "intrinsic" quantity). Measurements on FETs from the perspective of device physics have been considered elsewhere [7,8]. This section is restricted to techniques especially useful for in-process evaluation. These parameters are also relevant to the device user.

In-process measurements are usually made using a probe station that allows small probes to be placed on device pads. Several practical considerations are crucial to obtaining accurate results. The contact resistance resulting from the use of probes can and usually must be taken into account, as described in section 18.3.1. This contact resistance is a function of the pressure applied between the probe and the device pad — it should be consistent from device to device. Too low a pressure will result in inconsistency. Too great a pressure can result in bent or dulled probes, or sliding probes which damage devices. Probes should be examined often and replaced or resharpened as needed.

Because modern devices such as FETs have very high gain, they have a strong tendency to oscillate. The extended wiring found in probe stations is usually an ideal external circuit to support low frequency oscillations. It is therefore useful, and often essential, to take precautions such as the use of shielded cable and/or placement of bypass circuitry at each probe to short out low frequency oscillations. Even with these precautions, oscillatory effects may be observed that are sensitive to grounding, wire movement, or physical contact of a probe by a hand.

Many measurements are also sensitive to external illumination (the light generates electron-hole pairs). The lighting associated with the microscopes used on probe stations can be a particularly powerful source of illumination. It should be switched off during actual measurements unless there is a specific reason not to do so (such as evaluating light sensitivity).

Resistance in the source circuitry during measurements of FET parameters can lead to significant errors. Section 18.3.1 explores this effect. The remaining subsections address various FET measurements.

18.3.1 Source Resistance Effects

Two important quantities used to characterize FETs are the saturated current, I_{dss}, and the transconductance, g_m. The apparent value of these quantities can be significantly altered by the presence of resistance on the source side of the device. Resistance of even one ohm, for example, in the measurement system (on the source side) can be significant when probing high current FET devices. This will be considered in more detail after basic equations describing the effect are derived. Consider the FET and the resistance, R_s, shown in Figure 18.8. Following general usage, this quantity will be referred to as "source resistance" in this subsection — it refers to resistance anywhere in the source circuit (see below). However, after this subsection, the term "source resistance" will be used in another, more specific manner. In that context (see next subsection) it refers to source resistance in the device and does not include resistance in external circuitry. This variation in usage is not as confusing as it seems; the context is usually clear. Again, this subsection uses the more general meaning.

Referring to Figure 18.8, source-drain current flows through the device and also through the source resistance (the common source configuration is assumed). A voltage, V_s, appears across the resistance because of this current flow ($V_s = IR_s$). V_s is positive with respect to ground (the drain bias voltage is positive). An external gate voltage, V_e, is applied between the gate and ground. However, the effective gate voltage is the potential difference between the gate and the FET source, V_i. V_i is more negative than V_e and hence, less source-drain current flows than would occur in the absence of the resistance. The transconductance is also

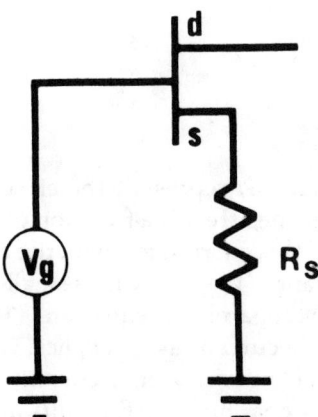

Figure 18.8 Parasitic resistance in the source circuit of an FET. This resistance can affect measurement of some FET parameters.

affected. If g_{me} and g_{mi} refer to the external and intrinsic transconductances (with and without R_s), then

$$g_{me} = dI/dV_e, \quad g_{mi} = dI/dV_i \qquad (18.13)$$

and

$$V_e = V_i + IR_s \qquad (18.14)$$

From (18.14) we have

$$dV_e = dV_i + dI\, R_s$$

or

$$\frac{1}{g_{me}} = \frac{1}{g_{mi}} + R_s$$

which can be rearranged to give the standard expression

$$g_{mi} = \frac{g_{me}}{1 - g_{me} R_s} \qquad (18.15)$$

So the effect of the source resistance is to make the measured transconductance, g_{me}, smaller than the actual transconductance, g_{mi}.

This effect will also lower the source-drain current. If I_i represents the current with $R_s = 0$, I_e the current with $R_s \neq 0$, and dI the difference, then

$$I_i = I_e + dI$$

and

$$dI = \frac{dI}{dV_i}, \quad dV_i = g_{mi} I_e R_s$$

and hence,

$$I_i = I_e (1 + g_{mi} R_s) \qquad (18.16)$$

Equations (18.15) and (18.16) represent the effect of the source resistance. The magnitude of the effect is a function of the magnitude of the source-drain current. If this current is small, the extra voltage dropped across the source resistance ($V_s = IR_s$) will be small and have little effect. This is not immediately apparent in equations (18.15) and (18.16) because they do not include current as an explicit variable; the transconductance is the variable. However, a small current (whether due to small gate width or due to bias near pinch-off) results in a small transconductance. Therefore, small gate-width FETs will be less sensitive to the effect than will large gate-width FETs, such as power FETs. To illustrate this difference, consider two FETs fabricated side-by-side on the same slice, one having ten times the gate-width of the other. The smaller gate-width device might be present for in-process measurements. The saturation current of the larger device should be ten times that of the smaller device. But unless the above effect is considered, measurements of the saturation current will not correspond to that ratio. Assume the external measuring system to have a total source resistance of one ohm (including cables, probe, and contact resistance between the probe and the source pad of the FET). Assume the (real) transconductance of the FETs to be 40 mS and 400 mS. Then by equation (18.16) the ratio I_i/I_e of the two devices would be 1.04 and 1.4 respectively. Clearly, measurements made on large gate-width FETs can be very misleading if the source resistance of the measuring system is not considered. Similar considerations apply to the transconductances.

Exactly the same considerations may be applied to the effect of internal parasitic source resistance in the FET. That source resistance includes the contact resistance of the source ohmic contact plus the resistance of the GaAs material between the ohmic contact and the gate metal. Therefore, there are at least three different transconductances, or saturation currents, that could be considered. First, the externally measured one that includes the effects of the measuring system's resistances. Second, the transconductance in which "ground" is placed at the FET's source pad. Third, the transconductance that would occur if the FET contained no internal parasitic source resistance. This latter quantity is referred to as the "intrinsic" transconductance and is the one used in the equivalent circuit model described in section 3.2. It also indicates the FET's maximum performance if parasitic source resistance could be reduced to zero. The second case above is the "extrinsic" transconductance. It is the one most meaningful to the circuit designer using the device, and best represents the parameters of the complete physical device.

"Probe resistance" usually is taken to mean all resistance in the measuring system up to the pad the probe is placed on. Therefore, it includes the contact resistance between the probe and the pad, the resistance of the probe itself, the cable, and the instrument to which it is connected. Measuring this resistance usually consists of placing two probes on the same metal pad, as close together as possible, but without the two probes actually touching each other. The total contact resistance is determined by the voltage required to draw a certain current (i.e., the slope of the *IV* trace on a standard curve tracer). Half this amount is ascribed to each probe.

A more accurate method is to use three probes. Three probes are usually needed anyway for characterizing FETs (source, drain, and gate probes). The total resistance is measured for each of the three possible pairings of the probes. This gives three equations in three unknowns and the probe resistance of each can be determined. In the ideal case, all would be equal. In practice, they rarely are. In fact, probe wear can change the values daily. If high accuracy is required, such measurements should be made often. As discussed above, highly accurate knowledge of the source resistance is more important when measuring high current devices than when measuring small current devices. If computer controlled probing stations are used for such measurements, the above determination of probe resistance can be easily programmed and used before every slice if need be.

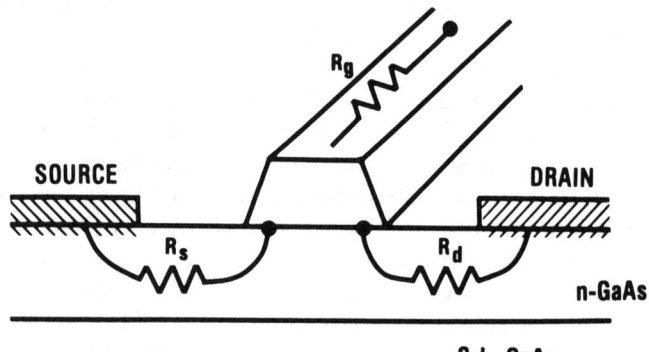

Figure 18.9 Diagram illustrating the definition of various parasitic resistances in the FET.

18.3.2 Parasitic Resistances of FETs

There are minor variations in terminology used in the literature. The definitions given below are adopted in this subsection (refer to Figure 18.9):

R_s = source resistance = the resistance between the source metal and the gate. It includes the ohmic contact resistance of the source. It does not include the metallic resistance of the gate stripe.

R_g = gate resistance = the metallic resistance of the gate stripe in the FET configuration (the distributed nature of the geometry results in the gate having an effective resistance that is 1/3 of the end-to-end resistance — see below).

R_{sg} = source-gate resistance = $R_s + R_g$

R_d = drain resistance = the resistance between the drain metal and the gate (analagous to source resistance).

R_{sg} may be measured by forward biasing the gate with respect to the source. The forward breakdown will have the form shown in Figure 18.10. At low current, the forward I-V is dominated by the exponential form typical of Schottky barriers (Chapter 12). At intermediate current,

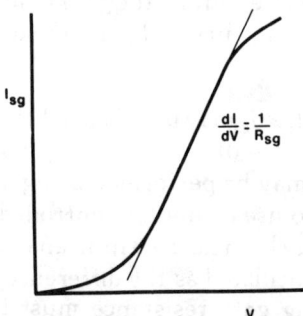

Figure 18.10 Forward gate I-V of an FET. The linear region is used to determine the sum of the gate and source resistance, R_{sg}.

the forward I-V is dominated by the material resistances. At high current, current saturation begins and dominates the IV characteristic. The middle region, the linear IV, therefore represents the total resistance of the source-gate path. (In this and all other measurements described subsequently, resistance associated with external circuitry and probes is assumed to be accounted for.)

R_s is measured in a slightly more involved manner. Assuming the source is the common or ground connection, voltages are impressed on the gate and drain which result in small to moderate forward gate current and very small (compared to the gate current) drain current. The forward gate current is increased in precise current steps. As indicated in Figure 18.11, the current reaching the drain is so small (compared to the source-gate current) that the drain voltage is essentially equal to the

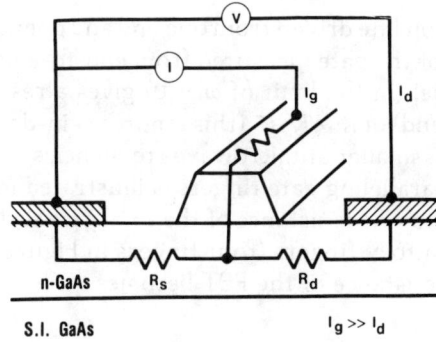

Figure 18.11 Diagram illustrating the method used to measure source resistance.

voltage between the source pad and the gate stripe. Hence, if the forward gate current is I_g, the drain current I_d, and the drain voltage V_d, then

$$R_s = dI_g/dV_d, I_g \gg I_d$$

(the requirement that $I_g \gg I_d$ may be relaxed if the measurement is made under the condition of $dI_d = 0$).

This measurement may be performed using a standard curve tracer. However, it is easier to use computer controlled instrumentation. The drain resistance, R_d, may be measured in an analogous manner. Knowing R_{sg} and R_s, R_g may be calculated as the difference.

Two issues involving gate resistance must be considered in these measurements. They are the difference between apparent gate resistance and end-to-end resistance; and the effect of multiple gate fingers connected in parallel. As illustrated in Figure 18.12, the source and gate (metal) resistances are not connected in series. Rather, the gate-source resistance connection is distributed. In practice, the gate may be thought

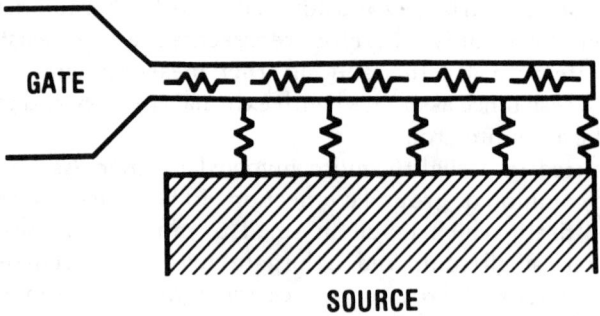

Figure 18.12 The distributed resistive connection between the gate and the source metalization.

of as a transmission line driven from one end and open at the other. If R_{ee} is the resistance of the gate measured from end-to-end, then the transmission line model (in the limit of $\omega \to 0$) gives a resistance (apparent from the driven end) of $R_g = R_{ee}/3$ (this ignores skin-depth effects which may make it even smaller at microwave frequencies).

The effect of paralleling gate fingers is illustrated in Figure 18.13. If the R_{ee} is the end-to-end resistance of the total gate width, and this total width is divided into N fingers (four fingers in Figure 18.13), then the equivalent gate resistance of the FET becomes

$$R_g = \frac{1}{3N^2}$$

where the factor of three arises as explained above, the factor of N^2 arises from paralleling N sections (or fingers). One factor of N comes from dividing the total width into N fingers; another from joining these N sections in parallel. Note that a special consideration arises in the "T" or "Pi" geometry of Figure 18.13(b). Probes must be placed on all the gate pads to properly measure gate or source-gate resistance as described above. Contacting only one of several gate pads will still allow accurate measurements of quantities such as I_{dss} or g_m. But quantities that involve current flow through the gate must be measured with contacts to all gate pads (or modifications made in the calculations). These considerations do not apply to the parallel finger geometry of Figure 18.13(a) — there is only one gate pad.

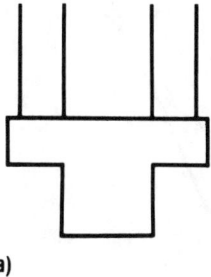

a)

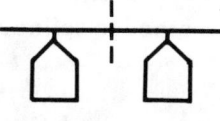

b)

Figure 18.13 The affect of paralleling gate fingers on gate resistance. Each configuration represents four parallel gate fingers — (a) is a parallel finger geometry, (b) is a "T" or "Pi" gate geometry.

18.3.3 Characterstic Voltages in FETs

Several voltages are of interest in characterizing FETs. The symbols are defined as follows:

V_{sat} = the saturation voltage (also called the knee voltage)
V_{bdgs} = the reverse breakdown voltage between gate and source
V_{bdgd} = the reverse breakdown voltage between gate and drain
V_{gf} = the forward gate breakdown voltage
V_{po} = the pinch-off voltage

Exact definitions of these quantities depend on the method of measurement. Each quantity is discussed below.

The saturation voltage is the voltage between source and drain at which current saturation occurs. As indicated in Figure 18.14, the saturation as it appears in a typical FET I-V curve is somewhat gradual. One useful definition is indicated in this Figure — the intersection of lines fitted to the linear and saturated parts of the trace. Another useful approach is to simply define V_{sat} as being some percent (such as 90%) of I_{dss} (which, of course, also depends on the precise definition of I_{dss}).

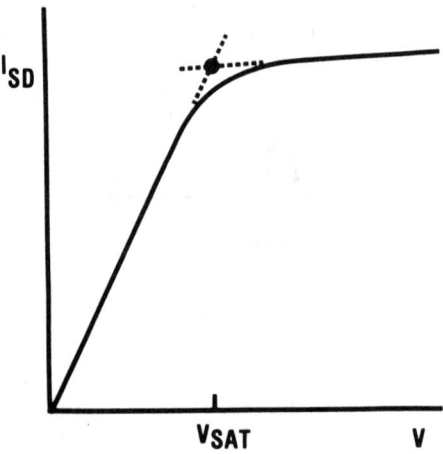

Figure 18.14 One possible method to define saturation voltage (or knee voltage).

The breakdown voltages are usually defined as the voltage required to draw a given current (per unit width of total gate periphery). Note that a definition based on extending a line fitted to the linear portion of the breakdown curve (Figure 18.10) until it intercepts the voltage axis is not reasonable — results depend critically on the resistance of various materials in the current path (including the gate metal). In this regard, the lower the specified current, the less the effect of parasitic resistance on the measurement.

Pinch-off voltage may be defined as the voltage required to reduce the source-drain current to a given percent (for example, 2%) of I_{dss}. This is the type of definition used on manufacturers' data sheets. It is a very easy method to implement. Of course, the definition also depends on the exact definition of I_{dss}. Other definitions are more appropriate for serious study of device physics [7].

18.3.4 Currents and Transconductance in FETs

The two currents of interest in FETs are as follows:

I_{dss} = the saturated source-drain current (V_g = 0)

I_{max} = the maximum source-drain current (under forward gate bias)

There are two common methods used to specify I_{dss}. The method used on manufacturers' data sheets is to define I_{dss} as the source-drain current at a specified source-drain voltage and with V_g = 0. If the I-V trace of the FET exhibits a maximum (see Chapter 3), I_{dss} may be defined as the source-drain current at that maximum (still specifying V_g = 0). Note that the gate must be grounded (held at V_g = O). If the gate is electrically floating, it wil acquire charge and the source-drain current will change in an ill-defined way.

I_{max} is determined by forward biasing the gate. This shrinks the depletion region and allows greater source-drain current to flow. The magnitude of forward voltage is limited by forward breakdown. Increasing the gate voltage in the positive direction above V_g = 0 will result in the source-drain current increasing until it reaches a maximum. But this is not a good method to determine I_{max}. As the forward gate voltage approaches breakdown, source-drain current increases very little. Device damage may occur. It is also inappropriate to define I_{max} as the current at a specified forward gate voltage. In that case, results will depend on the parasitic resistances (contact resistance in the ohmic, gate metal resistance, etc.). The preferable method is to specify the forward gate current (per unit gate width) and define I_{max} as the source-drain current at that forward gate current.

Transconductance is defined as

$$g_m = \frac{dI}{dV_g}$$

but practical measurements are generally made with finite steps in gate voltage. Hence, the measurement requires specification of the initial gate voltage, the gate voltage step, and the drain voltage at which the measurement is made (this is usually the drain voltage at which I_{dss} is defined). Because of the nonlinear behavior of source-drain current as a function of gate voltage, g_m will become less as pinch-off approaches. This effect also means that a smaller voltage step (beginning from the same initial voltage) will yield a higher transconductance. Transconductance is usually specified at V_g = 0 for a (negative) voltage step that typically is 0.25 to 1.0 volt. Sometimes, g_m is also specified at some fraction of I_{dss}, such as 20% (high transconductance near pinch-off is important for low noise FETs). Of course, transconductance is a function of the total gate width of the device, and so the width must also be given. Alternatively, g_m may be normalized to a unit gate width, usually 1 mm.

REFERENCES

[1] S.M. Sze, *Physics of Semiconductor Devices*, Second Edition. New York: Wiley, 1981.

[2] W.R. Runyan, *Semiconductor Measurements and Instrumentation*. New York: Wiley, 1981.

[3] J.D. Wiley and G.L. Miller, *IEEE Trans. Electron Devices*, 22, 1975, p. 265.

[4] W.C. Johnson and P.T. Panonsis, *IEEE Trans. Electron Devices*, 18, 1971, p. 965.

[5] R.A. Pucel and C.F. Krumm, *Electronics Letters*, 12(10), 1976, p. 242.

[6] K. Lehovec, *Appl. Phys. Lett.*, 25, 1974, p. 279.

[7] H. Fukui, *Bell Syst. Tech. J.*, 58, 1979, p. 771.

[8] H. Fukui, *Solid State Electronics*, 22, 1979, p. 507.

CHAPTER 19

DIAGNOSTIC TECHNIQUES

19.1 INTRODUCTION

It was noted in Chapter 1 that processing does not consist of following well-specified steps and obtaining guaranteed results. A great many things can go awry. Even in the well-established silicon processing industry, a process engineer will spend a good part of his time determining what has gone wrong. Such problems are much more common in the emerging GaAs processing technology. The small size of GaAs devices and their sensitivity to contamination make problems both likely and difficult to diagnose easily. But such problems must be understood and resolved rapidly; a process engineer must be a good detective. This is not a trivial matter. A process facility that cannot solve problems rapidly and efficiently will not be able to produce devices at adequate yield and cost.

The process engineer has two categories of tools to attack such problems. First, he must have a solid understanding of the basic principles of each process. This book is an introduction to these principles. Secondly, he should have available diagnostic tools and techniques to analyze materials — these are reviewed in this chapter. Sometimes, the first set of tools is sufficient. A good understanding of the fundamentals allows an educated guess as to the cause of a problem and its solution. A quick measurement or experiment can then confirm the diagnosis. Examples might include poor metal adhesion following a change in the cleaning process, or poor pattern definition in resist caused by a change in exposure light intensity. However, other problems are more difficult and the techniques reviewed in this chapter are enormously useful. A metalization that exhibits greater resistivity than usual can be examined for alloyed constituants and contaminants, which in general can be identified; GaAs material can be examined for crystallinity. Major silicon device manufacturers, knowing the value of diagnostic instrumentation, will have a sophisticated array of these tools available.

As treated in this chapter, the diagnostic techniques are divided into surface and bulk techniques. This division is somewhat arbitrary, but tends to follow general usage. In this sense, the surface techniques are those which are sensitive to the top 1000 Å of the material or less — some are sensitive only to the top 10 to 20 Å. Most of these techniques have been developed or popularized only within the last two decades, driven by the demands of the silicon semiconductor industry for improved understanding and control of surfaces, materials, and interface behavior of solid-state devices. The GaAs arena has only added to these needs. In total, there are more useful diagnostic and analytical techniques than any one institution could afford or use. The National Bureau of Standards has identified more than 75 techniques for surface analysis, at least 60 of which are involved with active research or application [1,2]. Although some of these use specialized, "homemade" equipment, many can be performed using commercially available instruments costing up to several hundred thousand dollars.

Some of these techniques are referred to by acronyms, the more popular of which are listed in Table 19.1. The scanning electron microscope (SEM) has not been listed or discussed here — it is so basic that it is difficult to conceive of any institution fabricating devices in any efficient manner without one. Large institutions will often have several SEMs. The most popular and useful surface techniques tend to be Auger electron spectroscopy (AES), scanning Auger microprobe (SAM), and x-ray photoelectron spectroscopy (XPS). Worldwide installations of these instruments outnumber the next most used surface techniques by a factor of about three [1,3]. The next group in popularity and usefulness includes secondary ion mass spectroscopy (SIMS) and low-energy electron diffraction (LEED). An excellent review of the surface techniques may be found in reference [3]. The most popular bulk techniques include the electron microprobe and its variations. These and other bulk analysis techniques are reviewed in references [4,5]. Much of the review literature is written in the context of silicon work, but all principles apply to GaAs also.

TABLE 19.1

Acronyms Of Common Diagnostic Techniques

	Surface Analysis Techniques
AES	Auger Electron Spectroscopy
EELS	Electron Energy Loss Spectroscopy
ISS	Ion Scattering Spectroscopy
HEED	High Energy Electron Diffraction (or RHEED: Reflection HEED)
LEED	Low Energy Electron Diffraction
SAM	Scanning Auger Microprobe
SIMS	Secondary Ion Mass Spectroscopy
SXAPS	Soft X-ray Appearance Potential Spectroscopy
UPS	Ultraviolet Photoelectron Spectroscopy
XPS	X-ray Photoelectron Spectroscopy
	Bulk Analysis Techniques
DLTS	Deep Level Transient Spectroscopy
EBIC	Electron Beam Induced Current
EMP	Electron Microprobe
NAA	Neutron Activation Analysis
RBS	Rutherford Backscattering
TEM	Transmission Electron Microscopy
XRF	X-ray Fluorescence

Fundamental characteristics of each technique include sensitivity, analytical spot size, and sampling depth. The sensitivity is the minimum amount of material the technique can detect (concentration, number of atoms, etc.). Analytical spot size is the resolution of the technique in the horizontal plane. This is essentially equivalent to the beam diameter of the probing particle or photon for some techniques, but is appreciably greater for other techniques because of scattering phenomena. The sampling depth is the depth into the material that the technique samples. There are fundamental trade-offs between sensitivity and sampling volume (composed of lateral resolution and sampling depth). A small sampling volume gives high resolution. But there will be fewer atoms in that volume from which to detect signals (photons, electrons, etc.). Hence, greater sensitivity requires less resolution. Nevertheless, there is a wide range of sensitivity displayed by the various techniques. The general range of these parameters for some of the more popular techniques is indicated in Figures 19.1-19.3 [6]. However, these indications

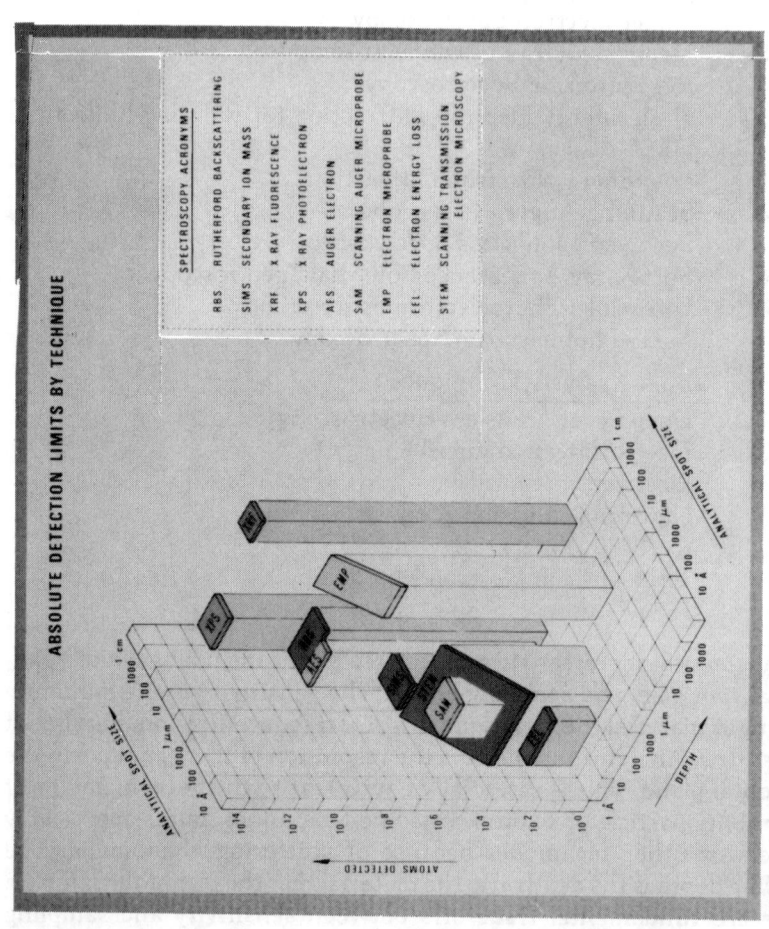

Figure 19.1 Absolute detection limits by technique [6].

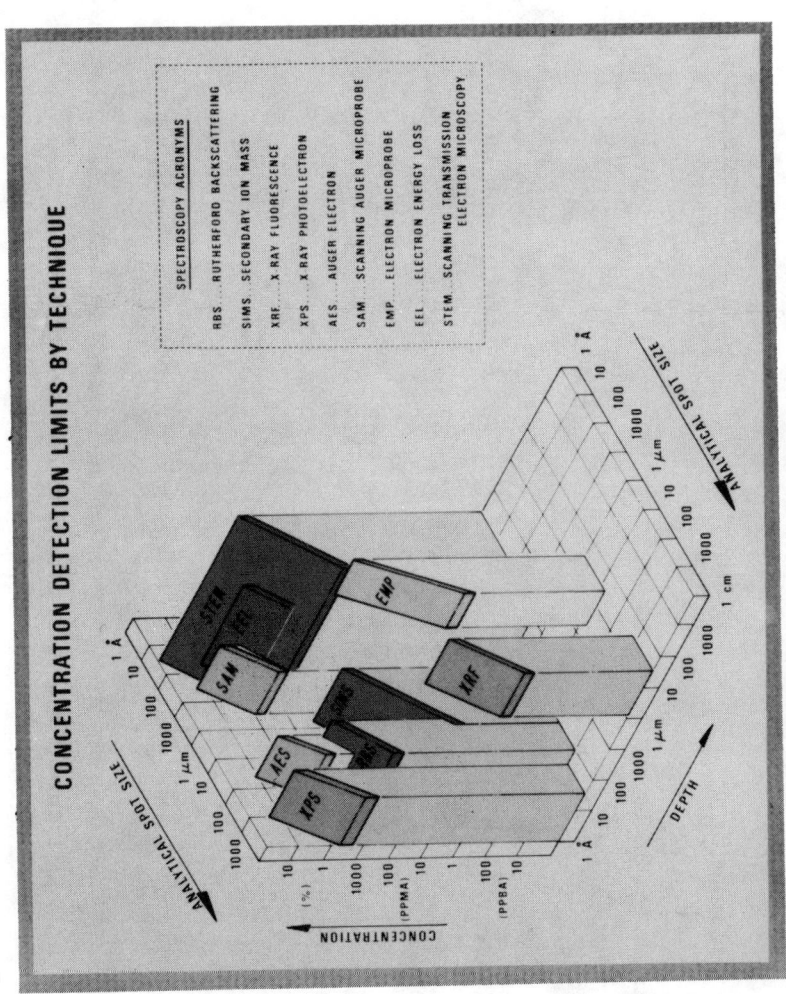

Figure 19.2 Concentration detection limits by technique [6].

390 GaAs Processing Techniques

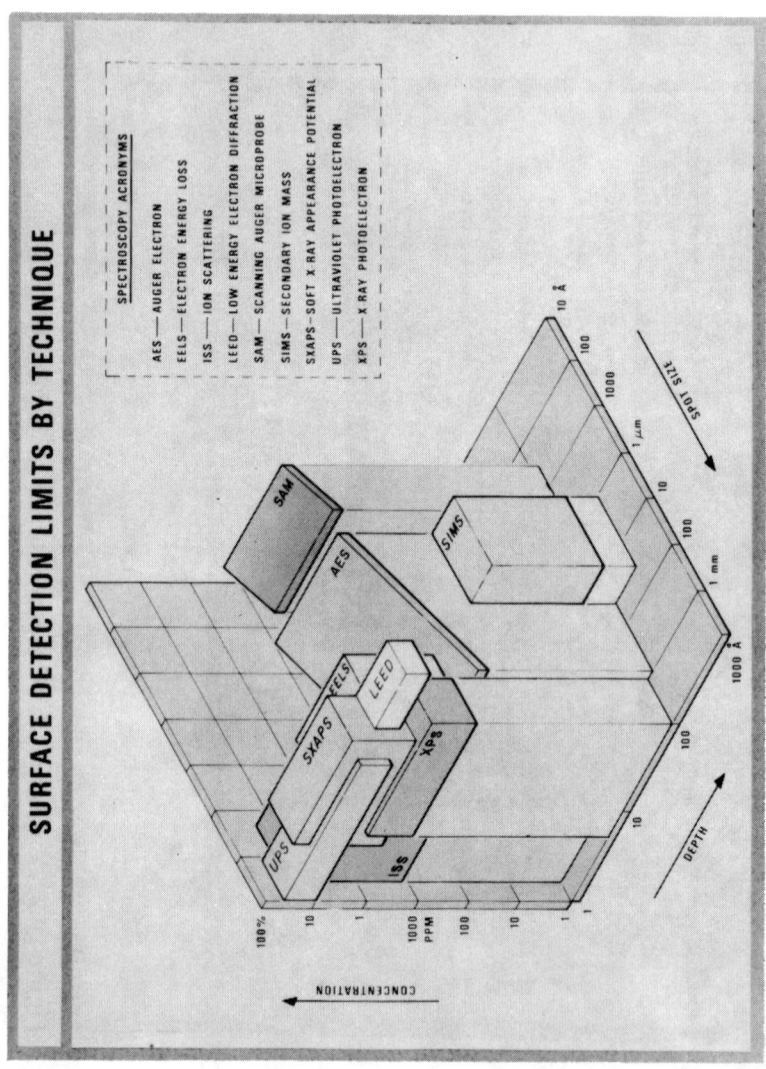

Figure 19.3 Surface detection limits by technique [6].

can only be general. Specific applications can depart from the indicated values. For example, sensitivity will depend on the element being detected, the material in which it is embedded, and the exact configuration of the instrument. Efficient use of these analytical techniques involves appreciably more information than this chapter will impart. Indeed, moderate length books could be written about some of the individual techniques. A good amount of skill and understanding is required to perform and interpret results of many of these measurements. Large institutions usually have persons knowledgeable in these analytical techniques and they essentially operate as a service organization to the institution. This chapter serves to introduce and summarize the more popular and useful techniques.

Most of the techniques involve electron, ion, or photon beams impinging on the sample. The resulting interactions and effects are then monitored by detecting other electrons, ions, or photons; or monitoring other parameters such as conductivity or capacitance transients. The general phenomena resulting from these beams are briefly reviewed in the remainder of this introductory section.

When an electron beam of 1 to 100 keV impinges upon a solid, the electrons interact with the electrons of the atoms comprising the solid, resulting in both elastic and inelastic events (interactions with the nuclei are essentially zero). A rich variety of interactions is possible, as indicated in Figure 19.4. The host atoms are excited to higher electronic states by inelastic collisions. There are a number of possible de-excitation mechanisms. One is decay by emission of a photon. Decays involving outer shell electrons typically result in optical photons, those involving inner shell electrons typically result in x-rays. In both cases, the absorption cross sections for the emitted photons are sufficiently small enough that the absorption length tends to be greater than the distance over which electron scattering occurs. Therefore, photons are emitted (through the surface of the material) from most of the electron scattering volume, which can be appreciably larger than the extent of the entering beam (Figure 19.4). However, another means of decay for excited host atoms is the ejection of an Auger electron — an event of great significance to material analysis. This process is indicated in Figure 19.5. An inner shell electron can decay to a lower energy level (vacated by ionization). Upon decay, the available energy can be transferred to another electron in the atom which gains sufficient energy to be ejected from the atom. It then has an energy equal to the difference in energy between the electron levels (of the initial electron transition) minus the escape energy. These electrons are named after the French physicist Pierre Auger (pronounced o-zjay) and typically range in energy from 20 significant characteristic for material analysis is that they have a mean path of only 10 to 30 Å. Thus, only those Auger electrons generated very near the surface will escape the material and be detected. This makes Auger analysis sensitive to only the top 30 Å of material and to the lateral extent of the impinging electron beam. The Auger process dominates in low atomic number materials, the x-ray process dominates in high atomic number materials, and both are equally likely around atomic number 33 (arsenic). Of course, in all the above cases (emission of an optical photon, x-ray, or Auger electron) the emission has an energy characteristic of the emitting atom.

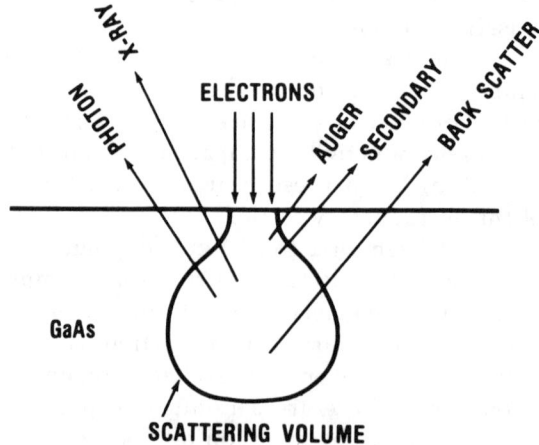

Figure 19.4 Interaction of an electron beam with a solid material.

Other electron collisions can result in backscattered electrons, secondary electrons (electrons emitted from the outer shells of host atoms), and transmitted electrons for sufficiently thinned materials.

Ion beams interact with solid surfaces in a matter characteristically different from that of electrons. There is some nuclear interaction (Chapter 2, section 2.5) and the substantial mass of ions can result in sputtering material from the sample. Below ion beam energies of about 300 keV, minimal sputtering occurs and the dominant result is ion implantation (section 2.5). At higher energies, typically 300 keV to 4 MeV, significant sputtering and backscattering phenomena occur. These can be used to characterize material constituents. Ion beams, as employed in material analysis, are generally not focused to submicron dimensions and hence, typically do not allow the lateral resolution that is possible with some of the electron techniques. However, some of the ion techniques (e.g., SIMS) are capable of very high sensitivity (see section 19.3).

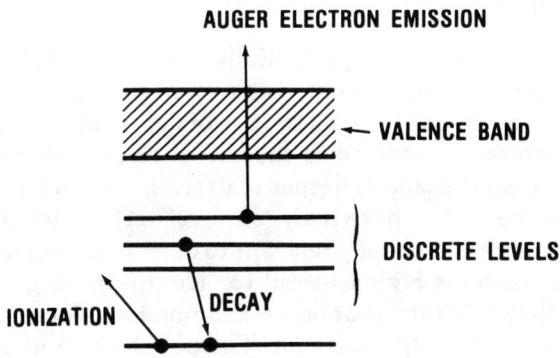

Figure 19.5 Auger electron emission during de-excitation of an atom.

X-rays can also be used as the impinging energy beam. There is the obvious disadvantage that such beams cannot be focused to small dimensions. However, techniques such as x-ray photoelectron spectroscopy (XPS) can establish chemical bonding information and so have a unique ability. Further, x-rays are also used for defect mapping (x-ray topography).

There are analytical techniques that involve still other particles, such as neutrons, and these will be discussed in the appropriate section. Some techniques used to study properties of bulk semi-insulating GaAs, such as photoluminescence or Fourier transform infrared spectoscopy, have been omitted. This chapter is basically restricted to a few of the more popular analytical methods that are useful to the process engineer. The discussion is certainly not meant to be exhaustive.

19.2 SURFACE ANALYSIS

As indicated in the previous section, the surface analysis techniques are somewhat arbitrarily defined as those that sample the top 1000 Å or less of material. Virtually all of these may be (and often are) used to profile samples by simultaneously sputtering material from the surface. This is especially effective with the Auger techniques because they sample only the top 10 to 30 Å of the material, and hence, are capable of great vertical resolution when combined with the sputtering procedure. Such procedures are used, for example, to obtain profiles (concentration as a funtion of depth) of the constituents under an alloyed ohmic contact, or to ascertain alloying affects (undesirable) under a Schottky metalization.

19.2.1 Auger Electron Spectroscopy (AES); Scanning Auger Microprobe (SAM)

Auger electron spectroscopy typically uses 1 to 5 keV electrons to generate Auger electrons. As noted above, the primary characteristic of the process is the limitation to 10 to 30 Å of depth into the material (and hence, also laterally restricted to the extent of the impinging electron beam rather than the larger, deeper scattering volume), making Auger spectroscopy one of the more truly "surface" techniques. As also noted above, the process is especially relevant to detecting low atomic number species, and hence, is highly useful for identifying organic materials (e.g., resist contaminants) that other common techniques (such as electron microprobe) may find difficult. The popularity and usefulness of AES has resulted in a number of commercial instruments, including those from JEOL (Peabody, MA), Perkin-Elmer (Eden Prairie, MN), Riber (Metuchen, NJ), and Varian (Palo Alto, CA). Detailed descriptions of some of these instruments have been reviewed in the literature [7]. Electron beams can be focused to very small diameters, but at some point this results in electron densities sufficiently large enough to cause surface modification or damage. For example, visible damage on fragile specimens can occur at 10^{24} e/cm^2, e-beam resists are altered near 10^{20} e/cm^2, and oxygen desorption from SiO_2 occurs near 10^{17} e/cm^2 [3]. The density impinging on the specimen is a function of beam current, spot size, and scan time. But greater densities result in a larger number of Auger electrons, and hence, improved sensitivity. These considerations have generally resulted in a compromise of 500 Å as the minimum spot size commonly used. The highly focused electron beam is often rastered over an area, hence, the name scanning Auger microprobe (SAM).

The Auger technique is often combined with sputter etching to allow profiling into the depth of the specimen. The basic principle is clear, but there are many practical complexities. Any sputtered surface tends to become rougher, especially if it has a crystalline structure. Knock-on and mixing effects become greater as sputter depth increases. Etch depth must be correlated to etch time and this correlation can be difficult, depending on substrate material, chemistry, and ion beam uniformity and angle. Such complexities make exact quantitative results (such as determining interface width between two materials) difficult. But qualitative results are often enormously useful.

The fundamental output of an Auger analysis is a spectrum of Auger electron energy. This spectrum can be correlated to atomic species and hence, the species identified. Many instrument vendors offer software packages that attempt to provide a quantitative analysis. Again, there are many complexities and subtleties involved and accuracy can be poor [8,9]; errors of 50% or greater can occur [3]. However, precision is usually good (repeatability when running sequential spectra), and such information is often valuable and the errors willingly tolerated.

19.2.2 Secondary Ion Mass Spectroscopy (SIMS); Ion Microprobe

This very sensitive technique uses an ion beam to sputter material from the surface of a sample. The resulting secondary ions (from the sample) are then analyzed using a mass spectrometer (the sputtered ions have a wide range of energy and must be electrostatically accelerated prior to entering the mass spectrometer). This process is referred to as secondary ion mass spectrometry. The term ion microprobe is also used, especially if the ion beam is kept small (2.5 – 10 μm) and rastered over the sample. Ion beam energy is generally in the range of 1 to 30 keV. Over 90% of the secondary ions are emitted from the outer two atomic layers of the sample, making the procedure very surface sensitive [4]. Mass spectra are continuously obtained as the sample is sputter etched. Total ranges of up to 10,000 Å are possible. But depth resolution and accuracy degrades with depth because of effects such as atomic knock-on and ion mixing. The highest sensitivities are obtained by analyzing large areas. Sensitivity can be very high, reaching below 10^{16} atoms/cm^3. SIMS is the only major technique capable of detecting dopant levels below about 10^{18} cm^{-3} and it can also detect light elements (which some other techniques cannot). Some problems arise from species that have the same charge-to-mass ratio. These can be generated in interactions which include background gases and/or the ion beam itself. Hence, high vacuum is desirable. Different ion sources can also prove useful. Typical ions used in SIMS include O_2^+, Cs^+, Ar^+, and O^-. Positive ions are generated using O_2^+ ion sources; negative ions are generated using O^- or Cs^+ ion sources. The sensitivity in any given application will depend on many factors, including the host material, the impurity atom, the quality of mass spectrometer, and the type of ion source. Given all these complexities, and the high vacuum requirement, ion microprobes (or SIMS instruments) are rather large, complex, and expensive instruments. It is also a destructive technique. But the sensitivity and high depth resolution make it an important analytical technique in semiconductor processing.

19.2.3 X-ray Photoelectron Spectroscopy (XPS); Ultraviolet Photoelectron Spectroscopy (UPS)

When photons impinge on a solid, atoms of the specimen absorb the photons and emit electrons (photoelectrons) which have energies characteristic not only of the parent atom, but also of neighboring atoms because of their effect on the atomic environment. Hence, the technique is capable of gathering chemical information in addition to atomic identity and concentration. If the impinging photons are in the ultraviolet energy range (4-50 eV), the technique is called ultraviolet photoelectron spectroscopy (UPS); if in the x-ray range, the technique is called x-ray photoelectron spectroscopy. The photoelectrons can only escape from the solid if they are generated near the surface, making XPS and UPS generally sensitive to less than the top 100 Å of the sample. The sample is not damaged in the process and there are no charging effects (because photons are used). Unlike the Auger process, which involves core electrons, these photoelectrons typically originate from outer s, p, or d atomic orbitals (XPS) or from the valence and conduction bands (UPS). These electrons (and their energies) are affected by the atomic environment. The chemical bonding effects are manifested in a perturbation, or chemical shift, of the energies characteristic of atomic transitions. In XPS, the shift is typically small enough to allow atomic identification, but great enough to be measured reliably and gain chemical information. However, UPS spectra typically have broad, overlaping peaks that are difficult to interpret. Hence, XPS is more relevant to most semiconductor diagnostic applications.

Although ultraviolet light can be focused easily, the more useful x-ray sources cannot. Apertures can be used, but XPS basically is restricted to low lateral resolution.

19.2.4 Electron Diffraction (LEED, RHEED)

Electron diffraction may be used to assess crystalline perfection near the surface of materials (diffraction arising from transmitted electrons is discussed in section 19.3). The diffraction patterns arising from the interaction of particles with crystal lattices are complex and will not be described here. An expert is needed to properly interpret results. The two basic uses of electron diffraction for surface characterization are indicated in Figure 19.6. Figure 19.6(a) shows backscattered electrons. In this case, the backscattering decreases as energy increases and so the electron beam energy must be kept low, typically tens or hundreds of electron volts (transmission electron microscopes typically use thousands of electron volts, even over 100,000 eV). The technique is referred to as low energy electron diffraction (LEED). Beam penetration is low and diffraction occurs only from the top few atomic layers, making the technique highly surface sensitive. Hence, surface damage or contamination (assuming the contaminant affects the surface) can be readily detected.

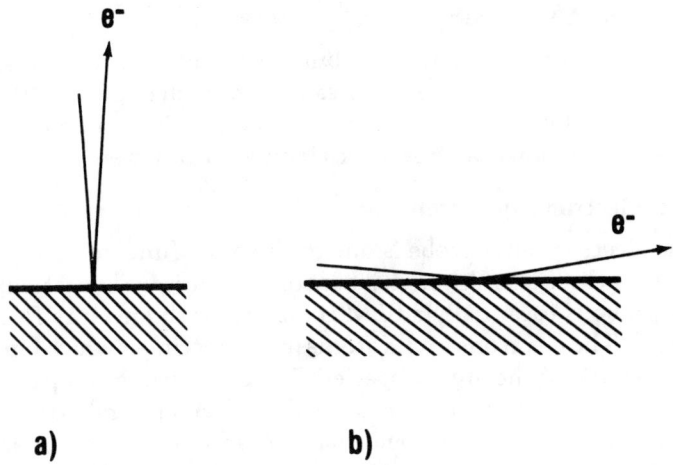

Figure 19.6 Electron diffraction configurations: (a) LEED, (b) RHEED.

Figure 19.6(b) illustrates another possibility. Higher energy electrons can be used if the grazing angle is very small. The forward scattered electrons provide the diffraction pattern. The technique is referred to as reflection high energy electron diffraction (RHEED). If the surface is highly smooth, the process will again be sensitive to only the top few atomic layers. However, surface roughness can result in a significant transmitted component through the peaks of the surface material.

19.2.5 Soft X-ray Appearance Potential Spectroscopy (SXAPS)

In this technique, an electron beam is directed onto the sample and the beam energy is ramped from 0eV to approximately 1 keV. Such electron excitation can result in x-ray emmission if the energy is great enough. As the ramped energy surpasses the threshold for a given element, it will begin to emit x-rays, and the energy of this threshold can be used to identify the element. The low energy electrons do not penetrate very deeply (< 100 Å) making this technique surface sensitive. Because of the low energy, there is not a complex x-ray spectrum from each element. Rather, there is only one major peak per element making interpretation easy. Nevertheless, there are some subtleties. For example, the spectrum will have some structure and often the derivative of the spectrum is taken to better identify the peaks. The process is non-destructive. The major disadvantage of the process is that it is insensitive to some elements. However, it is very sensitive to others, such as the rare earths.

19.3 BULK ANALYSIS

As noted in the introduction, bulk techniques have been somewhat arbitrarily defined as those which sample more than the top 1000 Å of the material. In this sense, all the surface techniques can be used for bulk analysis if combined with sputter etching (or other etching).

19.3.1 Electron Microprobe

The electron microprobe is one of the most fundamentally useful of the bulk techniques, although it has some significant limitations. The ion microprobe is essentially a SEM combined with an x-ray fluorescence detector. The electrons excite atoms in the specimen which emits x-rays characteristic of the atomic species. The depth being sampled is a function of electron energy, but is on the order of a micron. The x-ray detectors can be mounted on normal SEMs and are offered as accessories on many such instruments. Hence, cost is minimal (not including the SEM). The technique is very attractive because the SEM can be used to examine the sample and determine exactly the size (and on some instruments, the shape) of the area being sampled. The user sees the precise area being sampled. This area can be extremely small (limited only by the SEM's magnification). A typical procedure would be to focus on the questionable area (such as a particle or spot on a gate stripe) and then move over just off that area to obtain a "control" spectra. Some instruments can display the difference between the two spectra, aiding identification of contaminants.

There are essentially two types of x-ray detectors — energy dispersive (EDS, or energy dispersive spectoscopy) or wavelength dispersive (WDS). The EDS detector detects all wavelengths simultaneously and also has greater sensitivity than the WDS detector. The WDS method uses a crystal to direct x-rays of a given wavelength onto the detector. Some energy is lost in this process, resulting in the lower sensitivity. The EDS method would seem the best, being both faster and more sensitive. However, in some cases peaks in the x-ray spectra can overlap each other and the WDS method yields significantly improved resolution.

The method does have several limitations. First, as in all SEM work, the sample must be sufficiently conductive to prevent charging. Also, sensitivity is low. The SEM scan rate, spot size, and depth penetration all combine to result in a rather small effective volume from which x-rays are generated, resulting in a sensitivity of only 100 to 2000 ppm (depending on the element). Also, low atomic number species tend to emit Auger electrons rather than x-rays. In practice, detection of elements with mass below sodium can be difficult, making the technique insensitive to organics. WDS detectors tend to be able to detect lower mass elements than EDS detectors, although this has more to do with the presence of window materials than the actual ability of the detector. In some formats, carbon and oxygen can be detected, although with poor sensitivity. Nevertheless, the technique is non-destructive and allows rapid, multi-element surveys with high lateral resolution and visual correlation with the sampled area. Evaluation time is rapid and cost is low (add-on to an existing SEM). These features tend to make this technique one of the most useful for the process engineer.

19.3.2 X-Ray Fluorescence (XRF)

This technique is very similar to that of the electron microprobe (described above), except that x-rays instead of an electron beam are used to excite atoms and stimulate x-ray emission. As above, the technique is non-destructive, but it also avoids the charging problems that can restrict use of the electron microprobe. However, the lighter elements are again difficult to detect, the technique generally being insensitive to elements below aluminum. The determination of identity and concentration depends on the use of standards for calibration. This technique generally is capable of sampling to substantial depths, up to 30 μm. Use of x-ray for an excitation source also results in poor lateral resolution (lack of x-ray focusing). However, the large sampling volume can yield good sensitivity, typically between 1 – 10 ppm.

19.3.3 Deep Level Transient Spectroscopy (DLTS)

Deep level transient spectroscopy (DLTS) is very sensitive to deep levels in the bandgap of semiconductors, such as traps (Chapter 2). The general procedure is to use photons, electrons, or bias conditions to excite or populate such levels and then monitor capacitance (of a Schottky diode, for example) or current transients. There are a number of variations to the procedure, some requiring measurements over temperature. A rather large area is usually used, resulting in very high sensitivity. Detection limit can be as little as 10^{10} or 10^{11} cm^{-3}. The technique generally yields the energy and concentration of the deep levels; but identification can be difficult. A variation popular in GaAs work uses a Schottky diode. The diode is forward biased to populate the majority carrier traps in the depletion region. A reverse bias pulse empties the traps (above the Fermi level for n-type material). The charge released (expotential decay) is monitored by measuring the change in capacitance between two chosen, fixed times. That change in capacitance is monitored for different temperatures. This data is sufficient to calculate activation energies, concentrations, and capture cross sections. If an electron beam is used to scan over the sample, the technique is referred to as scanning deep level transient spectroscopy and does not yield as great a sensitivity. A photon beam can also be used to excite the states. The DLTS techniques are highly valuable in characterizing the electron traps in GaAs material.

19.3.4 Electron Beam Induced Current (EBIC)

This technique requires the presence of a Schottky diode, a p-n junction, or a MOS capacitor (not usually attempted on GaAs). An electron beam is scanned over the sample on or near the diode (or other active area) which collects the charge from the electron hole pairs generated by the e-beam. Any feature, such as a defect or inhomogeneity, that affects the generation or recombination of electron hole pairs will also affect the resulting current. If a CRT is used to monitor the detected current in synchronization with the scanning electron beam, the defect (or other) areas will be imaged. The EBIC ability can be easily and cheaply added to many SEMs. Lateral resolution of this technique is usually limited to 1 μm or greater.

19.3.5 Rutherford Backscattering (RBS)

Rutherford backscattering (RBS) is a technique in which impinging ions are backscattered from atoms in the host material. If E_o is the initial energy of the incident ion, the backscattered ion will have energy

$$E = [(M - m)/(M + m)]^2 E_o$$

where M and m are the masses of the target and incident ions respectively. This expression is for 180° scattering, other angles giving similar expressions (exactly 180° is not used). However, the ion also loses energy in traversing the host material, both incoming and outgoing. These energies can be determined (or are known and cataloged) for any host material. Hence, appropriate measurements can yield both the mass of the target atom and its depth in the sample. High accuracy requires a highly monoenergetic exciting ion beam and high energy resolution in the detector. Depth resolutions of approximately 100 Å are feasible. Maximum depth from which information can be obtained approaches 10,000 Å. Typically, 1 to 4 MeV helium ions are used for RBS. These beams are typically 100 μm to 1 mm in diameter; high lateral resolution is not attempted. Sensitivity is not great either, usually being no better than about 5×10^{18} cm^{-3}. The advantage of RBS is the ability to provide absolute concentration values that are insensitive to chemical effects in the substrate. It is quantitative to within about 5% without the use of standards. There are no charging problems, and the technique is sensitive to both elements and defects. For crystalline materials, the technique is especially useful if applied in a channeling direction. In this case it is very sensitive to atoms off lattice sites or other defects. As such, it may be used to assess crystalline perfection (before and after anneal, for example).

19.3.6 X-ray Topography

This technique uses x-ray diffraction to detect defects or other inhomogeneities in crystals. A slit and photographic plate are employed in a scanning process which records a permanent map of the slice on film. Usual diffraction exhibited by a perfect crystal will be locally perturbed by any departure from crystallinity and cause an image on the film (either darker or lighter from the general background). Defects such as stacking faults or dislocations are not directly observable (as they are using transmission electron microscopy) — only strain is revealed, not the cause of the strain. Also, lateral resolution is generally not great, a few microns at most (many microns is more typical). But defect distribution over entire slices can be obtained rapidly and nondestructively. The techniques can be performed in transmission, in which case the entire thickness of the slice is sampled. Reflection x-ray topography will generally sample only the top 10 to 30 μm of material.

19.3.7 Transmission Electron Microscopy (TEM)

Transmission electron microscopy (TEM) can be used to obtain detailed and direct information about crystalline imperfection. A thin section of the material must be prepared and electrons are directed through the sample and detected after transmission. Either a diffraction pattern or an image may be obtained. In conventional TEM (CTEM), a broad electron beam (1-50 μm) is used. In scanning TEM (STEM) a highly focused beam (as little as 2-5 Å) is used and the CRT images the detected signal. STEM has the advantage of ultra-high resolution (2-5 Å) and the ability to directly detect crystalline imperfections and defects. In conjunction with other detectors (secondary electron, X-ray, electron loss spectroscopy (ELS), etc.), it can analyze extremely small areas and detect as few as 1,000 atoms. However, it is a time-consuming and difficult technique — difficult both in sample preparation and image analysis. A skilled scientist is required to analyze and interpret many of the resulting images or diffraction patterns.

19.3.8 Neutron Activation Analysis (NAA)

Neutron activation analysis can provide highly sensitive analysis of some impurities but, unfortunately, has grave restrictions for use with GaAs material. Nevertheless, the technique is sensitive and valuable for some applications. In this procedure, a sample is exposed to high neutron flux (being placed in or near a nuclear reactor). The neutrons generate radioactive isotopes from the stable isotopes present in the sample. The most common nuclear reaction in this case is (n, γ) — a neutron is absorbed and a gamma ray emitted, resulting in an isotope one neutron heavier than the initial atom. Assuming the product nucleus is radioactive, it will decay and emit alpha, beta, or gamma radiation. The beta or gamma decay can be detected and measured using a multichannel analyzer, allowing determination of both the identity and amount of many elements. Although the process is expensive and time consuming, it has a sensitivity of 0.00001 to 0.01 ppma for over 40 elements. It is insensitive to elements lighter than sodium, including the important constituents of organic compounds, carbon, oxygen, and nitrogen. The difficulty in using this technique with GaAs material is the large interference that occurs from activation of heavy elements such as Ga and As. However, chemical means can be used to make an initial separation. Further, the technique can be very specific in that only materials present at the time of irradiation will be detected. No subsequent processing, cleaning, or handling will add artificial information.

REFERENCES

[1] C.J. Powell, *National Bureau of Standards Report*, NBSIR 75-945, Washington, D.C.: U.S. Dept. of Commerce, 1977.

[2] A.L. Robinson, *Science*, 191, 1976, p. 1253.

[3] T.J. Schaffner, in *VLSI Electronics: Microstructure Science*, Vol. 6, N.G. Einspruch, editor. New York: Academic Press, 1983, p. 497.

[4] G.B. Larrabee, in *VLSI Electronics: Microstructure Science*, Vol. 2, N.G. Einspruch, editor. New York: Academic Press, 1981, p. 37.

[5] J.A. Keenan and G.B. Larrabee, in *VLSI Electronics: Microstructure Science*, Vol. 6. New York: Academic Press, 1983, p. 74.

[6] Courtesy of T.J. Schaffner, Materials Characterization Laboratory, Texas Instruments, Inc., Dallas, Texas. A version of Figure 19.3 also appears in reference [3].

[7] C.A. Evans, Jr. and R.J. Blattner, *Semicond. Int.*, Nov. 1980, p. 109.

[8] J.M. Morabito and P.M. Hall, *Scanning Electron Microscopy/1976*, II-TRI, Chicago, IL, 1976, p. 221.

[9] P.H. Holloway, *Scanning Electron Microscopy/1978*, SEM Inc., Chicago, IL, 1978, p. 361.

INDEX

Airbridge, *see* bridge
Alignment marks (for e-beam), 153
Alloy, 236
Aluminum-gold interactions, 286
Angle evaporation (for gate), 282
Anneal (to active ion implant), 50
Anodic etching, 119
Assisted liftoff, 276
Auger electron, 391
Auger electron spectroscopy (AES), 394
Back side gating, 214
Bake (resist), 39
Bandgap, 29
Barrier height, 230, 267
Bilevel resist, 146
Blanket exposure, 142
Breakdown voltage, 266, 381
Bridge, 70, 334
Built-in voltage, 58, 262

Capacitance of gate, 58
 of Schottky barrier, 263
Capacitance-voltage measurement, 264, 365
Capacitors (monolithic), 306
Chlorobenzene method (with resist), 145
Chromium-doped GaAs, 39
Clean room, 87
Contact photolithography, 128
Contact resistance, 226, 228
 measurement, 241, 248
Coplanar transmission line, 317
Critical current density, 289
Cutoff frequency, 67
Debye length, 369
Deep level transient spectroscopy (DLTS), 399
Deep ultra-violet (DUV) exposure, 129
De-ionized water, 97

Depletion mode FET, 64
Depletion region, 58, 260
Descum, 141
Develop (resist), 142
Dielectric (capacitor), 297, 312
Dielectric constant, 27
Die separation, 351
Diffusion, 287
Diffusion limited etching, 102
Diodes (microwave), 74
Direct broadcast satellite (DBS), 2
Discrete device, 4
Dislocation density, 40
Distributed element, 303
Doping concentration, 4
Doping profile, 365
Drain (of FET), 61
Drift mobility, see mobility
Edge profile modification (resist), 145
Electroless plating, 332
Electron beam induced current (EBIC), 400
Electron beam lithography, 150
Electron diffraction, 396
Electron microprobe, 398
Electroplating, 329
End resistance, 247
Energy bands, 27
Enhancement mode FET, 65
Epitaxy, 42
Equivalent circuit (of FET), 65
Etch pit density, 40
Evaporation (of metal), 289
Field effect transistor (FET), 61
  Configuration, 67
  Logic (or digital), 72
  Low noise, 70
  Power, 68
  Recessed gate, 69
First-level metal, 5, 294
Four point probe, 364

Gallium arsenide (GaAs) material properties, 22
Gate (of FET), 61, 69, 272
Gate metal, 271
Gate resistance, 378
  distributed nature of, 380
  parallel fingers, 380
Gunn diode, 78
Hall mobility, see mobility
Handling of slices, 93
Heterojunction devices, 79
High electron mobility transistor (HEMT), 80
Horizontal Bridgman GaAs, 36
Humidity control, 90
I_{dss} (drain-source saturation current), 63, 383
I_{sat} (saturation current), 62
I_{max} (maximum current), 63, 383
Ideality factor, 231, 267
IMPATT diode, 75
Inductors, 315
Interdigitated capacitors, 308
Ion beam lithography, 162
Ion implantation, 47
Ion implant isolation, 218
Ion milling, 198
Ion microprobe, 395
Junction field effect transistor (JFET), 61
Kinetically assisted chemical reaction, 184
Kinetically limited etching, 102
Lapping, 343
Lattice configuration, 21
Liftoff, 126
Light assisted plating, 333
Liquid encapsulated Czochralski (LEC) GaAs, 38
Liquid phase epitaxy (LPE), 43
Loadline, 64
Looping, 62

Index 405

Low energy electron diffraction (LEED), 396
Lumped element, 303
Masks (for photolithography), 126
Mass transport limited etching, 102
Maximum frequency of oscillation (f_{max}), 67
Mercury emission lines (for photolithography), 137
Mercury probe, 369
Mesa etching, 214
Mesa orientation, 216
Metal-insulator-metal (MIM) capacitor, 307, 310
Metal-organic chemical vapor deposition (MOCVD), 45
Metal-semiconductor FET (MESFET), 61
Metalization properties, 285
Microfaceting, 105
Microstrip line, 317
Microwave integrated circuit (MIC), 5
Miller indices, 23
Mobility, 18, 33
  Drift, 371
  Hall, 369
Modulation transfer function (MTF), 133
Molecular beam epitaxy (MBE), 46
Monolithic microwave integrated circuit (MMIC), 5
Multilevel resist, 145
Negative resist, 125, 143
Neutron activation analysis (NAA), 401
Nickel overlay (for ohmics), 234
Noise figure, 70
Ohmic metalization, 234

Optical stepper (lithography), 130, 131
Orienting the slice, 216
Parasitic resistances (of FET), 378
Permeable base transistor (PBT), 73
Phased array radar, 2
Pinchoff voltage, 63, 382
Plasma etching, 187
Plated heat sink, 70
Plug bar, 153, 359
Positive resist, 125, 136
Process flow, 6
Projection lithography, 130
Proximity effect, 154
Proximity printing (lithography), 128
Reaction rate limited etching, 102
Reactive ion beam etching (RIBE), 197
Reactive ion etching (RIE), 192
Recessed gate, 69, 272
Reflection high energy electron diffraction (RHEED), 396
Resist, 125
Resistors (monolithic), 319
Resistivity, 2
Reticle, 130
Runout, 129
Rutherford backscattering (RBS), 400
SAINT process, 280
Saturated drift velocity, 19
Saturation voltage, 381
Sawing, 352
Scanning Auger microprobe (SAM), 394
Scattering parameters, 66
Schottky barrier, 230
Schottky diode, 58
Scribing, 351
Scumming, 141

Secondary ion mass spectroscopy (SIMS), 395
Second-level metal, 5, 300
Selective ion implantation, 221
Self-aligned gate, 69
Sheet conductivity (measurement), 364
Sheet resistance, 3
Skin depth, 288
Soft x-ray appearance spectroscopy (SXAPS), 397
Solvents & solvent cleaning, 93
Source (of FET), 61
Source-gate resistance, 378
Source resistance, 378
  effect on measurements, 374
Specific contact resistance, 226
Spinning resist, 136
Sputtering, 289
Standing wave pattern (in resist), 137
Static electricity, 92
Surface limited etching, 102
Thermal conductivity of GaAs, 27
Thermal expansion of GaAs, 26
Thermionic emission, 230
Thinning of GaAs substrate, 343
Threshold voltage, 72
Transconductance, 63, 383
Transfer length, 228
Transmission electron microscopy (TEM), 401
Transmission lines, 376
Travelers, 361
Trenching, 202
Trilevel resist, 147
Tunneling, 230
Ultraviolet photoelectron spectroscopy (UPS), 395
Unit cell of GaAs lattice, 22
Vapor phase epitaxy (VPE), 44
VARACTOR, 60

Vertical FET (VFET), 73
Via (front to back of substrate), 70, 346
Voltages, characteristic (of FET), 381
Water content of air, 91
Wulff plot, 107
X-ray fluorescence, 399
X-ray lithography, 158
X-ray photoelectron spectroscopy (XPS), 395
X-ray topography, 401

SUNIL V. HATTANGADY